Muhammad Yasir Ali, Nisar Ur-Rahman, Saeed Ahmad and Udo Bakowsky (Eds.)

Dosage Form Development

Also of Interest

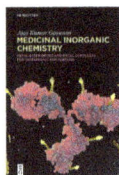

Medicinal Inorganic Chemistry.
Metal-Based Drugs and Metal Complexes for Therapeutic Applications
Ajay Kumar Goswami, 2024
ISBN 978-3-11-142609-9
e-ISBN 978-3-11-142613-6

Drug Delivery Technology.
Herbal Bioenhancers in Pharmaceuticals
Prashant L. Pingale (Ed.), 2022
ISBN 978-3-11-074679-2
e-ISBN 978-3-11-074680-8

Active Pharmaceutical Ingredient Manufacturing.
Nondestructive Creation
Girish K. Malhotra, 2022
ISBN 978-3-11-070282-8
e-ISBN 978-3-11-070284-2

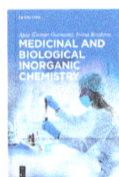

Medicinal and Biological Inorganic Chemistry
Ajay Kumar Goswami, Irena Kostova, 2022
ISBN 978-1-5015-2455-4
e-ISBN 978-1-5015-1611-5

Dosage Form Development

Liquid, Solid and Semi-Solid Dosage Forms

Edited by
Muhammad Yasir Ali, Nisar Ur-Rahman, Saeed Ahmad
and Udo Bakowsky

DE GRUYTER

Editors

Dr. Muhammad Yasir Ali
Department of Pharmaceutics
Faculty of Pharmaceutical Sciences
GC University Faisalabad
P.O. Box 38000, Faisalabad, Pakistan
myasirali@gcuf.edu.pk

Prof. Nisar Ur-Rahman
Department of Pharmacy
COMSATS University Islamabad
Abbottabad Campus
Islamabad 22060, Pakistan
college.430@gcuf.edu.pk

Prof. Saeed Ahmad
Department of Pharmaceutical Chemistry
The Islamia University of Bahawalpur
Bahawalpur 63100, Pakistan
rsahmed_iub@yahoo.com

Prof. Udo Bakowsky
Institute for Pharmaceutical Technology
and Biopharmaceutics
Philipps University of Marburg
Robert-Koch-Str. 4
35037 Marburg, Germany
ubakowsky@aol.com

ISBN 978-3-11-143801-6
e-ISBN (PDF) 978-3-11-143810-8
e-ISBN (EPUB) 978-3-11-143825-2

Library of Congress Control Number: 2026930151

Bibliographic information published by the Deutsche Nationalbibliothek
The Deutsche Nationalbibliothek lists this publication in the Deutsche Nationalbibliografie;
detailed bibliographic data are available on the internet at http://dnb.dnb.de.

© 2026 Walter de Gruyter GmbH, Berlin/Boston, Genthiner Straße 13, 10785 Berlin
Cover image: apomares/E+/Getty Images
Typesetting: Integra Software Services Pvt. Ltd.

www.degruyterbrill.com
Questions about General Product Safety Regulation:
productsafety@degruyterbrill.com

Contents

List of contributors

Abd-Ur-Rehman Khan
Department of Pharmaceutical Chemistry
GC University Faisalabad
Faisalabad
Pakistan

Abid Mahmood
Department of Pharmaceutical Chemistry
GC University Faisalabad
Faisalabad
Pakistan

Abida Sabir
Department of Pharmacy Practice
GC University Faisalabad
Faisalabad
Pakistan

Abu Bakar, Asghar
Department of Pharmaceutics
Faculty of Pharmaceutical Sciences
GC University Faisalabad
Faisalabad
Pakistan

Adeeba Ishaq
Department of Pharmacognosy
GC University Faisalabad
Faisalabad
Pakistan

Adeeba Ishaq
Department of Pharmacognosy
GC University Faisalabad
Faisalabad
Pakistan

Aisha Sethi
Department of Pharmaceutics
GC University Faisalabad
Faisalabad
Pakistan

Ali Abbas
Department of Pharmacology
GC University Faisalabad
Faisalabad
Pakistan

Ali Moghadam
Faculty of Agricultural Sciences
Shiraz University
Shiraz
Iran

Amna Shakeel
Department of Pharmacognosy
GC University Faisalabad
Faisalabad
Pakistan

Amna Shakeel
Department of Pharmacognosy
GC University Faisalabad
Faisalabad
Pakistan

Arshad Mahmood
College of Pharmacy
Al Ain University, Abu Dhabi, UAE

Asia Naz Awan
Department of Pharmaceutical Chemistry
University of Karachi
Karachi
Pakistan

Asiya Farheen
Department of Pharmaceutical Chemistry
University of Karachi
Karachi
Pakistan

Ayesha Akbar
Department of Pharmaceutical Chemistry
GC University Faisalabad
Faisalabad
Pakistan

https://doi.org/10.1515/9783111438108-203

Ayesha Fatima
Bakhtawar Amin College of Pharmaceutical
Sciences
Bakhtawar Amin Medical and Dental College
60000 Multan
Pakistan

Ayesha Saleem
Department of Pharmaceutical Chemistry
GC University Faisalabad
Faisalabad
Pakistan

Bisma Islam
Department of Pharmacognosy
GC University Faisalabad
Faisalabad
Pakistan

Bisma Islam
Department of Pharmacognosy
GC University Faisalabad
Faisalabad
Pakistan

Daulat Haleem Khan
Department of Pharmacy
Lahore College of Pharmaceutical Sciences
Lahore
Pakistan

Eisha Mashkoor
Department of Pharmacology
GC University Faisalabad
Faisalabad
Pakistan

Fatima Sajjad
Department of Pharmaceutics
GC University Faisalabad
Faisalabad
Pakistan

Ghazala Ambreen
Sargodha College of Medical Sciences
Sargodha
Pakistan

Ghulam Abbas
Department of Pharmaceutics
GC University Faisalabad
Faisalabad
Pakistan

Hafiza Arwa Nadeem
Department of Pharmacology
GC University Faisalabad
Faisalabad
Pakistan

Hammad Afzal
Department of Pharmaceutics
GC University Faisalabad
Faisalabad
Pakistan

Hania Ahmad
Department of Pharmacy Practice
GC University Faisalabad
Faisalabad
Pakistan

Humaira Gul
Department of Pharmacology
GC University Faisalabad
Faisalabad
Pakistan

Ijaz Ali
Department of Pharmacognosy
GC University Faisalabad
Faisalabad
Pakistan

Muhammad Ijaz
College of Health and Agricultural Sciences
University College Dublin
Irland

Imran Tariq
Eshelman School of Pharmacy
University of North Carolina
Chapel Hill
USA

Javeria Batool
Department of Pharmacognosy
GC University Faisalabad
Faisalabad
Pakistan

Javeria Batool
Department of Pharmacognosy
GC University Faisalabad
Faisalabad
Pakistan

Kashf Ul Iman
Department of Pharmaceutical Chemistry
GC University Faisalabad
Faisalabad
Pakistan

Khadeja Arshad
Department of Pharmacy
Iqra University
Islamabad
Pakistan

Khurram Waqas
Institute of Pharmaceutical Sciences
University of Veterinary and Animal Sciences
Lahore
Pakistan

Kifayat Ullah Khan
College of Pharmaceutical Sciences
Soochow University
Suzhou
China

Laraib Zeeshan
Department of Pharmaceutics
GC University Faisalabad
Faisalabad
Pakistan

Malik Saadullah
Department of Pharmaceutical Chemistry
GC University Faisalabad
Faisalabad
Pakistan

Maria Manan
Department of Pharmacology
GC University Faisalabad
Faisalabad
Pakistan

Mehma Meraal
Department of Pharmacy Practice
GC University Faisalabad
Faisalabad
Pakistan

Muhammad Arsal Rafiq
Department of Pharmaceutical Chemistry
GC University Faisalabad
Faisalabad
Pakistan

Muhammad Asjad Ur Rahman
Department of Pharmacognosy
GC University Faisalabad
Faisalabad
Pakistan

Muhammad Fayyaz
Department of Bioengineering
The Grainger College of Engineering
University of Illinois Urbana-Champaign,
Champaign
IL, USA

Muhammad Hammas
Department of Pharmacology
GC University Faisalabad
Faisalabad
Pakistan

Muhammad Hashim
Department of Pharmacology
GC University Faisalabad
Faisalabad
Pakistan

Muhammad Ijaz
School of Veterinary Medicine
College of Health and Agricultural Sciences
University College Dublin
Dublin
Ireland

Muhammad Irfan
College of Pharmacy
University of Sargodha
Sargodha
Pakistan

Muhammad Mehboob ur Rehman
Department of Pharmaceutics
GC University Faisalabad
Faisalabad
Pakistan

Muhammad Yasir Ali
Department of Pharmaceutics
GC University Faisalabad
Faisalabad
Pakistan

Mulazim Hussain Asim
College of Pharmacy
University of Sargodha
Sargodha
Pakistan

Nisar ur Rahman
Department of Pharmacy
COMSATS University
Abbottabad
Pakistan

Rabia Jabeen
Department of Pharmaceutics
GC University Faisalabad
Faisalabad
Pakistan

Rabia Munir
Department of Pharmaceutics
GC University Faisalabad
Faisalabad
Pakistan

Romana Riaz
Bakhtawar Amin College of Pharmaceutical
Sciences
Bakhtawar Amin Medical and Dental College
60000 Multan
Pakistan

Sabahat Abdullah
Department of Pharmaceutical Chemistry
University of Karachi
Karachi
Pakistan

Sadia Hakim
Department of Pharmaceutics
GC University Faisalabad
Faisalabad
Pakistan

Saeed Ahmad
Department of Pharmaceutical Chemistry
Faculty of Pharmacy
The Islamia University of Bahawalpur
Bahawalpur
Pakistan

Sana Inam
Department of Pharmaceutics
Faculty of Pharmaceutical Sciences
GC University Faisalabad
Faisalabad
Pakistan

Shakib Kazmi
Department of Chemistry
School of Science
Loughborough University
England

Shams ul Hassan
Department of Pharmaceutics
GC University Faisalabad
Faisalabad
Pakistan

Sufyan Iqbal
Department of Pharmaceutics
GC University Faisalabad
Faisalabad
Pakistan

Syeda Hijab Zahra
Department of Pharmacy Practice
GC University Faisalabad
Faisalabad
Pakistan

Sajid Ali
Department of Chemistry
University of Uppsala
Uppsala
Sweden

Tanzeela Awan
Bakhtawar Amin College of Pharmaceutical
Sciences
Bakhtawar Amin Medical and Dental College
60000 Multan
Pakistan

Udo Bakowsky
Department of Pharmaceutics and
Biopharmaceutics
Faculty of Pharmacy
Philipps University Marburg
Marburg
Germany

Umair Amin
Department of Pharmaceutics and
Biopharmaceutics
Philipps University Marburg
Marburg
Germany

Umaira Maqbool
Department of Pharmacy Practice
GC University Faisalabad
Faisalabad
Pakistan

Usman Saleem
Department of Pharmaceutics
Faculty of Pharmaceutical Sciences
GC University Faisalabad
Faisalabad
Pakistan

Uzma Saher Department of Pharmacy
The Women University Multan
Multan
Pakistan

Zeenat Arshad
Department of Pharmacy Practice
GC University Faisalabad
Faisalabad
Pakistan

Zermina Rashid
Department of Pharmacy
The Women University Multan
Multan
Pakistan

Part 1: **Basics of dosage form design**

Tanzeela Awan, Muhammad Yasir Ali, Ayesha Fatima, Fatima Sajjad,
Ayesha Akbar, Ayesha Saleem, Kashf Ul Iman, Romana Riaz,
Nisar ur Rahman

1 History of dosage form development

1.1 Introduction

Humans have used plants as medicines since ancient times because early people found that consuming certain plants made them feel content, happy, relaxed, or at peace. These humans used the parts of plants like leaves, stems, roots, flowers, fruits, and bark to treat various ailments. These parts are either swallowed or applied externally to heal the wounds and cuts. Leaves and animal fats are applied externally to aid in healing cuts and scrapes. Sometimes herbs were used by setting them on fire and breathing in the released vapors. Eventually, combinations of herbs would be employed, and they would have been added to substances like oils, fats, and honey [1].

Drugs, in the form of vegetation and minerals, have existed as long as humans. Human disease and the instinct to survive have led to their discovery through the ages. The use of drugs, crude though they may have been, undoubtedly began long before recorded history, for the instinct of primitive man to relieve the pain of a wound by bathing it in cool water or by soothing it with a fresh leaf or protecting it with mud is within the realm of belief. From experience, early humans would learn that certain therapies were more effective than others, and from these beginnings came the practice of drug therapy. Among many early races, disease was believed to be caused by the entrance of demons or evil spirits into the body. The treatment naturally involved ridding the body of supernatural intruders. From the earliest records, the primary methods of removing spirits were using spiritual incantations, the application of noisome materials, and the administration of specific herbs or plant materials [2].

Over time, these unprocessed materials underwent progressively more processing before being utilized. To extract the goodness, they could be cooked. They could be ground up, diluted with something else, or dissolved in a solvent. To enhance the taste, one could add something sweet, something aromatic to enhance the scent, or a dye to enhance the color. This led to the development of the last dose form, administered orally, topically, or through a body opening. This chapter lists a few of the numerous dosage forms used to treat or at least reduce the physical ailments of humans, both internally and externally [3].

https://doi.org/10.1515/9783111438108-001

1.2 Dosage form development eras

We can discuss the history of dosage form development by dividing it into different eras.

1.2.1 Mesopotamian era

In ancient Mesopotamia, the gods are believed to have guided the people in all matters for over 2,000 years. The Sumerian goddess, Gula, directed the people about the art of medicine. She guided the physicians to cure many diseases; however, these health issues were alleged to be caused by supernatural powers. In the old Babylonian period (2000–1600 BCE), the medical profession was well-developed.

The asu and asipu were two terms used to describe different healers in ancient Mesopotamia. The main difference between the two types is that the **asipu** depends specifically on supernatural prayers for the treatment of diseases, while the **asu** is treated by balms and herbal drugs to cure the diseases.

In the same era, antiseptics were prepared by combining alcohol, honey, and resins. The surgical procedures were also more developed compared to other regions. Teall writes: "In the treatment of all wounds, there are three critical steps: washing, applying a plaster, and binding the wound" [6]. The Mesopotamians realized that fast wound healing and prevention from infection could be achieved only by practicing hygiene procedures, i.e., dealing wounds with clean water and clean hands. A combination of beer and hot water was used by physicians to sanitize hands and wounds (Fig. 1.1).

1.2.2 Egyptian era

Modernized medical practices, especially in orthopedics, dentistry, and surgeries, evolved in the era of ancient Egypt (3300 BCE to 525 BCE) [4]. The use of malachite as medicine and eye paint was also discovered in the same era (4000 BCE) [5]. The physicians advised the use of more than one drug to treat most of the ailments. The mouth, rectum, vagina, skin, and nose were the five major routes of administration. So, different dosage forms were used such as pills, cakes, ointments, eye drops, gargles, suppositories, fumigations, and baths [6].

For developing these dosage forms, some materials with potential therapeutic activity were used as vehicles, such as beeswax, which was used for its binding property in ointments and other skin applications. Some other edibles like honey, beer, wine, water, and milk were also used as vehicles in pharmaceutical preparation [7].

Fig. 1.1: Depiction of a physician preparing an elixir.

1.2.3 Greek and Roman era

From Egypt, the art of medicine came to Greece, where it was further refined by Hippocrates and his descendants. Greeks involved science in mythological beliefs and started thinking about the scientific and logical explanations of diseases and their treatments. Greek medicine had a great influence on the Egyptian era. However, they also believed in the gods of healing. They used to treat the disease through sleep, fasting, hygiene, and diet to correct the balance of the body. Proper diagnosis and treatment of the disease were the main achievements of the Greeks. They have also developed some advanced surgical techniques. Hippocrates was considered as "Father of Medicine" [8].

In this era, several dosage forms were developed and are being used, including pills, powders, ointments, gargles, and eye ointments. Enemas were also introduced in this period, where a greasy liquid was injected into the rectum through a funnel or animal horn. For the relief of pain, they used hot fermentation, such as hot cotton sponges (soaked in hot water) or roasted millet folded in a warm cloth. At that time, cerates were used for external applications. It was an old and harder form of poultice and used to retain heat at a specific body area. Another dosage form they developed was terra sigillata, which was a greasy mineral clay disc, used for many therapeutic purposes.

Treacles were the most significant medicines used in that era against venoms and stings, present-day known as an antidote. These were made using numerous plants and honey. The specially made preparations with chemicals, plant materials, and animal materials were only used by people living in big cities and the army [1].

1.2.4 Chinese era

The Chinese have a long history of using traditional medicines as external medicines. According to the history recorded in an old Chinese reference book called Shan Hai Jing, a yellow irrigation bath was used for the treatment of scabies. Topical formulations like lotions, paints, PEI, liniments, and baths have been mentioned in Zhou Li Tian Gong for skin problems during changing seasons. The Chinese were also familiar with other external dosage formulations like pastes, powder, ointments, smoke agents, and iron, as mentioned in Huang Di Nei Jing. Specific formulations for a particular drug were also elaborated in certain reference books, for example, Canon of Medicine suggested making pills for Bezoar and boiling Angelica before use. The external governance theory was in its initial stages at that time [9].

Subsequently, in traditional Chinese books, more than 10 different kinds of drug dosage forms were described, including a variety of pills (such as honey pills, bitter pills, and medicine pills), pastes, powders, ointments, pillows, creams, sachets, and washes. Several types of suppositories were also developed in the Chinese era. All dosage forms were also well-documented in the Chinese books. Pills were the first most common and widely used dosage form, paste came in the second place, and then came the powders. Suppositories were also developed and used for the first time in the Chinese era. Those ancient suppositories served as the foundation for the development of other forms of suppositories, like ear bolts and vaginal and nasal suppositories [10].

During the Sui and Tang Dynasties, the dosage forms were practiced as required by the individual patients. Prescription development and choice of the dosage form in a clinical setting were important milestones that were achieved by the Chinese.

1.2.5 Arab medicine

Avicenna wrote two books that had several formulas, a section on poisons, and instructions on developing remedies. It is believed that Avicenna introduced the custom of gilding and silvering pills. The sugar used by the Arabs came from sugarcane, which was first grown in India and then spread to Persia, Cyprus, Sicily, and Spain. Among the sweet dishes they brought were syrups, conserves, confections, electuaries, and juleps. Sugar was boiled with liquor to make syrups. To preserve them beforehand, flowers, herbs, roots, peels, or fruits were combined with sugar to conserve them. To make confections, combine the powdered and dried components with honey or syrup until the mixture has the consistency of a thin electuary. Electuaries were thicker and more prone to fermentation or candying than confections, depending on their thickness. Clear, sweet drinks known as "Juleps" were typically a pleasant way to deliver medication. To improve the taste of medications, the Arabs also added flavors like orange and lemon peel, and rose water [3].

1.2.6 Medieval medicine

After the Romans left in the early fifth century AD, there were three main kinds of medicine practiced in Britain: household medicine, herbal medicine, and monastic medicine, which was used by monks and was based on old books and prayers; and practiced by Anglo-Saxon healers, called leeches. Many of the later recipes were written in books called leechbooks.

Magic was very helpful in many treatments, besides using plants, animal body parts, and their waste. The leech books have preparations like ointments, poultices, plasters, and fomentations. Animals were mixed with water, ale, or wine, and people breathed in the vapors. Animal parts were used to burn, and the patient sat over the smoke [3].

1.2.7 Early modern medicines

The introduction of printing in the fifteenth century gave people easy access to pharmacopeias by making them widely available. The *Pharmacopoeia Londinensis*, which came out in 1618, was the first to do this work for the entire country (England) instead of just a small area. Only a few doctors knew Latin well, so the texts were difficult to understand. Many books about herbs and medicines printed in English became more accessible, making it easier for people to learn about medicine. This was the first time people had learned about medication [3].

The tablets were introduced in the nineteenth century when a base made up of lactose was utilized to develop tablet triturate. To make tablets in a specific shape, metallic molds with a lower plate having pegs in it and an upper plate with holes in it were used. A hard paste of the ingredients was made and fixed into the plates and then allowed to dry. These tablet triturates were readily dissolved sublingually [1].

1.2.8 Novel dosage forms

One important development in the second half of the twentieth century was the introduction of dosage forms that either delayed the release of the drug they contained or else released it slowly over time. Tablets could be given an enteric coating to prevent their disintegration in the stomach if they irritate the lining, or be made from combining granules that would disintegrate at different rates in the same tablet to produce sustained-release tablets. In the same way, hard capsules could be filled with small drug pellets designed to dissolve at different rates to make a sustained-release capsule. Implants are pellets designed to be inserted under the skin where they gradually release a drug over a while. They have been used to administer steroids and sex hormones. Transdermal patches can now be applied to release drugs for pain relief, hormone replacement therapy, or to help reduce the craving for nicotine, the drugs being absorbed gradually through the skin [3].

1.2.9 The first dispensary

Before the days of the priestcraft, the wise man or woman of the tribe, whose knowledge of the healing qualities of plants had been gathered through experience or handed down by word of mouth, was called upon to attend to the sick or wounded and prepare the remedy. The art of apothecary originated from the preparation of medicines. The art of the apothecary has always been associated with mystery. The medical practitioners were believed to have a connection with the world of spirits and thus performed as intermediaries between the seen and the unseen. The belief that a drug had magical associations meant that its action, for good or for evil, did not depend upon its natural qualities alone. The compassion of a god, the observance of ceremonies, the absence of evil spirits, and the healing intent of the dispenser were individually and collectively needed to make the drug therapeutically effective. Because of this, the tribal apothecary was the one to be feared, respected, trusted, sometimes mistrusted, worshipped, and revered, for it was through his potions that spiritual contact was made, and upon that contact, the cures or failures depended [11].

Throughout history, the knowledge of drugs and their application to disease has always meant power. In the Homeric epics, the term "pharmakon" (Gr.), from which our word "pharmacy" was derived, connotes a charm or a drug that can be used for good or for evil. Many of the tribal apothecary's failures were probably due to irrational prescriptions, underdosing, overdosing, and even poisoning. Successes may be attributed to experience, mere coincidence of appropriate drug selection, natural healing, inconsequential effect of the drug, or placebo effects, that is, successful treatment due to psychological rather than therapeutic effects. Even today, placebo therapy with inert or inconsequential chemicals is used successfully to treat individual patients and is a routine practice in the clinical evaluation of new drugs, in which subjects' responses to the effects of the actual drug and the placebo are compared and evaluated [12].

As time passed, the art of the apothecary combined with priestly functions, and among the early civilizations, the priest-physician became the healer of the body as well as of the soul. Pharmacy and medicine were indistinguishable in their early history because their practice was the combined function of the tribal religious leaders.

Early Drugs: Because of the patience and intellect of the archaeologist, the types and specific drugs used in the early history of drug therapy are not as indeterminate as one might suspect. Numerous ancient tablets, scrolls, and other relics as early as 3000 BC have been uncovered and deciphered by archaeological scholars to the gratitude of historians of both medicine and pharmacy [13]. Perhaps the most famous of these surviving artifacts is the Ebbers papyrus, a continuous scroll some 60 feet long and a foot wide, dating to the sixteenth century BC. This document, now preserved at the University of Leipzig, is named for the noted German Egyptologist Georg Ebbers, who discovered it in a mummy's tomb and partly translated it during the last half of the nineteenth century. Since then, many scholars have participated in the translation of the document's challenging hieroglyphics, and although they are not unanimous in their interpretations, there is little doubt that by 1550 BC, the Egyptians were using some drugs and dosage forms that are still used today. The text of the Ebbers papyrus is dominated by drug formulas, with more than 800 formulas or prescriptions being described and more than 700 drugs mentioned. The drugs are chiefly botanical, although mineral and animal drugs are also noted. Such botanical substances are acacia, castor bean (from which we express castor oil), and fennel [2].

1.3 History of dosage form

Animal use in traditional folk medicine dates back to thousands of years, with ancient civilizations like the Egyptians, Greeks, and Chinese using various animal-derived substances for medicinal purposes. Early dosage forms included raw animal parts, such as organs, skins, bones, and secretions, which were consumed orally or applied topically.

1.3.1 Tinctures and extracts

Over time, traditional healers developed techniques to extract bioactive compounds from animal tissues, creating concentrated forms of medicine. Tinctures and extracts were prepared using methods like maceration, percolation, or distillation, resulting in potent remedies. The use and development of extracts and tinctures are discussed below in detail.

1.3.1.1 Herbal teas/mixtures/extracts

Tea's history is inextricably linked to that of botany and herbal medicine, as it originated in China and was considered to have therapeutic benefits. According to legend, the Chinese emperor Shennong, who was credited with inventing Chinese medicine, drank the first cup of tea around 2737 BC. Shennong was sitting beneath the shade of a *Camellia sinensis* tree, heating water to drink, when dried leaves from the tree floated into the pot, altering the color of the water. Shennong tested the infusion and was impressed with its flavor and therapeutic qualities.

A more catastrophic Indian tradition credits the Buddha with discovering tea. During a journey to China, he swore to meditate continuously for 9 years, but fell asleep. Outraged by his weakness, he severed his eyelids and tossed them to the ground. Where they fell, the first tea tree grew, with eyelid-shaped leaves.

Regardless of the reality behind the tales, tea has been an important part of Asian culture for millennia. The first known treatise on tea is "Ch'a Ching," or "The Classic of Tea," penned by the Chinese scholar Lu Yu. The book discusses tea's mythical roots, as well as its horticultural and medicinal virtues, and provides extensive instructions on tea preparation and etiquette. This was a highly regarded ability in China, and being unable to prepare tea effectively and elegantly was considered a disgrace. Tea was regarded as a therapeutic beverage until the late sixth century. Tea drinking became very popular during the Tang dynasty, which lasted from the seventh to the eleventh century. Different preparations evolved, with more oxidation yielding darker teas ranging from white to green to black. Other plant compounds were added, such as onion, ginger, citrus, and peppermint, with various infusions thought to have specific medical benefits. Tea was eventually used as a beverage rather than just for therapeutic purposes [14].

Tea was introduced to Europe in the late sixteenth century, during the Age of Discovery, a period of considerable abroad adventure. Natural philosophers found several new plants, which they gathered and employed as remedies or for human nourishment. Stimulant herbs like coffee, chocolate, tobacco, and ginseng piqued people's attention. Europeans learned about the therapeutic properties of plants from locals. However, Asians were skeptical that tea's curative virtues would benefit Europeans' health, arguing that the medicinal value was exclusive to Asians [15].

Portuguese traders were the first to return from China with tea (known to them as "Cha," from Cantonese slang). However, the Dutch were the first to commercially import tea, which swiftly gained popularity throughout Europe. Tea arrived in Britain in the seventeenth century, and its popularity may be traced back to Catherine of Braganza, a Portuguese princess and tea addict who married Charles II. Her love of tea made it popular at court and among the upper classes. For many years, due to hefty taxes, tea was only available to the rich. In the eighteenth century, an organized criminal network for tea smuggling and adulteration grew. Other plant leaves were utilized instead of tea leaves, and a convincing color was obtained by combining ingredients ranging from deadly copper carbonate to sheep's waste [16].

In Britain, tea was marketed as a medication when it was first brought. Thomas Garraway, the owner of Garraway's coffee establishment in London, said that tea would "make the body active and lusty" but also ". . . remove the obstructions of the spleen . . ." and was "very good against the Stone and Gravel, cleaning the Kidneys and Ureters." Cornelius Decker, a Dutch doctor, enthusiastically advised tea intake, advising eight to ten cups per day, and claimed to drink 50 to 100 cups himself. Samuel Johnson, another doctor noted for indulging in excessive tea drinking, was said to have eaten as much as sixteen cups at one tea party and was a staunch supporter of the health advantages of tea. In 1730, Thomas Short conducted several studies on the health effects of tea and published the findings, saying that it had healing capabilities for scurvy, dyspepsia, chronic dread, and sadness. However, the health effects of tea were debated, and by the mid-eighteenth century, allegations that tea was harmful to health were circulating. Wealthy benefactors were concerned that excessive tea consumption among the lower classes would promote weakness and depression. One French doctor cautioned that drinking too much tea would cause the body to overheat, resulting in disease and perhaps death. John Wesley, an Anglican preacher, denounced tea for its stimulating characteristics, claiming that it was damaging to both the body and the soul, causing a variety of neurological illnesses. Wesley even guided on how to handle the unpleasant scenario of having to decline an offered cup of tea [17].

Jonas Hanway, an English traveler, considered that tea consumption posed a risk to the nation, resulting in poor worker health. He was especially worried about the effect on women, claiming that it made them less attractive. Arthur Young, a political economist, opposed tea because of the time lost during tea breaks. He criticized the fact that some members of the working class would drink tea instead of eating a hot meal at midday, thereby reducing their nutritional intake: tea replaced the traditionally working-class drink of home-brewed beer, which had a higher nutritional value than tea; tea contains no calories because it lacks milk and sugar. Thomas Short, a Scottish doctor, stated that tea caused terrible illnesses and believed that people would spend more money on tea than food. In actuality, the working class frequently purchased extremely inexpensive grades of tea or utilized tea leaves from richer households.

Tea regained popularity because philanthropists recognized the importance of tea drinking in the temperance campaign and offered tea as an alternative to alcohol. Many coffee houses and cafes began in the 1830s as alternatives to taverns and inns, while tea rooms and tea shops gained popularity and fashion beginning in the 1880s. Today, tea is the most popular beverage in the world. Tea is estimated to account for 40% of the British population's daily fluid consumption. According to a study conducted at Harvard University Medical School, tea may have health benefits. Tea contains polyphenols, which are particularly abundant in green tea, and has anti-inflammatory and antioxidant properties that could prevent damage caused by elevated levels of oxidants, including damage to artery walls, which can contribute to cardiovascular disease. However, these effects have not been explicitly investigated in people, and it's possible that tea drinkers just live better lives. However, there is no solid proof that tea has any meaningful impact on health [18].

Herbal teas are infusions produced from an herbal blend of leaves, seeds, and/or roots of various plants and hot water [19], which are often used for their medicinal and energizing characteristics, including the ability to promote calm [20]. Herbal teas can aid with gastrointestinal troubles and enhance the immune system. Distinct plants may have unique medical effects; for example, herbal teas are well-known for their relaxing characteristics and ability to decrease blood pressure [19].

Since ancient times, people have been using herbs for experimentation and practice. Herbal tea is derived from the Latin word "herba," which means "plant." Herbs are referred to as "yao chao cha" or "hua chao cha" in Chinese, which means "herbal and flower tea." According to historical documents, the Sumerians were the first to employ plants 5,000 years ago. A thousand years later, records show that China and India were also employing herbalists. About 2,000 years ago, a classic Chinese herbal book, *Shen long Ben Chao Jing*, recorded 365 various herbs that the Chinese employed in their daily lives. One of the herbals listed was tea; however, the more current definition of herbals excludes tea's distinctive properties [21].

Most herbals are caffeine-free and lack the antioxidants EGCG and theanine found in tea. Humans have been utilizing herbs for millennia, discovering several that are beneficial and safe to take, gradually transitioning from eating them to drinking them like tea. Popular herbs include chrysanthemum, jasmine, mint, goji berries, osmanthus, rose, lavender, bitter melon, licorice, and "qian ri hong," a Chinese herbal that is included in one of our presentation teas. Herbs may be utilized in a variety of applications, including baths, salves, oils, and cookery. For this reason, herbals are quite popular all around the world [22].

Herbal tea is an essential aspect of tea culture since it complements tea consumption and is used to maintain health and provide healthcare. The green tea production method produces the majority of non-fermented teas. There are also semi-fermented (oolong tea) and fermented teas (black tea) [23]. Non-fermented tea is prepared from plant materials that are heated soon after harvest, mechanically coiled and compacted, and then dried to preserve the color and natural components [24, 25]. Semi-fermented tea is produced primarily by harvesting, drying, rolling, additional drying, screening, and baking [26]. Catechin is the primary biochemical component of fermented tea, accounting for roughly 20% of its dry weight. Condensation during fermentation by polyphenol oxidase converts catechins to theaflavins and thearubigins [27, 28]. The fermenting step of fermented tea manufacture affects its distinct scent and color, with theaflavins contributing to its orange-red color [29].

Non-fermented tea: Most herbal teas are non-fermented and made by picking, withering, blanching, rolling, and drying. Drying the delicate buds of *Castanopsis lamontii* yields herbal tea. This remedy is utilized in southwest China to relieve breath and prevent mouth irritation [30]. *Combretum micranthum* G. Don is commonly used in West African traditional medicine as a "long-life herbal tea" or "plant to heal." Traditional therapists commonly employ these herbs to treat renal disorders [31]. *Jasonia glutinosa* D.C. (Asteraceae), sometimes called rock tea, is a popular medicinal plant in the Mediterranean region. The flower stems are used to make herbal drinks that aid with digestive issues [32]. Chamomile tea is a highly popular herbal tea produced from the dried flower heads of Compositae plants (*Matricaria recutita* L., *Chamomilla recutita*

L., and *Matricaria chamomilla*). Traditionally, it is utilized for medicinal purposes due to its anti-inflammatory properties [33]. Yacon (*Smalanthus sonchifolius*) is an indigenous Andean plant farmed as a crop and used as a traditional medicine for diabetic, intestinal, and renal ailments [34].

Fermented tea: Fermented tea is made by selecting, withering, rolling, fermenting, and drying herbs, making up just a small percentage of herbal tea production. Rooibos is a famous herbal tea made from *Aspalathus linearis* plants native to South Africa. Fermented herbal teas include polyphenols, which have been linked to several health benefits [35, 36]. Honeybush tea (*Cyclopia intermedia*) is harvested, dried, and then fermented [37]. Vine tea, prepared from fermented leaves of *Ampelopsis grossedentata*, has been a traditional herbal tea and folk medicine in southern China for centuries [38]. Vine tea is processed similarly to traditional black tea manufacturing methods. Fermented tea is made by selecting, withering, rolling, fermenting, and drying herbs, making up just a small percentage of herbal tea production. Rooibos is a famous herbal tea derived from *A. grossedentata*'s fragile branches and leaves are kneaded into strips. After fermentation, they are fried and dried to form tea-shaped strips with a mild, sweet, and sour flavor.

1.3.1.2 Tinctures

Before tinctures, apothecaries were just shops for storing crude drugs like roots, stems, and seeds. Maceration was used initially to prepare the extracts, generally in acidic media, and mostly, vinegar was used as a solvent. But Raimundus Lullus (1235–1315) was the first to introduce the tincture in the history of pharmacy by preparing it through maceration of different drugs by using spirits. Paracelsus was known for the development and upgradation of dosage forms. Different colored tinctures were displayed in the front windows of the pharmacy for decoration purposes [39].

1.3.2 Solutions and colloids

Solutions, syrups, elixirs, emulsions, and suspensions are different forms of homogeneous liquid dosage forms that have been utilized long ago. Even today, this drug design has been extensively suggested for children and elderly patients because of the ease of administration and dosage adjustment according to weight or age. However, this dosage form needs a proper device that can measure the dose. Additionally, these formulations also need taste-masking agents or sweeteners, and preservatives.

Solutions are also used for Intravenous routes to overcome the difficulty of swallowing and gastrointestinal problems. Some solutions can also be used to fulfill the purpose of nutrition, either through the oral or parenteral route, depending on the condition of the patient. Sir Christopher was the first to introduce the Intravenous nutrition term by injecting wine. With the advancement of technology, new devices and more stable parenteral products are being developed to improve the bioavailability and efficacy of drugs. Some examples of solutions that serve the purpose of nutrition are electrolyte solutions (KCl and NaCl), micronutrient solutions (glucose and dextrose), and macronutrient solutions (fats, lipids, and proteins) [40].

1.3.3 Poultices

Our first thought when considering any injury is infection and its potential consequences. Antibiotics, which were first developed by Alexander Fleming in the 1920s, revolutionized the field of medicine by providing a dependable means of treating a wide range of bacterial infections. We are really lucky to have lived over the previous 100 years when a variety of medicines have been developed to prevent and cure various diseases. However, a new age is upon us, and with it, fewer and fewer alternatives. Since we relied on antibiotics, bacteria have evolved more swiftly than humans can create new ones, since they are basic creatures. Scientists are being compelled to review the methods we have employed historically as antibiotic resistance emerges as one of our most pressing contemporary issues. If a practice has endured for hundreds or thousands of years, it could represent a novel approach to management that offers us a safety net or substitute for the use of antibiotics.

An age-old treatment known as a poultice is said to pull out "pus" to clear infection or to calm and minimize inflammation. A poultice that had developed from centuries of recipes passed down via generations of families, farriers, religious healers, and herbal healers born without any understanding of microorganisms and aseptic procedures was recommended until the 1920s and the discovery of antibiotics [41].

At approximately 400 BC, Hippocrates started what we know Hippocrates, who lived around 400 BC, is credited with starting what is now known as modern medicine and physiology. Galen, who lived around AD 160, reinforced Hippocrates' ideas, which became fundamental to the management of wounds and abscesses. One of their fundamental conclusions was that a wound would exude and produce a sloughy discharge before healing; they independently concluded that a yellow discharge was a normal process critical to wound healing; in reality, they were observing the inflammatory phase of healing, the natural process where white blood cells arrive at the wound site and break down bacteria, devitalized tissue, and debris.

The two ancient physicians had different interpretations of these observations. After the dead and devitalized tissue was removed, the healing process would continue. He proposed that debridement and cleansing with boiling water could facilitate this discharge, which he described as a natural healing process that resulted in the formation of healthy tissue. Debridement is the physical removal of dead and devitalized tissue, akin to lancing an abscess and extracting its contents. Galen saw this procedure as well; however, he reached a somewhat different conclusion. He argued that wounds should be encouraged to excrete this purulent fluid because he thought that exudate and slough were actively promoting healing [42].

1.3.4 Clay minerals

One of the greatest libraries in history, dating from between 668 and 627 BC, was found in Mesopotamia in the middle of the 1800s. Almost 800 of the 30,000 cuneiform-written clay tablets that make up the library have medical themes, some of which are pertinent to plastic surgery. They handle congenital deformities and wound healing [43].

When used as active ingredients or excipients in pharmaceutical preparations, spa treatments, and beauty therapy medications, clay minerals can have positive effects on human health. Man has utilized clay for sustenance and healing since the beginning of time. There is more evidence supporting the usage of therapeutic earth in Mesopotamia and Ancient Egypt, including the use of Nubian dirt as an anti-inflammatory and the use of mud materials for corpse mummification. Lemnos Earth mud components were utilized as cicatrizers, antiseptic cataplasms, and remedies for snake bites during the time of the Ancient Greeks. Aristotle and Hippocrates were two of the scholars who created categories of therapeutic earth. The majority of these substances are clays, referred to by various names based on natural clays that have been utilized to mend skin diseases since the earliest written history. They could disclose an antimicrobial mechanism that offers a low-cost remedy for different skin ailments [44]. We were alerted to the therapeutic use of Fe-smectite-rich French green clay for the treatment of Buruli ulcer, a necrotizing fasciitis (or "flesh-eating" illness) brought on by *Mycobacterium ulcerans*. French green clays and other clays are commonly used for healing. "Healing clays" must be distinguished clearly from the clays that we have found to be antibacterial. Despite not being antibacterial, many clays' extremely adsorptive qualities may help treat a range of illnesses [45].

1.3.5 Ointments and salves

External applications of animal-based remedies led to the development of ointments, salves, and balms for treating skin conditions, wounds, and inflammations. Animal fats, oils, and extracts were combined with herbal extracts or other ingredients to create topical preparations. Ointments are oil-based, while creams are water-based topical preparations and are less greasy than ointments, e.g., cold cream.

A Greek physician, Galen, introduced the mixing of herbal drugs and other compounds to form a proper dosage form. He was also known as the "Father of Pharmacy." His most famous product was a cold cream, the ingredients of which were pretty similar to the ones used these days. Medicated plasters for local application on the skin were also seen in ancient Chinese medical history, and these were the predecessors of today's transdermal patches.

A Muslim Persian physician, Avicenna (Ibn Sina) (AD 980–1037), first used the principle that some drugs can be absorbed through the skin. In *The Canon of Medicine*, he suggested that the drugs used for topical applications can be either soft or hard. According to him, the soft part could be absorbed through the skin, while the hard part of the medication could not cross the skin barrier. He also explained that when the drug is absorbed through the skin, it is not only beneficial for local issues on the skin but also affects the underlying tissues and joints and can also be absorbed into the blood for systemic effects. Hence, this way he formulated many drugs for systemic effect that could not be administered orally. A plaster containing sulfur and tar was one of his famous topically used formulations, which was used to treat sciatica pain. The mixture was applied on the back of the paper and then applied to the skin.

1.3.6 Plasters (plaisters)

Plasters were used to increase the contact of the medicament to the skin to increase bioavailability. Medicated plasters were prepared by incorporating drugs into some bases, like resin, fat, or beeswax. The shape of plasters depends on the site of the application. These plasters were applicable after gently heating them in a water bath or flame. Some examples are Belladonna to stop milk secretion, cantharides for counter irritation on blisters, and mercury for inflamed joints or glands, while resin plasters were used to support joints or splints. In 1950 commercial supply of plasters was discontinued and has been replaced by patches like nicotine patches or estrogen patches.

1.3.7 Patches

The *German Pharmacopeia* (1872) listed 28 plaster formulations that used various plant-based adhesive materials along with medicaments. The purpose of these drugs was to produce systemic effects. For example, peppermint oil was used to treat stomach problems, and plasters obtained from leaves of *Atropa belladonna* were used to treat TB and cancers. However, there were also some systemic poison cases reported after the use of belladonna patches.

In the twentieth century, it was realized that the skin was permeable to lipid-soluble substances but not to water, as in the early 1900s, many systemic poisoning cases were reported because of topical substances. A strict check is needed to prepare and apply patches to the skin. Various topical preparations in the form of patches were ready for systemic absorption, for example, transdermal nicotine patches used for smoking cessation, nitroglycerine patches for angina treatment, scopolamine patches for motion sickness, clonidine patches for hypertension, and estradiol patches for female hormone replacement.

1.3.8 Tablets

The development of pills and tablets as dosage forms allowed for easier administration and standardized dosing in traditional folk medicine. Animal-derived ingredients were ground into fine powders and mixed with binders to form compressed tablets or coated pills.

Pills or tablets have been in use for thousands of years. In 4000 BC, medicines were prepared in liquid rather than solid form. The evidence of pills has been found on the papyrus in the Egyptian era, where different types of pills were made for ease of administration and to mask the bitter taste of the drugs. These pills were made up of medicinal ingredients mixed with bread dough, honey, or grease, and then small balls like tablets were made by hand. In ancient Greece, these pills were called *katapotia* (which means something to swallow), and by the Romans, these were known as *pilula,* from which we derived the present-day name, pill. As pills were hand-made, they had a large size, which made them difficult to swallow; therefore, efforts were made to make the process easier, and efforts have been made to make them easy to swallow, such as by coating them with slippery materials.

Sugar-coated tablets originated in France in the 1830s and quickly expanded to other parts of the world. Traditionally, sugar coating consisted of a mixture of gum and sugar, which led the pharmacists to develop various other coatings. In addition to sugar and gum, gold and silver leaf, magnesia, starch, licorice powder, lycopodium, and gelatin were also used for coating purposes, and later the list went on. One example of the coating was pearl coating, where the tablets were coated with talc, which made them look like pearls. Even though talc is water-repellant and insoluble in body fluids, this coating practice remained popular among the manufacturers [46].

William Brockedon invented a die to compress powders into the shape of tablets and lozenges [47]. Hence, in 1844, he made the first compressed tablets of potassium bicarbonate and sodium bicarbonate without excipients or adhesives. He took this idea from manufacturing pencil lead, where the lead was compressed. However, these tablets faced the issues of poor solubility and disintegration due to their hardness. Some other researchers used the same technique, but in 1872, John Wyeth and co-workers upgraded this compression machine to a hand-operated machine, which also reduced production costs [48].

1.3.8.1 Enteric coating tablet

Another method, which was started in the nineteenth century, was to cover/coat the tablets with gold or silver. However, this coating also protected the tablets from being dissolved in the GI tract. In the 1800s, sugar coating and gelatin coating were introduced along with gelatin capsules. In 1884, a German dermatologist, Unna, brought the concept of keratinized pills by coating the salicylic acid and iron chloride pills with keratinized reagent and demonstrated the stability of these pills against acidic medium and better bioavailability in the small intestine. The concept of enteric coating can be found in the pacific medical and surgical journal, demonstrating that collodion coating can save pills from the stomach's acidic environment [46].

In the nineteenth century, some ancient oral dosage forms were not considered pharmaceutical materials. The idea of coating pills came from Arabs who used gum mucilage as a coating material. While Avicenna used silver and gold for coating pills, this technique remained popular till the nineteenth century. Keratinized tablets remained popular till 1950; later, it was proved from studies that their effectiveness was being compromised. However, the site of disintegration for tablets coated with keratin was unpredictable [49]. Keratin was replaced by salol, phenyl salicylate, and a blend of salol and tolu to overcome the problems of keratin coating [50]. In 1938, studies discovered that salol hydrolyzed in the alimentary canal and phenol and salicylic acid by-products were formed, which had some biological actions. In 1935, it was revealed that tolu is indigestible and has no therapeutic effect [46].

Dr. Weyland, in 1897, invented enteric coating by hardening the capsule with formaldehyde and named it the "glutoid" capsule. In the nineteenth and twentieth centuries, the area of pharmacist interest was the development of the most efficient and effective dosage form. This struggle led to the discovery of almost 34 coating substances in the USA, like fats and fatty acids, cellulose nitrate, shellac, wax, and cellulose acetate coating in 1906, 1928, 1930, 1931, and 1937, respectively [49], while Germans in the 1930s were working on polymers that could be synthesized. Almost 34 enteric coating substances and 12 auxiliary substances were mentioned by Thompson and Lee in 1938. The 10th edition of *Remington's Practice of Pharmacy* had a list of enteric coatings like keratin, salol, and tolu; shellac, casein, and stearic acid, and synthetic compounds PVA (polyvinyl alcohol), CMC (carboxy methyl cellulose), etc. In the 13th edition of Remington, along with a list of coating materials, some references supporting and disproving the activity of coating materials were also mentioned. Sixty coating materials have references in their favor, while 24 coating materials have unsatisfactory references, but some materials have mixed references. For example, CAP (cellulose acetate phthalate) has 13 references in its favor, while 3 were against it and 5 had mixed illustrations [46].

These diverse references produced the need to develop a test to check the effectiveness of the coating material. In 1930, an old method was coating a tablet with a mixture of calcium sulfide, methylene blue, starch, and sugar. So, if the tablet cracked in the stomach, it would generate hydrogen sulfide, while intestinal activity on the coating material produced blue urine, and an inert coating would be eliminated unchanged. In the USA, Alec Williams used the coating technique to develop the time-release dosage form in the 1950s and 1960s [51].

1.3.8.2 Controlled-release dosage form

Pharmacists' interest in developing dosage forms created the need to expand their knowledge in the pharmacokinetic (PK) field of drugs to gain information on the site of action, drug release, drug elimination, and the effect of the biological environment on the drug's PK [52, 53]. Discoveries and knowledge in the field of biopharmaceutics helped prove the safety, efficacy, and efficiency of dosage forms. So, a dosage form or a treatment to be successful must have a therapeutic concentration in blood or tissues, and a longer half-life to be absorbed properly; similarly, proper disintegration and dissolution require particular excipients and particle size.

The concept of distribution and 100% bioavailability of the drug with the intravenous route of administration opened a new field to delay the dissolution of the drug until it reaches the basic media of the small intestine. Delivery of medicine from one state to another was also a challenge for pharmacists in 1958. To capture the market, businessmen of pharmaceutical industries indulged in new formulations of delaying, repeating, and sustaining the action or blood level of therapeutic entities. About 180 pharmaceutical products with delayed-release properties were introduced in the market in 1959, while only 20 products were from competent research industries. These novel formulations generated 4% revenue for pharmaceutical industries, which was almost 87 million dollars [54].

Timed-release dosage form was further divided into following classes; one was the drug that had two doses in single doses one dose released immediately after intake and the other after a specified time of interval, to reduce the frequency of dosing known as repeat action dosage form; second formulation was designed to release the active drug at a slow rate for prolonged action of the drug; Third dosage form had a dose to be released immediately to provide therapeutic effect following sustained-release pattern [55, 56].

Opium has historically been the primary ingredient in opioid medications. Since ancient times, opium has been derived from the *Papaver somniferum* L. plant, also known as the opium poppy [57]. The two most common applications of poppies in traditional medicine were as a sedative and an analgesic. Many innovative medicinal therapies with prolonged or sustained-release (SR) capabilities were introduced in the late 1940s and early 1950s. Israel Lipowski first developed medicines that reduced the frequency of administration for immediate-release dosage forms known as sustained-release medicines [58]. These medications are very helpful through the rectal route for those who cannot take medicines orally, such as infants, children, and those who cannot swallow drugs. Researching historical sources reveals the existence of similar treatments that contain opium for addicts, which were first developed by a Persian physician, "Imad," in the sixteenth century [59].

The purpose of the search for novel dosage forms was to find a solution for fluctuating concentrations of drugs in blood and to reduce the chances of overdosing/multiple dosing. During this era of pharmacokinetic properties development, pharmacists and chemists didn't synthesize new drugs or compounds but tried to change the formulation of drugs. Fantus in 1918 coated the granules to mask the taste of the drug. Afterward, Boyd Welin in 1939 used the coating concept and suggested that different thicknesses of coating would dissolve the drug at different rates, and the release of the drug could be controlled that way.

In an attempt to resolve the problem of oral dosage form, Smith, Kline, and French Laboratories developed the first Spansula with the help of three pharmacists and one biochemist. "Spansule" is patented by Rudolph Blythe. He used the technique of Lipowski and coated the drug in different thicknesses, in addition to an uncoated amount of drug, to release the dose immediately after administration. In 1949, the idea was put forward to put the medicine in honeycomb structures of cellulose derivatives that were water soluble and permeable to hydrophilic fluids; this idea was even older than coated beads in capsules. Then, in 1956, coated pills with a time-release mechanism were compressed to give the dosage form of a tablet rather than enclosed in capsules [46].

1.3.8.3 Multilayered tablet/bilayer tablet

Compressed tablets were upgraded to layered tablets in which the active drug blended with alcohols and waxes was placed in the center of the tablet, then the drug was released slowly while the outer layer had therapeutic moiety and excipients. A new advancement was introduced that was to put the medicament in multiple layers, and each layer would release the drug one after another. Then SFK prepared a tablet that used both techniques of sustained release and immediate release.

In 1953, a multilayered tablet was developed, where one layer contained the time-released agents, and the other provided the immediate release of the drug. Pills up to six layers were prepared to achieve long-term therapeutic effects with a single dose administration. In 1960, a tablet with three layers was introduced with an outer layer of tasteless material, which was released immediately in the oral cavity. Then there was a layer of sour taste which indicated to the patients to swallow the drug, and then there was a core made up of such material that is suitable to be absorbed in the gastrointestinal tract for a longer duration of action [60].

1.3.9 Capsules

Capsules were first described as a dosage form in a papyrus in about 1500 BC, where it has been written as an effective dosage form for medicines along with many others. The next reference for capsules was found in 1730, when de Pauli, a pharmacist, formed a capsule from starch to mask the bitter taste of the anti-gout medicine. After almost 100 years, in 1833, two French pharmacists applied for a patent for capsules that they called "bladders made of gelatin." This formulation was successful very soon, and many other pharmacists tried to improve the structure and palatability of the capsule over many years. Later in 1846, a two-piece hard gelatin capsule was formulated by Jules Cesar Lehuby. These capsules were cylindrical with two compartments that fitted on each other. The diameter of the two semi-cylinders was slightly different so that they could easily fit on each other [61].

In 1875, Parke, Davis, and Company manufactured these gelatin capsules on a large scale in the USA. The same idea was used to make a unit dose of cachet by mixing water and flour to make the top and bottom pieces of a drug container that can be digested. Stanislas Limousin, a pharmacist from France, introduced cachets by sealing the drug between wafers and then soaking it in water to make it a soft and easy-to-swallow product. Due to the difficulty in swallowing, lamels made up of gelatin were invented by Savory and Moore in 1870. Afterward, the shapes of these lamellae were modified into discs and square shapes.

1.3.10 Aerosols

Some traditional healing practices involved the use of animal parts in inhalations and fumigations for respiratory ailments and spiritual purposes. Animal-derived substances, such as resins, glands, or dried tissues, were burned or vaporized for their therapeutic or aromatic properties.

Asthma and chronic obstructive pulmonary disease (COPD) are now treated with inhalation therapy, which was first developed around a century ago. Other applications, including pulmonary hypertension, cystic fibrosis, vaccinations, and malignancies, have effectively used inhalation techniques in recent years. A medicine can be delivered to the lungs using one of three methods: nebulizers, dry powder inhalers (DPIs), or pressurized metered-dose inhalers (pMDIs). pMDIs are important, especially when quick delivery and effects are required; nevertheless, their use has recently been restricted due to the bans on propellant systems. Drops and pumps are used for nasal drug delivery because of their good bioavailability, fast absorption, drug stability, and reducing gastrointestinal adverse effects and first-pass metabolism in the liver [62, 63].

Since the mid-1990s, dry powder inhalation (DPI) systems have become increasingly significant, and they are currently the most widely used dose form for pulmonary disease prevention. DPI systems consist of tailored particles or an interactive powder blend that is administered via a device component. Cartridges (multidose-containing devices), blisters (monodose-containing devices), and capsules (also containing monodose) are used with the help of simple external devices. In both developed and developing countries, capsule-based DPI systems are becoming more and more common because of their low cost and ease of production [64].

1.3.11 Parenteral dosage form

The intravenous route of administration is almost more than 350 years old technique. In 1628, when William Harvey discovered the circulatory system, it gave insight into the new route to administer drugs and nutrients, which Christopher Wren initially trialed. In 1930, this mode of administration was developed, but clinically it was available in 1960 [65]. The parenteral feeding and blood transfusion idea was put forth by Lower in 1662 before the Royal Society [66]. Lower performed a successful experiment of transfusing sheep blood to a young man. William Courten followed the steps of Wren and Lower and injected 1 g/kg dose of olive oil into a dog, which died due to respiratory collapse more probably because of fat deposition in the lungs [67]. Modified dosage form was an idea that evolved from such failed experiments. In the nineteenth century, severe bleeding in patients with sepsis and outbreaks like cholera further valued the IV route of administration [68]. Latta was the first to effectively infuse the electrolytes in dehydrated electrolyte-deficient patients during an epidemic of Cholera (1831–1832). Edward Hodder took one step forward and tried to infuse the suspension form of fat, as in whole milk. Out of three participants, two patients survived while one died. Hodder concluded that the dose of milk was not sufficient for unfortunate patients. Later experiments showed that the infusion of milk had deleterious effects as well, and the method was painful, so abandoned after trials by other researchers [69].

In Vienna, Menzel and Perco substituted the IV route for fat administration with the subcutaneous route in dogs. Paul Friedrich (1904) performed the same successful trials on human beings, but later the nutrition infusion method was abandoned due to associated pain [70]. Fat emulsions were introduced by the US and Japanese researchers in the twentieth century to overcome the problems related to milk infusion [71]. The use of different proportions of fats and amino acids for longer periods resulted in an increase in weight and faster wound healing in infants, while rapid closure of fistula in adults. Although researcher from the USA corresponded to these experiments due to the unavailability of fats in their country, they substituted fats with others with high doses of micronutrients like amino acids, and glucose and named it the glucose system. The Swedish fat system was a combination of lipid emulsion and glucose. Once the fact that emulsions are safe for infusion and have promising nutritive and therapeutic effects was established, till now, emulsions have under the subject of new advancements to deal with the challenges of previous dosage forms. Nano-emulsions and microemulsions are new expansions of old two-phase liquid dosage form systems [72].

1.3.12 Modern dosage forms

With ongoing research, it has been identified that some medications/drugs do not work properly until they are fully bioavailable, efficient with fewer side effects, non-degradable by the gastric environment, and incompatible. The emergence of these issues led to the development of modified dosage forms that are more efficient, bio-available, and compatible. For example, coated tablets are formulated to bypass the stomach's acidic environment, and slow and sustained-release tablets/suspensions were formed to keep a constant level of drug in the plasma for an extended time. Delayed and extended-release dosage forms were used to reduce the dose frequency of the drug to increase medication adherence. Plasters are being replaced by patches in the modern era to overcome the side effects of the oral dosage form. Parenteral routes are now more pain-free and dosage forms with the highest bioavailability for patients who cannot take oral dosage forms.

Until 1950, the approved dosage forms were comprised of capsules, tablets, and liquid dosage forms like syrups, emulsions, elixirs, and suspensions. A tremendous transformation of dosage forms took place in the last 70 years. The first 12-h sustained-release tablets were approved by the US Food and Drug Administration in 1952. To improve medication safety and efficacy as well as patient adherence, the FDA has approved several innovative dosage forms, including digital dosage forms, liposomes, polymeric nanoparticles, digital dosage forms, pressurized metered dose inhalers (pMDIs), transdermal patches, microparticles, and drug-device combination products [73, 74].

During the drug development phase, the choice between developing a unique formulation and a standard dosage form is decided based on many considerations, including the physicochemical and biological qualities of the drug candidate. The drug's safety and efficacy must be guaranteed by the delivery method. For example, if prepared in traditional dose forms like tablets, capsules, or solutions, poorly soluble pharmacological compounds like cyclosporine would not result in the intended therapeutic response. Therefore, cyclosporine is sold as a microemulsion in the commercial sector [75].

A proper dosage form development is necessary: a) to achieve maximum therapeutic effects from the drug; b) to minimize the difficulty of administration; c) to control the dosage of a given medicine; d) to decrease the chances of side effects; and e) to target specific organs or areas. The history of dosage forms started with the evolution of humans, and newer innovative dosage forms are being added continuously.

References

[1]   Jackson WA From electuaries to enteric coating: A brief history of dosage forms. Making Medicines: A Brief History of Pharmacy and Pharmaceuticals. 2005:203.

[2]   Allen L, Ansel HC Ansel's pharmaceutical dosage forms and drug delivery systems: Lippincott Williams & Wilkins; Philadelphia, PA, USA., 2013.

[3]   Jackson WA, Anderson S From electuaries to enteric coating: A brief history of dosage forms. Making medicines: A brief history of pharmacy and pharmaceuticals: Pharmaceutical Press London; London, UK., 2005.

[4]   Nunn JF Ancient Egyptian medicine: University of Oklahoma Press; Norman, Oklahoma, USA, 2002.

[5]   Žuškin E, Lipozenčić J, Pucarin-Cvetković J, Mustajbegović J, Schachter N, Mučić-Pučić B, et al. Ancient medicine-a review. Acta Dermatovenerologica Croatica. 2008;16(3):149–157.

[6]   Metwaly AM, Ghoneim MM, Eissa IH, Elsehemy IA, Mostafa AE, Hegazy MM, et al. Traditional ancient Egyptian medicine: A review. Saudi Journal of Biological Sciences. 2021;28(10):5823–32.

[7]   Leake CD The old Egyptian medical papyri: Lawrence, Kan., University of Kansas Press; Lawrence, Kansas, USA., 1952.

[8]   Santacroce L, Bottalico L, Charitos IA Greek medicine practice at ancient Rome: The physician molecularist Asclepiades. Medicines. 2017;4(4):92.

[9]   Chen P, Xie P. History and development of traditional Chinese medicine: IOS Press; Amsterdam, Netherlands, 1999.

[10]   Zuo J, Park C, Kung JYC, Bou-Chacra NA, Doschak M, Löbenberg R Traditional Chinese medicine "pill", an ancient dosage form with surprising modern pharmaceutical characteristics. Pharmaceutical Research. 2021;38:199–211.

[11]   Kay E A history of herbalism: Cure, cook and conjure. 2022. p. 224.

[12]   Loudon IS The origins and growth of the dispensary movement in England. Bulletin of the History of Medicine. 1981;55(3):322–42.

[13]   Masic I, Skrbo A, Naser N, Tandir S, Zunic L, Medjedovic S, et al. Contribution of Arabic medicine and pharmacy to the development of health care protection in Bosnia and Herzegovina-the First Part. Medical Archives. 2017;71(5):364.

[14]   Franklin-Jeune S More than just a cup of tea: Florida Atlantic University, Honors College; Jupiter, Florida, USA, 2012.

[15]   Weir TT. Superfoods: Cultural and scientific perspectives: Springer; Cham, Switzerland, 2022. p. 141–55.

[16]   Zhang LT. Moral foods: The construction of nutrition and health in modern Asia. University of Hawai'i Press; Honolulu, 2020. pp. 201–220.

[17]   Bond TJ The origins of tea, coffee and cocoa as beverages. Teas, Cocoa and Coffee: Plant Secondary Metabolites and Health. 2011:1–24.

[18]   Khan N, Mukhtar H Tea and health: Studies in humans. Current Pharmaceutical Design. 2013;19 (34):6141–47.

[19]   Ravikumar C Review on herbal teas. Journal of Pharmaceutical Sciences and Research. 2014;6(5):236.

[20]   Zielinski AAF, Haminiuk CWI, Alberti A, Nogueira A, Demiate IM, Granato D A comparative study of the phenolic compounds and the in vitro antioxidant activity of different Brazilian teas using multivariate statistical techniques. Food Research International. 2014;60:246–54.

[21]   Sun M, Shen Z, Zhou Q, Wang M Identification of the antiglycative components of Hong Dou Shan (Taxus chinensis) leaf tea. Food Chemistry. 2019;297:124942.

[22]   Cooper R, Morré DJ, Morré DM Medicinal benefits of green tea: Part II. Review of anticancer properties. Journal of Alternative & Complementary Medicine. 2005;11(4):639–52.

[23]   Heck CI, De Mejia EG Yerba Mate Tea (Ilex paraguariensis): A comprehensive review on chemistry, health implications, and technological considerations. Journal of Food Science. 2007;72(9):R138–R51.

[24] Knor F, Vellosa J, Beltrame F, Pereira V Determination of phenolic compounds and antioxidant activity of green, black and white teas of Camellia sinensis (L.) Kuntze. Theaceae Revista Brasileria de Plantas Medicinais. 2014;16:490–98.

[25] Turan N, Kennedy JF Phytochemicals as bioactive agents. Carbohydrate Polymers. 2002;47(4), 394

[26] Yang J, Liu RH The phenolic profiles and antioxidant activity in different types of tea. International Journal of Food Science & Technology. 2013;48(1):163–71.

[27] Asil MH, Rabiei B, Ansari RH Optimal fermentation time and temperature to improve biochemical composition and sensory characteristics of black tea. Australian Journal of Crop Science. 2012;6 (3):550–58.

[28] Wilson T, Temple NJ Beverage impacts on health and nutrition: Springer; Cham, Switzerland, 2016.

[29] Samanta T, Cheeni V, Das S, Roy AB, Ghosh BC, Mitra A Assessing biochemical changes during standardization of fermentation time and temperature for manufacturing quality black tea. Journal of Food Science and Technology. 2015;52:2387–93.

[30] Gao Y, Wang J-Q, Fu Y-Q, Yin J-F, Shi J, Xu Y-Q Chemical composition, sensory properties and bioactivities of Castanopsis lamontii buds and mature leaves. Food Chemistry. 2020;316:126370.

[31] Kpemissi M, Potârniche A-V, Lawson-Evi P, Metowogo K, Melila M, Dramane P, et al. Nephroprotective effect of Combretum micranthum G. Don in nicotinamide-streptozotocin induced diabetic nephropathy in rats: In-vivo and in-silico experiments. Journal of Ethnopharmacology. 2020;261:113133.

[32] Valero MS, Berzosa C, Langa E, Gomez-Rincon C, Lopez V Jasonia glutinosa DC (rock tea): Botanical, phytochemical and pharmacological aspects. Boletin Latinoamericano y del Caribe de Plantas Medicinales y Aromáticas. 2013;12(Supl 6):543–57.

[33] McKay DL, Blumberg JB A review of the bioactivity and potential health benefits of chamomile tea (Matricaria recutita L.). Phytotherapy Research: An International Journal Devoted to Pharmacological and Toxicological Evaluation of Natural Product Derivatives. 2006;20(7):519–30.

[34] Sugahara S, Ueda Y, Fukuhara K, Kamamuta Y, Matsuda Y, Murata T, et al. Antioxidant effects of herbal tea leaves from yacon (Smallanthus sonchifolius) on multiple free radical and reducing power assays, especially on different superoxide anion radical generation systems. Journal of Food Science. 2015;80(11):C2420–C9.

[35] Arries WJ, Tredoux AG, de Beer D, Joubert E, De Villiers A Evaluation of capillary electrophoresis for the analysis of rooibos and honeybush tea phenolics. Electrophoresis. 2017;38(6):897–905.

[36] Zhang K-X, Tan J-B, Xie C-L, Zheng R-B, Huang X-D, Zhang -M-M, et al. Antioxidant effects and cytoprotective potentials of herbal tea against H 2 O 2-induced oxidative damage by activating heme oxygenase1 pathway. BioMed Research International. 2020; 2020:7187946.

[37] McKay DL, Blumberg JB A review of the bioactivity of South African herbal teas: Rooibos (Aspalathus linearis) and honeybush (Cyclopia intermedia). Phytotherapy Research: An International Journal Devoted to Pharmacological and Toxicological Evaluation of Natural Product Derivatives. 2007;21 (1):1–16.

[38] Ye L, Wang H, Duncan SE, Eigel WN, O'Keefe SF Antioxidant activities of Vine Tea (Ampelopsis grossedentata) extract and its major component dihydromyricetin in soybean oil and cooked ground beef. Food Chemistry. 2015;172:416–22.

[39] Raubenheimer O History of maceration & percolation. American Journal of Pharmacy. 1910;82:32–42.

[40] Shamsuddin AF Brief history and development of parenteral nutrition support. Malaysian Journal of Pharmacy (MJP). 2003;1(3):69–75.

[41] Lambert WG The twenty-one "Poultices". Anatolian Studies. 1980;30:77–83.

[42] Doehne E, Schiro M, Roby T, Chiari G, Lambousy G, Knight H, editors. Evaluation of poultice desalination process at Madame Johns' Legacy, New Orleans. Proceedings of the 11th international congress on deterioration and conservation of stone, 15–20 September 2008, Torun, Poland; 2008.

[43] Mazzola RF, Foss CB Plastic surgery: An illustrated history: Springer Nature; Cham, Switzerland, 2023.

[44] Stojiljković ST, Stojiljković MS, editors. Application of bentonite clay for human use. Proceedings of the IV advanced ceramics and applications conference; Paris, France, 2017: Springer.

[45] Williams LB, Haydel SE Evaluation of the medicinal use of clay minerals as antibacterial agents. International Geology Review. 2010;52(7–8):745–70.

[46] Helfand WH, Cowen DL Evolution of pharmaceutical oral dosage forms. Pharmacy in History. 1983;25(1):3–18.

[47] Bueno AG, Nozal RR Innovation vs. tradition: The election of an European way toward pharmaceutical industrialisation, 19th-20th centuries. Anales de la Real Academia Nacional de Farmacia. 2010;76(4):459–478.

[48] Kebler L The tablet industry–its evolution and present status–the composition of tablets and methods of analysis. Journal of the American Pharmaceutical Association. 1914;3(6):820–48.

[49] Cook EF, Martin EW Remington's practice of pharmacy (10th edition). 1951. p. 1615.

[50] Tbompsont HO, Lee C History, literature, and theory of enteric coatings. Journal of the American Pharmaceutical Association. 1945;34(5):135–38.

[51] Williams A Sustained release pharmaceuticals, 1969: William Andrew; Norwich, New York, USA, 1969.

[52] Wagner JG History of pharmacokinetics. Pharmacology & Therapeutics. 1981;12(3):537–62.

[53] Mohsin S, Ahmed S, Ahmed I Sustained release of captopril from matrix tablet using methylcellulose in a new derivative form. Latin American Journal of Pharmacy. 2011;30(9):1696–701.

[54] Lazarus J, Cooper J Absorption, testing, and clinical evaluation of oral prolonged-action drugs. Journal of Pharmaceutical Sciences. 1961;50(9):715–32.

[55] Notari RE Biopharmaceutics and pharmacokinetics: An introduction. Marcel Dekker; New York, 1975.

[56] Aslam Z, Akhter KP, Ahmad M, Aamir MN, Naeem M, Ali MY Preparation of modified-release tramadol tablets and drug release evaluation using dependent and independent modeling approaches. Latin American Journal of Pharmacy. 2012;31(10):1417–21.

[57] Armstrong SC, Wynn GH, Sandson NB Pharmacokinetic drug interactions of synthetic opiate analgesics. Psychosomatics. 2009;50(2):169–76.

[58] Park K Controlled drug delivery systems: Past forward and future back. Journal of Controlled Release. 2014;190:3–8.

[59] Soleymani S, Zargaran A A historical report on preparing sustained release dosage forms for addicts in medieval Persia, 16th century AD. Substance Use & Misuse. 2018;53(10):1726–29.

[60] Mcdermott CB Multi-layered pill or tablet with indicating lamination. Google Patents; 1960.

[61] Podczeck F, Jones BE Pharmaceutical capsules: Pharmaceutical press; London, UK, 2004.

[62] Brox W, Zande H, Meinzer A Soft gelatin capsule manufacture. EP Patent 0649651; 1993.

[63] Djupesland PG Nasal drug delivery devices: Characteristics and performance in a clinical perspective – A review. Drug Delivery and Translational Research. 2013;3:42–62.

[64] Siew A Wearable injection devices address delivery challenges. BIOPHARM INTERNATIONAL. 2018;31 (4):10–13.

[65] Darn H, Mark R The architecture of Christopher Wren. Scientific American. 1981;245(1):160–75.

[66] Keynes G The history of blood transfusion. Blood Transfusion. 1949:1–40.

[67] Courten W, Sloane H, II. Experiments and observations of the effects of several sorts of poisons upon animals, &c. Made at Montpellier in the Yeats 1678 and 1679, by the late William Courten Esq; communicated by Dr. Hans Sloane, RS Secr. Translated from the Latin MS. Philosophical Transactions of the Royal Society of London. 1712;27(335):485–500.

[68] Olah K, Keane D Abell I Professional attainments of Ephraim McDowell. Bulletin of the History of Medicine 1950; 24: 161–67. 2. Abramovits A. Joseph Asherman, 60th anniversary celebration. Harefuah 1950; 38: 11. 3. Abramson DJ. Charles Bingham Penrose and. Ginecologia (Bucharest). 1976;24:97–8.

[69] Hodder E Transfusion of milk in cholera. Practitioner. 1873;10(14):14–16.

[70] Menzel A Ueber die Resorption von Nahrungsmetteln von Unterhautzellgewe. Wiener Medizinische Wochenschrift. 1869;25:753.

[71] Yamakawa S, Nomura T, Fujinaga I Zur Frage der Resorption des emulgierten Fetts vom Dickdarm aus. The Tohoku Journal of Experimental Medicine. 1929;14(2–3):265–74.

[72] Preeti SS, Malik R, Bhatia S, Al Harrasi A, Rani C, et al. Nanoemulsion: An emerging novel technology for improving the bioavailability of drugs. Scientifica. 2023;2023(1):6640103.

[73] Lee PI, Li JX Evolution of oral controlled release dosage forms. Oral Controlled Release Formulation Design and Drug Delivery: Theory to Practice. 2010:21–31.

[74] Ali MY, Tariq I, Sohail MF, Amin MU, Ali S, Pinnapireddy SR, et al. Selective anti-ErbB3 aptamer modified sorafenib microparticles: In vitro and in vivo toxicity assessment. European Journal of Pharmaceutics and Biopharmaceutics. 2019;145:42–53.

[75] Rahman Z, Xu X, Katragadda U, Krishnaiah YS, Yu L, Khan MA Quality by design approach for understanding the critical quality attributes of cyclosporine ophthalmic emulsion. Molecular Pharmaceutics. 2014;11(3):787–99.

Uzma Saher, Muhammad Yasir Ali, Zermina Rashid, Hammad Afzal,
Muhammad Hammas, Abu Bakar Asghar, Nisar ur Rahman,
Saeed Ahmad

2 Physicochemical and pharmacokinetic basis of dosage form

2.1 Introduction

The dosage form can be described as the physical form of a drug or chemical compound in which it is manufactured or administered for medicinal purposes in a particular amount, usually called a dose. Commonly available dosage forms in markets
are tablets, pills or capsules, pure powders or solid crystals, liquid syrups, drinks or
suspensions, liquid aerosols or inhalers, liquid injectables, or some plant-based herbal
products. Whatever the dosage form used, the administration route depends on the
type of the dosage form [1]. The physicochemical properties, such as aqueous solubility, lipophilicity, and physical and chemical stability of substances, affect their pharmacokinetics and fate in our bodies [2]. Hence, for developing a stable dosage form
and more effective drug delivery with improved therapeutic efficacy, it is important
to thoroughly understand the physicochemical properties of the active pharmaceutical ingredients (APIs) and pharmaceutical excipients and their interactions with the
drug's pharmacokinetics.

2.2 Physicochemical basis of dosage form

"Physicochemical" refers to the combination of physical and chemical properties.
From the development of a formulation until its administration to the patient, the
physicochemical properties of the API and the excipients play a pivotal role. Both
physical (such as particle size of the drug, its surface area, and partition coefficient)
and chemical (pH value and solubility) properties play significant roles in the production process of the pharmaceutical formulation and the biological performance of the
drugs so keen consideration is required to make the formulation remain stable in
whole stages of drug life [3, 4]. The physical properties, such as the particle size, significantly affect the powders' flowability during production. Similarly, chemical reactions can lead to the loss of the API or excipients in a dosage form and resulting in the
production of toxic substances that can be harmful if administered to the patient. The
prior knowledge of any such possible chemical reactions occurring during and/or
after the production of a pharmaceutical product is crucial to prevent the loss of the
product's integrity and for the patient's safety [5].

https://doi.org/10.1515/9783111438108-002

2.2.1 Physical characteristics

2.2.1.1 Solubility

Solubility is a solute to dissolve in a single solvent or a mixture of two or more solvents. Both the solute and solvent can be a solid, liquid, or gas; however, the solvent is found in higher proportion volumetrically as compared to the solute in a solution, but some exceptions do exist. Usually solute is solid and the solvent is liquid [6].

In pharmaceutical drug development solubility of the drug is one of the key aspects as it decides various aspects of the dosage form and its biological performance. Although the oral route of drug administration is the most preferred one, owing to higher patient compliance and lower production cost, the bioavailability of the orally administered drugs is the main challenge. Drugs need to be solubilized in the surrounding media prior to absorption across the membranes to the systemic circulation [7, 8].

Among all the solvents, water is found most useful solvent in pharmaceutical product development. Water is considered an attractive solvent due to its non-toxic, biocompatible, and easily accessible nature. Due to its high dielectric constant and ability to dissolve a wide range of ionizable substances, water can be an excellent solvent for many hydrophilic drugs. However, this property can also have some disadvantages and restrictions because it can also dissolve a variety of undesirable impurities found in pharmaceutical products. Another major concern is the possibility of microbial growth when water is used as a solvent in pharmaceutical formulations because it provides a favorable medium for many microorganisms that could harm the components of the product or have toxic effects when administered to a patient. However, there are a number of strategies to mitigate this issue, such as adding a suitable preservative to the pharmaceutical formulation to stop microorganisms from growing in the media [9].

Even though water is a desirable solvent for use in pharmaceutical products, many substances or drugs have low solubility in water. This makes it challenging to formulate these substances in aqueous media because they may precipitate even in the presence of little solvent evaporation, pH changes, or other variables. Several techniques, including the use of cosolvents, pH regulation, the use of solubilizing agents, and the modification of molecular structure, can be employed to increase the solubility of such medicinal compounds [10].

Effect of pH: Since many drug compounds have the characteristics of weak acids or bases, the pH of the solution plays an important role in estimating how soluble the medication is in the solvent. The Henderson-Hasselbalch equation can be used to predict a substance's solubility at a given pH if the pK_a of the weak acid or base is known:

$$pH = pK_a - \log\frac{[HA]}{[A^-]}$$

where [HA] is the protonated form of the substance at the given pH, and [A⁻] is the concentration of the deprotonated form of the substance at the given pH. Therefore, half of the substance will exist in the protonated form and the other half in the deprotonated form when the pH is equal to the pK_a of the material. In general, for a weak base, when the pH of the solution is below the pK_a of the weak base, it causes the weak base to ionize, resulting in high solubility in the solvent. Conversely, for a weak acid, if the pH of the solution is above the pK_a of the weak acid, it causes the weak acid to exist in an ionized state, which has a higher solubility as compared to the unionized state.

Although adjusting the pH of the solution solubility of a drug substance in a pharmaceutical formulation can be improved, there are several limitations. For instance, the stability of the API and the excipients is significantly affected by the pH. Research has shown that variations in pH levels can initiate chemical reactions in solutions, leading to modifications in the molecular structure of many drugs. As a result, the ideal pH needed for the drug to be stable in the solution is frequently different from the optimum pH needed to have maximum solubility of the drug substance in the solution [11].

Another drawback of using pH control to increase solubility is that many drug formulations must have their pH values compatible with the pH of the application site. For instance, for ocular preparations, a large pH difference between the pH of the ocular membrane and the pharmaceutical formulation meant to be applied on the membrane can damage or irritate the ocular membrane. Similarly, the pH of the pharmaceutical formulations meant to be administered parenterally should be appropriately adjusted to prevent any irritation or damage owing to extreme pH variation with that of the site of the application.

The pH of the formulation also impacts the absorption and bioavailability of the formulation. Sometimes, the pH value, which is ideal for the appropriate solubility of the drug in solution, may affect the absorption and other attributes of the drug. Therefore, pH of the pharmaceutical must be carefully and properly controlled to achieve the best solubility without affecting the stability of the formulation, absorption, and incompatibility at the site of its application [12, 13].

Effect of cosolvents: The term "cosolvent effect" describes the potential for dissolving a material with low solubility in aqueous solvents by adding another solvent having miscibility with the aqueous solvent. Although the use of cosolvents in pharmaceutical formulation can greatly boost the solubility of drugs or substances that are poorly soluble in a given solvent, there are a number of toxicity-related restrictions. In general, ethanol, glycerol, and propylene glycol are the most often used solvents with water since they have a solid safety record and can improve the solubility of a variety of drugs that would otherwise have low solubility in water alone [14–16].

Effect of solubilizing agents: The solubility of poorly aqueous soluble substances can be increased by adding solubilizing agents to solutions [17]. Surface-active agents, or surfactants, are the most commonly used solubilizing agents. By forming different types of micelles – structures with both a hydrophobic and a hydrophilic portion – surfactants can cause a variety of substances to dissolve in aqueous solutions. The hydrophilic part of the surfactant molecules, pointing outward, interacts with the surrounding aqueous medium to enable the hydrophobic component to be solubilized and dissolved in the solution through the micelles. The hydrophobic substance is hidden within the inner hydrophobic part of the surfactant molecules.

Spherical-shaped micelles are the most commonly created type of micelle structure, while other, more complicated liposomes can also form. It has also been demonstrated that liposomes and related structures have a greater capacity to address problems associated with a drug's low water solubility [8]. Numerous more cutting-edge nanocarriers have been created to improve the drug's solubility and enable site-specific delivery of the medication to the affected body part [16, 18–20].

The physical characteristics of solid APIs and excipients employed in pharmaceutical formulations have particular importance due to their significant effect on the production process of the pharmaceutical formulation and the biological performance of the drugs. Solid-state materials can generally be classified as crystalline or amorphous based on the arrangement of their molecules [1, 21, 22].

2.2.1.2 Crystal habit

Substances are known to be present with a highly defined, ordered, and repeatable arrangement of atoms and molecules in their internal structure, called as crystal habit. Weak intermolecular forces, such as van der Waals contacts, ionic bonds, and hydrogen bonding, hold the molecules or atoms together [23]. During crystallization, substances transform from a liquid into a solid called a crystal. The material to be crystallized is typically present in a solution as a solute. Crystallization occurs in a supersaturated solution when the solute is added to the solution until it reaches the maximum amount that can dissolve; at that point, the solute precipitates and crystals form [24]. For making a solution supersaturated to make crystals, the amount of solvent is evaporated so that the solute concentration increases relative to the solvent [25].

2.2.1.3 Polymorphism

Different substances can crystallize in different ways, causing the crystals to take on distinct behaviors. The stirring pattern, the solvent used in the crystallization process, and the kind of impurities that exist in the crystallization medium are a few examples of conditions that can change and cause rearrangement of the molecules and atoms in different orientations. Repetition in the organized structures is also seen in the crystal's varied molecular and atomic arrangements. The term "ability to crystallize into different polymorphs" refers to substances that show this potential to have different crystals based on the parameters of the crystallization process [26, 27].

Polymorphism is a very challenging characteristic of substances, and it can affect the bioavailability of drug formulation. Many substances, including API and excipients, exhibit polymorphism, which is why this property should be controlled and characterized; otherwise, significant consequences can result [28]. For instance, an antibacterial drug, chloramphenicol palmitate, exhibits two polymorphs having different rates of absorption after administration. Polymorph B is more bioavailable as compared to polymorph A, so formulation is made keeping in view this character also [28].

2.2.1.4 Hydrates

Some of the molecules of solvent get trapped during the crystallization process within the crystals. When the solvent is water, the resultant crystals are known as hydrates, and if any organic solvent is used instead of water, then the crystals are called solvates [29–31].

Such substances that are crystallized in the form of hydrates are named according to the number of water molecules trapped within their crystals, like monohydrates (one water molecule with the substance), dihydrates (two water molecules), and so on [32]. Hydrates have different rates of dissolution as compared to the same substance without water, known as an anhydrous form. Because this entrapped water effect substantially affects the properties of the substance [33]. Generally, hydrates have a slower rate of dissolution as compared to anhydrous forms because the trapped water molecule makes a hydrogen bond with a lattice of crystals, making it difficult to dissolve and providing strength to its structure [34].

Sometimes it happens that hydrates exhibit more solubility as compared to the anhydrous form because there are fewer sites to interact with water during the process of dissolution. A classical example of such drugs is theophylline hydrate which shows a faster solubility than the anhydrous form. Here, the water molecules trapped within the crystal structure either disrupt or weaken the hydrogen bonds of a substance in the crystal lattice, thus weakening the crystal structure [35, 36].

2.2.1.5 Solvates

Solvates are also termed pseudopolymorphs. As discussed earlier, how solvates are formed in the crystallization process by the organic solvents, which are the major types of solvents usually used in drug formulations. The only difference between solvates and hydrates is the prediction of the number of molecules of solvents trapped in crystals is unpredictable or difficult to predict. The reason is the ratio of solvates to crystallizing substance is not a whole number as in the case of hydrate for instance it may be less than 1 but whatever the ratio is it should be under consideration during the formulation development process because residues of organic solvents may impact the safety of drug formulation, or it can increase the side effects of drugs [37, 38].

Solvates also affect the rate of dissolution of drug formulation in the same way as hydrates do, because whether it is unpredictable or difficult to know the exact number of solvent molecules in the crystallization process still impacts the strength of the crystal lattice, especially during dissolution; if it is stronger, it will be hardly solubilized. Consequently, the bioavailability of the drug is affected by using different solvates, and so is the case with the absorption rate of the drug [39]. Glibenclamide, a well-known antidiabetic drug isolated as solvates of pentanol and toluene, exhibits faster solubility and dissolution rate than the anhydrous solvate forms [37, 40, 41].

2.2.1.6 Amorphous solids

When molecules of a solid substance are not arranged in a repeating ordered structure, it is referred to as an amorphous solid. Generally, due to the rapid solidification process amorphous state is formed rather than a crystalline state. However, the crystalline solid can be transformed into its amorphous form by applying mechanical or thermal energy. The amorphous form of a particular substance has higher aqueous solubility, higher dissolution rate, and lower chemical stability in comparison to its crystalline counterparts [42, 43]. Moreover, amorphous substances have no sharp melting points and exhibit a glass transition temperature (T_g). When the temperature is below T_g, the material exhibits a glassy state where it becomes brittle, and when the temperature is above Tg, the material is in a rubbery state where it exhibits flexible properties. The T_g of the substance can be lowered by the addition of plasticizers, which enhance the mobility of the molecules [44, 45].

2.2.1.7 Particle size

It is crucial to reduce the size of solid components while developing drugs and formulating the existing ones. Reduced particle size has an impact on a variety of solid properties, including bulk density, powder flow, and a host of other properties during formulation development. It also has an internal impact on the body, where it increases the rate of dissolution and, ultimately, increases drug absorption and bioavailability by increasing the drug's surface area [46–48]. Even medications with low water solubility benefit from size reduction, as it speeds up the dissolution process. For instance, oral griseofulvin has low bioavailability due to its low water solubility; however, after particle size reduction, griseofulvin's surface area increased, increasing its bioavailability. However, it may cause an increase in the toxicity of certain drugs, for instance, reduction in the particle size of nitrofurantoin leads to increased bioavailability, leading to an instant rise in toxicity of the drug [49]. The filling procedure used in the manufacturing of tablets and capsules is also greatly impacted by the particle size. For example, filling machines employ volumetric filling to control the amount of powder when preparing tablets or capsules. This influences the consistency of the drug content [50]. Thus, particle size plays a vital role in both the production process and the drug's behavior inside the body; therefore, it is necessary to have a thorough knowledge and understanding of the drug's particle size.

Many techniques have been developed for the analysis of particle size; the majority of these techniques are used to calculate the equivalent diameter of the particles, which is a hypothetical diameter and is meant to approximate the true diameter of a particle. The irregularly shaped particles are often considered approximate spheres, and the equivalent diameter is the diameter of these hypothetical spheres. Therefore, depending on the method used to measure it, a particle may have more than one equivalent diameter. Each method used in particle size study has advantages and limitations of its own, as well as a range of particle diameters that can be estimated [4, 35, 38]. Apart from that, the field of medicine is currently shifting toward goods based on nanotechnology, where particle size is crucial [51]. It is a fact that the reduction of the particle size to the nanoscale significantly changes its physicochemical characteristics, which in turn modifies its biological function. Numerous medication delivery carrier systems based on nanotechnology are said to function better in biological systems than in traditional systems [52, 53].

2.2.1.8 Wettability

Wettability is the term used to describe the tendency of a liquid to adhere to the surface of a solid material. When developing pharmaceutical goods, a substance's wettability is crucial since it influences various biopharmaceutical processes, such as the drug's dissolution. The contact angle (θ), which is the angle formed by a solid's surface and the liquid spreading across it, is used to determine a substance's wettability. The amount of wettability of a substance can be determined by looking at its θ value. Values of θ near 180° show low or no wetting of the substance, whereas values near 0° indicate a significant or complete wetting of the substance [41]. The wettability of a solid material to a liquid is determined by the forces of attraction between and among molecules. When the attractive intermolecular interactions between the molecules of the solid substance and the molecules of the liquid are equal to or larger than the intramolecular forces between the liquid molecules, spreading of the liquid and complete wetting is preferred, and θ degree is observed. On the other hand, when the intramolecular forces between the molecules of the liquid are noticeably greater than the intermolecular forces between the molecules of the solid substance and the molecules of the liquid, the spreading of the liquid will be an unfavorable process, and high θ values will be observed [54, 55].

2.2.1.9 Hygroscopicity

Hygroscopicity is the ability of a substance to absorb and hold onto moisture from the surrounding atmosphere. According to their propensity to absorb water from their surroundings, solid materials differ in their hygroscopicity. A variety of intermolecular interactions, including ion-dipole interactions and hydrogen bonding with particular functional groups found in the solid material, can cause water molecules to become adsorbed on its surface [56]. The stability and efficacy of pharmaceutical products can be considerably impacted by the presence of water molecules resulting from the inclusion of a hygroscopic component in the formulation. This is because water can hydrolyze a variety of common functional groups present in several medication molecules. Therefore, it's critical to identify hygroscopic substances in pharmaceutical formulations and choose appropriate excipients to eliminate or reduce any potential negative effects of water molecules on the medication [57, 58].

2.2.2 Chemical characteristics

The chemical attributes of any drug are crucial in the development process of any pharmaceutical product. The chemical reactivity of the drug dictates how stable it will be in a given dosage form. Because drug degradation in pharmaceutical products can result in a reduction in the potency and therapeutic benefits of the drug, it is necessary to characterize the potential for unfavorable reactions that a drug can experience [59]. Drug deterioration can also result in pharmaceutical items that are aesthetically unappealing since the product may alter in color or smell as a result of the drug's chemical breakdown. Therefore, the chemical stability of the pharmaceutical formulation should be rigorously evaluated and verified to prevent the degradation of the drug or the generation of dangerous chemicals as a result of chemical reactions in the pharmaceutical formulation [60].

The API in the pharmaceutical formulation comes across several sources that may induce chemical degradation of the drug at different stages. For example, the water present in the surrounding atmosphere can cause several functional groups of drugs to hydrolyze. Similarly, oxygen present in the air can oxidize the API or the excipients, compromising the integrity of the product. In addition, light is considered another source that can degrade a variety of drugs; thus, precautions must be made when handling and storing these pharmaceuticals. Chemical deterioration can also be caused by variations in the pH of the solution since these changes can set off several reactions that would not ordinarily occur.

The reactions triggered by pH shifts can be intramolecular; for instance, the acidic pH can enable an intramolecular reaction to take place in penicillin and yield inactive compounds. The pH variations can also potentially induce an interaction between the API and an excipient, or between two or more excipients, that normally does not occur under the formulation's original pH. Since the pH greatly varies between different sections of the gastrointestinal tract, it is important to understand how gastrointestinal pH can affect a drug's chemical stability when administered orally. For example, many drugs can degrade in the extremely low (acidic) pH of the stomach's media or the intestine's alkaline media. It is crucial to understand the drug's chemical stability at various pH levels [61].

It should be considered that a given drug undergoes many reactions that lead to the API's degradation, which complicates the analysis. The effects and mechanisms of the various possible reactions that a drug may experience are covered in the sections that follow.

2.2.2.1 Hydrolysis

Hydrolysis is described as a reaction between a substance and water, which results in the break of a bond in the substance by employing a water molecule. A hydrolyzable drug can contain a variety of functional groups, such as lactones, amides, esters, and amides. Different functional groups hydrolyze at different rates. For instance, due to electronic effects, ester hydrolyzes at a rate that is typically higher than amides with identical structures. The rate of hydrolysis is also influenced by steric effects. Large substituents surrounding the amide or ester functional groups have been shown to significantly slow down the hydrolysis rate. Esters hydrolyze to produce the appropriate carboxylic acid and alcohol compounds. One such anesthetic agent is procaine, which hydrolyzes to produce these chemicals [62]. Amides can be hydrolyzed to produce the corresponding carboxylic acid and amine compounds; this process can affect both acyclic and cyclic amides or lactams. Procainamide, cinchocaine, and acetaminophen are a few examples of drugs with amide functional groups. All of these pharmaceuticals can hydrolyze at their amide bond and produce the corresponding carboxylic acid and amine [63]. The rate of amide hydrolysis is increased by raising the electrophilicity of the carbonyl carbon; therefore, the presence of electron-withdrawing groups on the alpha (α)-carbon atom is thought to be one of the key factors regulating the rate of amide hydrolysis. Steric hindrance surrounding the amide linkage is also a significant issue since hydrolysis of more hindered amides is less likely than that of less hindered amides. Cyclic amides (lactams) are also highly susceptible to hydrolysis. Among the antibiotics with a lactam ring are those in the beta-lactam class, which hydrolyze at even faster rates than acyclic amides. Since these cyclic structures have more ring strain than acyclic or less strained amides, they have higher overall reactivity and a higher rate of hydrolysis, which is thought to be the cause of this high rate of hydrolysis [64].

2.2.2.2 Oxidation

Oxidation, another very frequent drug degradation process, usually proceeds with the molecular oxygen found in air. The reaction may occur in the long-term storage of the medicine or during the process of manufacturing. A drug's ability to oxidize is mostly dependent on the presence of functional groups that are susceptible to oxidation. For instance, methyldopa, an alpha-adrenergic receptor agonist, can be oxidized to create the corresponding quinine metabolites [38, 65].

2.2.2.3 Photochemical degradation

Another typical degradation pathway for many drugs is photochemical degradation. In comparison to other degradation pathways like hydrolysis or oxidation, the photochemical breakdown of medicines is thought to be far more complex. In general, multiple degradation products are produced from the same drug substance due to photochemical degradation. Additionally, the presence of oxygen might enable the process to continue via distinct routes, producing diverse compounds. The photochemical reactions are more unpredictable and can occur during the manufacturing, storage, or administration of the medication. The calcium channel blocker nifedipine is an example of a drug that degrades photochemically; in the presence of light, it undergoes a dehydrogenation reaction [66–68].

2.2.2.4 Isomerization

A chemical reaction in which a stereoisomer or a constitutional (geometrical) isomer of a material transforms into another isomer is known as an isomerization reaction. The isomers (including stereoisomers) might have significantly different biological activity and toxicity profiles, therefore it is crucial and discover the pharmacological compounds that can undergo isomerization reactions under that can undergo isomerization reactions under various conditions [69]. Tetracycline is a drug that can undergo epimerization in an acidic environment; this process entails the interconversion of the stereogenic center at carbon 4 to produce the 4-primer of tetracycline, which is not only inactive but can also have harmful effects [70].

2.3 Pharmacokinetic basis of dosage form

Pharmacokinetics is the combination of two Greek words, drug (*Pharmakon*) and its movement (*kineticos*), so pharmacokinetics deals with the study of the movement of the drug inside the body starting from its absorption, distribution, metabolism and ending on excretion, as a whole, we can use the word "ADME" of a drug is its kinetics. In addition to ADME, pharmacokinetic studies also involve the toxic, corresponding pharmacologic, and therapeutic responses in the living body, either man or the animal (American Pharmaceutical Association, 1972). Pharmacokinetics research encompasses various aspects such as bioavailability assessments, the impact of physiological and pathological circumstances on drug distribution and absorption, drug interaction analysis, clinical forecasting, and the utilization of pharmacokinetic parameters to customize drug dosage and subsequently deliver optimal drug therapy [71].

2.3.1 Absorption

The process by which a drug moves from the point of administration to the measurement site (often blood, plasma, or serum) is known as absorption. When developing new medicines, one of the primary goals of the pharmaceutical industry is to maximize the bioavailability of drugs, especially the ones taken orally, because the drugs taken parenterally reach the bloodstream instantly, so they can be easily bioavailable. At the same time, there are many hurdles for orally administered drugs to reach the bloodstream and so at the site of action [72]. Bioavailability measures the amount of the active ingredient absorbed from the pharmaceutical form and the rate at which it enters the bloodstream. This suggests that the molecule must pass through one or more cellular membranes to enter the bloodstream. The amount of medication that enters the general circulation through the oral route depends on numerous variables. It has long been believed that a drug's physicochemical characteristics primarily influence its oral bioavailability.

Physicochemical properties may include the molecular weight, the pK, and the lipophilicity as characterized by the octanol/water partition coefficient (log P). However, it is now understood that focusing solely on these factors is overly straightforward. In fact, for several compounds, there is no relationship at all between log P(o/w) and the proportion absorbed or oral bioavailability [8, 16, 73]. When taking a drug orally, the amount that enters the bloodstream naturally depends on how much is absorbed by the intestinal epithelium, but there are additional variables that might affect the drug's bioavailability. These include the drug's solubility [72], which highlights the formulation's significance, the pH in the gastrointestinal tract, gastric emptying, gastrointestinal transit, and other interactions like those with food or mucus. Furthermore, there are some scenarios in which the product is excreted from the body even before reaching systemic circulation, and this happens by gastrointestinal lumen degradation and/or enterocyte metabolism during the passage. Most of the compounds have high bioavailability by taken orally, but some drugs have low bioavailability and/or significant interindividual heterogeneity in their absorption. When administering them orally, e.g., anticancer drugs, get poorly absorbed after oral administration. Similarly, therapeutic peptides are not highly bioavailable when taken orally; however, in this instance, this is due to significant metabolism in the small intestine and stomach lumen. The pharmaceutical industry has several critical goals, including lowering bioavailability variability, generating new oral medications, and increasing drug bioavailability to overcome these problems. To accomplish these goals, one must have a solid understanding of the physiological and molecular mechanisms involved in drug absorption, in addition to knowledge of the physicochemical characteristics of the drugs. Therefore, it is critical to have accurate predictive models since it is valuable to ascertain the mechanism of the intestinal absorption of possible drugs early in the drug development procedure. These models ought to provide quantitative data on absorption mechanisms, the impact of enhancers on absorption, the function of formulation, and studies on metabolism or preferred targeting to the best absorption sites. These models include not just in situ and in vivo models, but also in vitro ones.

2.3.1.1 Factors affecting drug absorption

The oral bioavailability of medications can be affected by extrinsic or intrinsic variables, which include physiological and physical aspects. There is a physical order to the way the active substance dissolves and its dosage form disintegrates. A drug's ability to dissolve and become soluble largely depends on its formulation, which affects this process in several ways. Some of them may include the active ingredient's chemical nature, particle size, and crystal structure, as well as the characteristics of the coatings and excipients that are used in the formulation. Drug dissolution is an essential step; regardless of the mucosal absorption pathway, a medication must dissolve to be absorbed; hence, for poorly soluble pharmaceuticals, the dissolution phase may be the limiting factor [73, 74]. Because pH affects both the ionization of weak bases and the breakdown of solid forms, it also plays a complex role in gastrointestinal absorption. Most of the drugs are weak acids or weak bases, and unless transport carriers are present, only the non-ionized fraction – that is, the most lipophilic form – crosses biological membranes. Additional physicochemical interactions, such as a drug's contact with mucus, also contribute to reducing bioavailability.

Mucus is a complex, viscous gel that coats the gastrointestinal tract's epithelial surface and is secreted by ciliated cells. Mucus's physicochemical characteristics affect how well drugs are absorbed. Its primary roles include preventing tissue water loss to preserve cellular integrity, shielding the mucosa from mechanical harm and foreign objects, and acting as a lubricant to facilitate the passage of intestinal contents. Mucus may also affect how well a medicine absorbs, since the first obstacle the active ingredient must overcome to reach the mucosa is mucus. Through the formation of ionic or hydrogen bonds, mucus can slow down or limit the absorption of specific molecules. The intestinal transit and stomach emptying are two physiological processes that affect absorption. The length of time a medication is in contact with the gut surface will influence its overall absorption. In turn, intestinal transit and stomach emptying will have an impact on this. The rate of stomach emptying might vary depending on several factors. Drugs such as metoclopramide, moderate physical activity, being in a prone position, and hunger are all known to speed up emptying. Some drugs, for instance, amphetamines, morphine, anticholinergics, and other drugs, as well as other circumstances, including emotion, pain, and strenuous activity, all reduce the stomach emptying time [75]. Substances that slow down peristalsis can also affect intestinal transit, extending the period that medications stay in the intestine and facilitating absorption. Conversely, elements that increase transit result in the opposite effects. Food and other extrinsic variables might also affect absorption. Certain physiological changes are brought about by eating, such as an increase in biliary secretion or a decrease in stomach emptying. Certain foods can change how drugs work in the pharmacokinetic system. For instance, calcium alters how first-generation tetracyclines are absorbed. It has been discovered more recently that ferrous salts reduce the absorption of numerous antibiotics. However, some medications – like fenofibrate – show enhanced absorption when taken with food. The concurrent administration of another medication has the potential to alter the effects of this one as well. There are various reasons why a drug coadministered with another medication may alter the drug's absorption. First, in the gastrointestinal lumen, this connection causes a change in the drug's diffusible form. Tetracycline with divalent or trivalent metallic derivatives could produce a non-absorbable complex. When two medications compete with one another for the same transport route, the absorption of both chemicals is sometimes reduced (phenytoin, for example, decreases the absorption of folic acid). As long as the solubility is adequate, the most significant rate-limiting factor is probably the absorption at the gut mucosal level, despite all these factors that may influence the bioavailability of a medication taken orally.

2.3.1.2 Mechanism of drug absorption

Size and hydrophobicity – that is, the speed at which tiny lipophilic substances move through the intestinal mucosa – are frequently thought to be the primary determinants of drug transport across this membrane. This is somewhat accurate, but it presents an oversimplified picture of how molecules move through intricate biological barriers. There are four ways that a drug can pass through the layer of epithelial cells [1]: Diffusion, which occurs passively across membranes. This is the pathway that low molecular weight lipophilic medicines are most likely to take [2]. Carrier-mediated transport, where small hydrophilic compounds use a carrier that is a saturable method that involves a particular interaction between the molecule and the transporter or carrier. There are two different types of carrier-mediated transport: passive and active. Bifunctional carriers are coupled to either Na^+ or H^+ to facilitate active carrier-mediated transport. This mechanism can operate against a substrate concentration gradient and needs metabolic energy. Conversely, substrate concentration drives passive carrier-mediated transport since it doesn't need metabolic energy. Certain drugs use these carrier-mediated pathways to enter cells. Antibiotics, e.g., cephalosporins and β-lactams, are the best-documented examples, which serve as the dipeptide transporter's substrates when propelled by protons. The carrier-mediated systems also facilitate the absorption of peptide mimetic ACE inhibitors, including captopril [3].

Endocytosis: For the uptake of macromolecules, the majority of mammalian cells exhibit this constitutive mechanism. Except for some specialized cell types, such as reticuloendothelial system cells, it requires metabolic energy and is often a gradual absorption mechanism that fuses endocytotic vesicles with lysosomes that have high levels of enzymatic activity. Certain receptors may be involved in endocytosis; vitamin B_{12} is absorbed by this method of absorption. It is also the process by which the intestinal epithelium may absorb large peptides and other macromolecules [4]. The paracellular transport/pathway is another way that molecules can get beyond the gut barrier. Drug absorption through paracellular transport, in which hydrophilic molecules traverse tight junctions and move between cells, is of great interest. Substances absorbed through this paracellular pathway may be amenable to modulation by absorption enhancers and should not be broken down by intracellular enzymes. Several products can "open" tight junctions, improving molecule absorption through the paracellular pathway. It should be highlighted, though, that the tight junctions only make up 0.1% of the intestine's surface area. P-glycoprotein, a transmembrane protein, is also present in intestinal epithelial cells. It is known to give tumor cells multidrug resistance by expelling a variety of drugs through energy [76]. Many normal tissues have been shown to contain the P-glycoprotein (P-gp) pump, and the small intestine's epithelium is one of its most abundant locations. P-gp is almost exclusively found on the brush border of the enterocyte's apical surface of the intestine, where it is responsible for pumping foreign substances back into the intestinal lumen from the cytoplasm. P-gp functions as a "flipase-like" drug efflux pump. This secretory drug-efflux pump is more likely to detect compounds with physicochemical characteristics that permit penetration across the apical and basolateral membranes. P-glycoprotein causes problems with drug absorption and is partly responsible for the low and inconsistent bioavailability of several medications. Several drugs, including cyclosporin, certain peptides, digoxin, fluoroquinolones, ranitidine, cytotoxic medications, and P-adrenoreceptor blockers, are secreted into the intestinal tract by P-gp-like pathways. P-gp may be more important than previously thought in regulating the oral bioavailability of numerous medication types [77].

2.3.2 Distribution

The reversible transport of an unmetabolized drug to and from the measurement location is known as distribution (typically blood or plasma). Any drug that disappears from the measuring location and does not reappear has been eliminated. Drug distribution in various parts of the body determines its efficacy or toxicity, and due to drug distribution in tissues, we do not find a direct relation between the plasma concentration of the drug and its observed effects. Drugs do not get distributed equally in all types of tissues, like the brain, muscles, and fat tissues; it all depends on their molecular structure and how much the drug gets distributed in them. Among all organs, the brain and testes have membrane barriers that make them special tissues because these barriers remarkably decrease the chances of drug distribution [78]. Drugs can be categorized as lipophilic or hydrophilic, based on their solubility in lipids and non-lipids; this factor also affects their distribution. Different other factors also affect the rate and extent of drug distribution some of them are drug-related for example tissue and/or organ perfusion by blood, how much the drug is bound to plasma proteins or tissues, and the permeability of membranes to allow the drug to pass through them while some are body-related for instance body water composition, fat composition, and the ailments body is suffering, these all factors decide the distribution of drugs. The physicochemical characteristics and chemical structures (i.e., the existence of functional groups) of a pharmacological molecule also affect the distribution of the drug.

2.3.3 Metabolism

The process of changing one chemical species into another by specialized enzymatic systems to get excreted from the body is called metabolism. Another term, "biotransformation," is also used in place of metabolism, which mostly occurs inside the liver. Some drugs are administered as their prodrug types, and they must be metabolized first to exhibit their therapeutic activity. Metabolites typically have little to no activity of the parent drug [79]. There are certain exceptions, though. Drugs that have therapeutically active metabolites include: Procainamide (Procan; Pronestyl) is an anti-dysrhythmic drug; *N*-acetyl procainamide is its active metabolite. Desmethyldiazepam is the active metabolite of 4-hydroxy propranolol diazepam (Valium), which is used to treat anxiety and tension symptoms. Propranolol HCl (Inderal) is utilized as a non-selective beta-antagonist. Metabolism of a drug can occur before its absorption, as given in the following discussion, or post-absorption metabolism can be seen in some drugs.

2.3.3.1 Preabsorption metabolism

The small intestine can metabolize drugs in different ways, even though most of the drugs taken orally get absorbed through the small intestine, which is why it is typically considered an absorptive organ for many drugs. Drugs can also be metabolized even before their absorption by the gut mucosa, mainly from two places. The first one is the lumen of the small intestine, and the second one is the brush border of the intestine. The lumen is a key location for amylases, lipases, peptidases, and carboxypeptidases. The pancreas provides the intestine with five peptidases: carboxypeptidases A and B, chymotrypsin, elastase, and trypsin, which are endopeptidases. These enzymes function cooperatively and are always present in mixed form. The luminal enzymes play a critical role in terms of quantitative analysis. Quantitative estimation of pancreatic enzymes produced differs and exhibits individual variation, but is expressed in grams. These enzymes are responsible for the instant metabolism of protein-nature drugs, and, for this reason, most of the time during formulation development, this factor is kept under consideration to minimize obstacles of enzymatic degradation of orally administered therapeutic peptides. Additionally, the lumen may contain enzymes from the gut flora, particularly in the lower part of the GI tract (large intestine). The bacterial inhabitants of the human gastrointestinal tract can have an impact on drug metabolism. Humans contain very few bacteria in the proximal intestine, but 10^{10} bacteria have been found in the large intestine, and 10^6 to 10^7 bacteria per gram in the distal small intestine. The field of microbially controlled drug delivery to the colon is experiencing a surge in research interest due to the realization that the enzymes of bacteria present in the human colon can hydrolyze prodrugs and other compounds into active therapies. The brush border of the intestine, which is the surface of the intestinal epithelium cells and is thought to have an area of more than 200 m^2, is the second site of preabsorptive metabolism. It is home to several peptidases, alkaline phosphatase, and sucrase. These brush border resident enzymes are responsible for the low bioavailability of peptide drugs. The ileum typically has lower brush border enzyme activity than the duodenum and jejunum, while the colon has very low membrane enzyme activities. It is easy to assess a drug's presystemic metabolism in vitro by culturing the molecule in various gastrointestinal preparations. These are prepared by homogenizing the tissues in buffer and washing away the contents of the stomach, intestinal tract, or colon with buffer at the pH appropriate for the activity of the enzyme. These media can be used to incubate drugs so that the molecule's breakdown and/or the metabolites' appearance can be assessed. These kinds of studies have produced valuable insights into the preabsorption metabolism of drugs.

2.3.3.2 Intracellular metabolism

Certain drugs' first-pass effect (metabolism before absorption in the blood) in the gut wall, particularly due to the action of the microsomal enzyme CYP3A4, may contribute to their low and variable oral bioavailability. The cytochrome P450 enzymes (phase-I enzymes) are mainly produced in the liver but are also present in extrahepatic organs like the intestine. It is now widely known that the cytochrome P450 isoform CYP3A4 exists in the human intestine and that it is the most prevalent kind of enzyme. It is absent from crypt cells and is exclusively present in the mature enterocytes that line the villi. Human enterocyte-derived CYP 3A4 seems to be physically and functionally identical to human liver-derived CYP 3A4. Although the CYP3A5 protein was recently discovered in enterocytes, its role in first-pass metabolism is yet unknown. The human duodenum has also been found to contain CYPlAl. According to recent findings, many of the same medications that interact with CYP3A4 have limited bioavailability because of the P-glycoprotein (P-gp) in the enterocyte brush boundary. P-glycoprotein and CYP3A4 may be functionally related, as it appears that the majority of drugs that are substrates for CYP3A4 are also transported by P-gp. It is true that P-gp and CYP3A4 colocalize in the small intestine, and that their genes are next to one another on the same chromosome. Oral bioavailability appears to be limited by their synergistic actions and coordinated regulation. According to Hunter and Hirst, the drug's reabsorption and P-gp recirculation out of the cell provide more access to the restricted amount of CYP3A4; as a result, a larger percentage will be metabolized, indicating a lower absorption of the drug [80]. There are several possible explanations for how P-gp affects the degree of CYP3A4 metabolism. Primary metabolites will be eliminated from within the enterocyte if CYP3A4-generated metabolites are superior P-gp substrates than the parent molecule. This will stop CYP3A4 from performing secondary metabolism and speed up the rate of primary metabolism. Phase I enzymes may play a significant part in detoxification and are also present in the gastrointestinal tract. Glutathione-*S*-transferase is present in the human small intestine in both the glutathione-*S*-transferase π and α forms, with the gastrointestinal tract having a higher concentration of the enzymes than the liver. The gut has the highest concentration of sulfotransferase activity among extrahepatic locations, and its acetyltransferase activity is half that of the liver [77].

2.3.4 Excretion

Excretion is the permanent elimination/removal of a drug from the body in its unmodified or unchanged chemical form. If the parent molecule is hydrophilic, otherwise. If the drug is lipophilic, it must be biotransformed into hydrophilic metabolites by the liver or any other suitable place to be excreted from the body. The liver and kidneys are the two main organs in charge of drug disposal. The kidney is the main organ where drugs that have undergone chemical changes are eliminated (also known as excretion), along with their metabolites. The main organ used for drug metabolism is the liver. Periodically, the lungs could play a significant role in the removal of high vapor-pressure chemicals (such as alcohol and gaseous anesthetics). Mother's milk is another possible medication elimination. It's not a major way for the woman to get rid of a drug, but it can still be taken in large enough doses to have an impact on the baby.

2.3.4.1 Renal excretion

When it comes to eliminating medicines and their metabolites, the kidneys are the most crucial organ. For 25% to 30% of medications, the primary route of elimination is renal excretion of unaltered pharmaceuticals. The three modes of renal excretion – glomerular filtration, tubular secretion, and passive reabsorption – make the kidneys the primary excretion pathway [81]. Unbound drugs get excreted through glomerular filtration (highly bound medications, including non-steroidal anti-inflammatory drugs, are not eliminated via glomerular filtration). Many drugs get excreted by tubular secretion, especially in cationic forms (like cimetidine) and anions (such as NSAIDs, penicillins, cephalosporins, and glucuronic acid conjugates). Net passive reabsorption occurs in the renal tubules for the non-ionized forms of weak acids and bases. The pH affects the passive reabsorption of weak electrolytes because the ionized versions of these substances are less permeable to the tubular cells. Weak acids are mostly ionized and eliminated more quickly and completely as the tubular urine becomes more alkaline. On the other hand, weak acid excretion and fractional ionization will decrease when the urine becomes more acidic. For weak bases, the effects of varying urine pH are the opposite. When treating drug poisoning, the urine may be appropriately alkalized or acidified to speed up the elimination of certain medications. For instance, using sodium bicarbonate to adjust the urine's pH is one way to treat aspirin poisoning [82]. Lipophilic drugs are usually passively reabsorbed usually the amount of tubular reabsorption can be significantly impacted by the pH of the urine. Drugs that are more lipophilic or less polar must first undergo metabolization to a more polar or water-soluble substance to be excreted.

2.3.4.2 Biliary excretion

In humans, biliary excretion plays a significant role in the removal of several drugs, especially lipophilic drugs and their metabolites. Lipophilic drugs can be biotransformed in the liver and then re-enter the bloodstream. After that, they can either be removed by bile or returned to the kidneys for excretion into urine. The gallbladder is where medications that are actively transferred from the liver into bile are kept. Drug excretion through stool may occur if the gallbladder releases bile into the GI tract. As an alternative, free drugs may be reabsorbed via the intestinal mucosa and reach the portal vein, which returns the drug to the liver, once the bile has reached the gut lumen. Enterohepatic recirculation is the term for this mechanism. Drug-related factors, including chemical composition, polarity, and molecular size, as well as liver-related factors like particular active transport sites inside liver cell membranes, are the elements that dictate clearance through the biliary route. It is possible for a drug that is excreted by bile to be reabsorbed from the digestive system or for a drug conjugate to be hydrolyzed by gut bacteria and release the original drug into the bloodstream. Enterohepatic circulation prolongs the pharmacological effects of both drugs and their metabolites; however, in humans, this phenomenon seems to have less quantitative significance than in other species. The interindividual variations in medication response seen in both healthy persons and patients with specific disorders may be related to biliary elimination. Drug elimination through this pathway will be influenced by cholestatic illness situations, where bile flow is normally lowered. This will raise the risk of drug toxicity. In cases of renal failure, bile may function as a different pathway for elimination; however, human research has not yet confirmed this. Because of the relative inaccessibility of the human biliary tract, there is a dearth of trustworthy knowledge addressing the biliary excretion of medicines in humans. Most research on medication excretion in human bile has been done on patients who have had surgery and are receiving drainage through a T-tube. This approach to bile collection is not optimal since enterohepatic circulation is partially disrupted, bile flow and composition are frequently drastically changed over the study period, and not all bile is collected [83].

2.3.4.3 Half-life of drugs

A substance's half-life is the amount of time or length required for its concentration in the body to decrease to half of its initial dose; this usually relates to the drug's elimination half-life. Knowing the half-life of a drug makes it easier to calculate its steady-state concentrations and excretion rates. Although the half-lives of drugs differ from each other, some drugs remain in the body for a longer time while others stay for a short time, but the one thing they all follow is: After one half-life has elapsed, 50% of the initial drug amount is excreted from the body. Simple differential equations can illustrate the well-established topic of pharmacokinetics, which is the study of how a drug's properties vary over time. First-order pharmacokinetics, in which drug clearance rates are proportionate to plasma concentrations, is often observed in the majority of therapeutically significant medicines. Certain medications, such as ethanol, on the other hand, exhibit zero-order elimination, in which the drug's concentration declines over time by a constant amount. It is also worthwhile to talk about the connection between the quantity of half-lives and the percentage of drugs removed. First-order pharmacokinetic medications all have specific quantitative constants that apply to all drugs, assuming a physiologically healthy individual, no drug-drug interaction, and no administration of extra drug after an initial dosage. For example, 90% of a given drug will have been eliminated after roughly 3.3 half-lives. Furthermore, after four to five half-lives, 94% to 97% of a drug will have been removed. It follows that a drug will be deemed removed when its plasma concentration falls below a clinically relevant concentration, which typically occurs after 4 to 5 half-lives. On the other hand, during an infusion, drug accumulation may achieve a steady state. After around 4 to 5 half-lives drug finally achieves a steady-state level when administered at regular intervals or in a consistent dosage (such as an infusion) and does not further accumulate in the body with repeated administrations. This situation results in achieving a balance between the rate of clearance and rate of infusion, consequently steady-state level gets approached in the body. The dosage, dosing interval, and clearance all affect the steady-state concentration's value. Despite being one of the first pharmacokinetic characteristics to be discussed in the medical community, half-life still causes difficulty for many doctors, including experienced ones. Numerous presumptions are required to understand the idea of half-life, such as the drug being metabolized by a single compartment, the system being completely first-order and free of any hepatic or renal limitations, and the system being isolated and devoid of any alternative metabolic pathways or drug-drug interactions. In a clinical context, this is rarely the case with patients who present with chronic renal disease or other illnesses, and who may be taking multiple medications that could interfere with one another. Additionally, the age of the patient plays a crucial role in estimating the precise half-life of a medication, especially in cases involving children and the elderly, because the metabolism of drugs in the elderly and children are very different from a middle-aged man and consequently half-life of drug or stay of the drug inside the body also varies greatly in this population. Half-life is a highly theoretical paradigm, making it difficult to apply in real-world situations and apply it to clinical decision-making as a tool. Consequently, to ensure safe and efficient pharmaceutical administration, medical professionals and students must take these realities into account when calculating half-lives. Several research works have endeavored to develop approaches that consider these subtleties in the treatment of illness depending on the unique pharmacokinetic characteristics of medications [84].

References

[1]   Allen LV, Jr. Dosage form design and development. Clinical Therapeutics. 2008; 30(11): 2102–11.
[2]   Deb PK, Al-Attraqchi O, Jaber AY, Amarji B, Tekade RK. Physicochemical aspects to be considered in pharmaceutical product development. In: Dosage form design considerations: Academic Press (an imprint of Elsevier); Cambridge, Massachusetts, USA, 2018. pp. 57–83.
[3]   York P. Design of dosage forms. Aulton's pharmaceutics the design and manufacture of medicines (6th ed): Aulton ME, Taylor KMG, Eds. Elsevier; Location: Edinburgh, UK, 2013. pp. 1–12.
[4]   Kesisoglou F, Wu Y. Understanding the effect of API properties on bioavailability through absorption modeling. The AAPS Journal. 2008; 10: 516–25.

[5]   Waterman KC, Adami RC. Accelerated aging: Prediction of chemical stability of pharmaceuticals. International Journal of Pharmaceutics. 2005; 293(1–2): 101–25.

[6]   Savjani KT, Gajjar AK, Savjani JK. Drug solubility: Importance and enhancement techniques. International Scholarly Research Notices. 2012; 2012:1–10.

[7]   Krishnaiah YS. Pharmaceutical technologies for enhancing oral bioavailability of poorly soluble drugs. Journal of Bioequivalence and Bioavailability. 2010; 2(2): 28–36.

[8]   Ali MY, Tariq I, Sohail MF, Amin MU, Ali S, Pinnapireddy SR, et al. Selective anti-ErbB3 aptamer modified sorafenib microparticles: In vitro and in vivo toxicity assessment. European Journal of Pharmaceutics and Biopharmaceutics. 2019; 145: 42–53.

[9]   Miller JM, Beig A, Krieg BJ, Carr RA, Borchardt TB, Amidon GE, et al. The solubility–permeability interplay: Mechanistic modeling and predictive application of the impact of micellar solubilization on intestinal permeation. Molecular Pharmaceutics. 2011; 8(5): 1848–56.

[10]   Chakraborty S, Khandai M, Sharma A, Patra CN, Patro VJ, Kumar Sen K. Effects of drug solubility on the release kinetics of water soluble and insoluble drug from HPMC based matrix formulations. Acta Pharmaceutica. 2009; 59(3): 313–23.

[11]   Vemula VR, Lagishetty V, Lingala S. Solubility enhancement techniques. International Journal of Pharmaceutical Sciences Review and Research. 2010; 5(1): 41–51.

[12]   Rawat S, D DV, P BS, S PR. Self emulsifying drug delivery system (sedds): A method for bioavailability enhancement. International Journal of Pharmaceutical, Chemical & Biological Sciences. 2014; 4 (3):479–494.

[13]   Völgyi G, Baka E, Box KJ, Comer JE, Takács-Novák K. Study of pH-dependent solubility of organic bases. Revisit of Henderson-Hasselbalch Relationship. Analytica Chimica Acta. 2010; 673(1): 40–46.

[14]   Jouyban A. Review of the cosolvency models for predicting solubility of drugs in water-cosolvent mixtures. Journal of Pharmacy & Pharmaceutical Sciences. 2008; 11(1): 32–58.

[15]   Jouyban-Gharamaleki A, Valaee L, Barzegar-Jalali M, Clark B, Acree W, Jr. Comparison of various cosolvency models for calculating solute solubility in water–cosolvent mixtures. International Journal of Pharmaceutics. 1999; 177(1): 93–101.

[16]   Ali MY, Tariq I, Ali S, Amin MU, Engelhardt K, Pinnapireddy SR, et al. Targeted ErbB3 cancer therapy: A synergistic approach to effectively combat cancer. International Journal of Pharmaceutics. 2020; 575: 118961.

[17]   Prajapati S, Patel L, Patel D. Studies on formulation and in vitro evaluation of floating matrix tablets of domperidone. Indian Journal of Pharmaceutical Sciences. 2009; 71(1): 19.

[18]   Devadasu VR, Deb PK, Maheshwari R, Sharma P, Tekade RK. Physicochemical, pharmaceutical, and biological considerations in GIT absorption of drugs. In: Dosage form design considerations: Academic Press (Elsevier); Cambridge, Massachusetts, USA, 2018. pp. 149–78.

[19]   Maheshwari RK, Chaklan N, Singh S. Novel pharmaceutical application of mixed solvency concept for development of solid dispersions of piroxicam. European Journal of Biomedical and Pharmaceutical Sciences. 2015; 1(3): 578–91.

[20]   Rathore GS, Tanwar YS, Sharma A. Fluconazole loaded ethosomes gel and liposomes gel: An updated review for the treatment of deep fungal skin infection. Pharmaceutical Chemistry Journal. 2015; 2(1): 41–50.

[21]   He X. Integration of physical, chemical, mechanical, and biopharmaceutical properties in solid oral dosage form development. In: Developing solid oral dosage forms: Academic Press (Elsevier); San Diego, California, USA, 2009. pp. 407–41.

[22]   Kawakami K. Modification of physicochemical characteristics of active pharmaceutical ingredients and application of supersaturatable dosage forms for improving bioavailability of poorly absorbed drugs. Advanced Drug Delivery Reviews. 2012; 64(6): 480–95.

[23]   Vippagunta SR, Brittain HG, Grant DJ. Crystalline solids. Advanced Drug Delivery Reviews. 2001; 48(1): 3–26.

[24] Hilfiker R, Blatter F, Raumer M. Relevance of solid-state properties for pharmaceutical products. Polymorphism: In the Pharmaceutical Industry. 2006; 1–19.

[25] Huang L-F, Tong W-QT. Impact of solid state properties on developability assessment of drug candidates. Advanced Drug Delivery Reviews. 2004; 56(3): 321–34.

[26] Purohit R, Venugopalan P. Polymorphism: An overview. Resonance. 2009; 14: 882–93.

[27] Pangarkar P, Tayade A, Uttarwar S, Wanare R. Drug polymorphism: An overview. International Journal of Pharmaceutical Technology. 2013; 5: 2374–402.

[28] Singhal D, Curatolo W. Drug polymorphism and dosage form design: A practical perspective. Advanced Drug Delivery Reviews. 2004; 56(3): 335–47.

[29] Yu LX, Furness MS, Raw A, Outlaw KPW, Nashed NE, Ramos E, et al. Scientific considerations of pharmaceutical solid polymorphism in abbreviated new drug applications. Pharmaceutical Research. 2003; 20: 531–36.

[30] Mohsin S, Ahmed S, Ahmed I. Sustained release of captopril from matrix tablet using methylcellulose in a new derivative form. Latin American Journal of Pharmacy. 2011; 30(9): 1696–701.

[31] Zaman M, Rasool S, Ali MY, Qureshi J, Adnan S, Hanif M, et al. Fabrication and analysis of hydroxypropylmethyl cellulose and pectin-based controlled release matrix tablets loaded with loxoprofen sodium. Advances in Polymer Technology. 2015; 34(3):21473.

[32] Khankari RK, Grant DJ. Pharmaceutical hydrates. Thermochimica Acta. 1995; 248: 61–79.

[33] Jurczak E, Mazurek AH, Szeleszczuk Ł, Pisklak DM, Zielińska-Pisklak M. Pharmaceutical hydrates analysis – overview of methods and recent advances. Pharmaceutics. 2020; 12(10): 959.

[34] Thirunahari S, Aitipamula S, Chow PS, Tan RB. Conformational polymorphism of tolbutamide: A structural, spectroscopic, and thermodynamic characterization of Burger's forms I–IV. Journal of Pharmaceutical Sciences. 2010; 99(7): 2975–90.

[35] Hajera K, Baig MS, Zaheer Z, Nazia K. Development and validation of RP-HPLC method for the estimation of gemifloxacin mesylate in bulk and pharmaceutical dosage forms. Asian Journal of Research in Chemistry. 2011; 4(11): 1685–87.

[36] Eddleston MD, Madusanka N, Jones W. Cocrystal dissociation in the presence of water: A general approach for identifying stable cocrystal forms. Journal of Pharmaceutical Sciences. 2014; 103(9): 2865–70.

[37] Kelley SP, Narita A, Holbrey JD, Green KD, Reichert WM, Rogers RD. Understanding the effects of ionicity in salts, solvates, co-crystals, ionic co-crystals, and ionic liquids, rather than nomenclature, is critical to understanding their behavior. Crystal Growth & Design. 2013; 13(3): 965–75.

[38] Idrees M, Rahman N, Ahmad S, Ali M, Ahmad I. Enhance transdermal delivery of flurbiprofen via microemulsions: Effects of different types of surfactants and cosurfactants. DARU Journal of Pharmaceutical Sciences. 2011; 19(6): 433.

[39] Docherty R, Pencheva K, Abramov YA. Low solubility in drug development: De-convoluting the relative importance of solvation and crystal packing. Journal of Pharmacy and Pharmacology. 2015; 67(6): 847–56.

[40] Suleiman M, Najib N. Isolation and physicochemical characterization of solid forms of glibenclamide. International Journal of Pharmaceutics. 1989; 50(2): 103–09.

[41] Stumpe MC, Blinov N, Wishart D, Kovalenko A, Pande VS. Calculation of local water densities in biological systems: A comparison of molecular dynamics simulations and the 3D-RISM-KH molecular theory of solvation. The Journal of Physical Chemistry B. 2011; 115(2): 319–28.

[42] Babu NJ, Nangia A. Solubility advantage of amorphous drugs and pharmaceutical cocrystals. Crystal Growth & Design. 2011; 11(7): 2662–79.

[43] Ohtake S, Shalaev E. Effect of water on the chemical stability of amorphous pharmaceuticals: I. Small molecules. Journal of Pharmaceutical Sciences. 2013; 102(4): 1139–54.

[44] Berthier L, Biroli G. Theoretical perspective on the glass transition and amorphous materials. Reviews of Modern Physics. 2011; 83(2): 587–645.

[45] Aslam Z, Akhter KP, Ahmad M, Aamir MN, Naeem M, Ali MY. Preparation of modified-release tramadol tablets and drug release evaluation using dependent and independent modeling approaches. Latin American Journal of Pharmacy. 2012; 31(10): 1417–21.
[46] Zhang Y, Johnson KC. Effect of drug particle size on content uniformity of low-dose solid dosage forms. International Journal of Pharmaceutics. 1997; 154(2): 179–83.
[47] Shekunov BY, Chattopadhyay P, Tong HH, Chow AH. Particle size analysis in pharmaceutics: Principles, methods and applications. Pharmaceutical Research. 2007; 24: 203–27.
[48] Tariq I, Mumtaz AM, Saeed T, Shah PA, Raza SA, Jawa N, et al. In vitro release studies of diclofenac potassium tablet from pure and blended mixture of hydrophilic and hydrophobic polymers. Latin American Journal of Pharmacy. 2012; 31(3): 380–87.
[49] Taylor KM. Particle size analysis. Aulton's Pharmaceutics E-Book: The Design and Manufacture of Medicines. 2017; 1(2): 140.
[50] Sandler N, Wilson D. Prediction of granule packing and flow behavior based on particle size and shape analysis. Journal of Pharmaceutical Sciences. 2010; 99(2): 958–68.
[51] Raval N, Maheshwari R, Kalyane D, Youngren-Ortiz SR, Chougule MB, Tekade RK. Importance of physicochemical characterization of nanoparticles in pharmaceutical product development. In: Basic fundamentals of drug delivery: Academic Press (Elsevier); Cambridge, Massachusetts, USA, 2019. pp. 369–400.
[52] Jhajharia R, Jain D, Sengar A, Goyal A, Soni P. Synthesis of copper powder by mechanically activated cementation. Powder Technology. 2016; 301: 10–15.
[53] Maheshwari R, Todke P, Kuche K, Raval N, Tekade RK. Micromeritics in pharmaceutical product development. In: Dosage form design considerations: (vol. I). Academic Press (Elsevier); Cambridge, Massachusetts, USA, 2018. pp. 599–635.
[54] Jiménez-castellanos MR, Zia H, Rhodes C. Mucoadhesive drug delivery systems. Drug Development and Industrial Pharmacy. 1993; 19(1–2): 143–94.
[55] Ali MY. Advanced colloidal systems for targeted chemotherapy: Elsevier; Location: Edinburgh, UK, 2019.
[56] Zhang R, Khalizov AF, Pagels J, Zhang D, Xue H, McMurry PH. Variability in morphology, hygroscopicity, and optical properties of soot aerosols during atmospheric processing. Proceedings of the National Academy of Sciences. 2008; 105(30): 10291–96.
[57] Allada R. Hygroscopicity categorization of pharmaceutical solids by gravimetric sorption analysis: A systematic approach. Asian Journal of Pharmaceutics (AJP). 2016; 10(04):S697–S705.
[58] Tereshchenko AG. Deliquescence: Hygroscopicity of water-soluble crystalline solids. Journal of Pharmaceutical Sciences. 2015; 104(11): 3639–52.
[59] Honmane SM. General considerations of design and development of dosage forms: Pre-formulation review. Asian Journal of Pharmaceutics (AJP). 2017; 11(03)S394–S401.
[60] Baertschi SW, Alsante KM, Reed RA. Pharmaceutical stress testing: Predicting drug degradation: CRC Press; Boca Raton, Florida, Pharmaceutical Press; London, United Kingdom, 2016.
[61] Kawabata Y, Wada K, Nakatani M, Yamada S, Onoue S. Formulation design for poorly water-soluble drugs based on biopharmaceutics classification system: Basic approaches and practical applications. International Journal of Pharmaceutics. 2011; 420(1): 1–10.
[62] Florence AT, Attwood D. Physicochemical principles of pharmacy. In: manufacture, formulation and clinical use: Pharmaceutical Press; London, United Kingdom, 2015.
[63] Testa B, Mayer JM. Hydrolysis in drug and prodrug metabolism: John Wiley & Sons; Weinheim, Germany, 2003.
[64] Mitchell SM, Ullman JL, Teel AL, Watts RJ. pH and temperature effects on the hydrolysis of three β-lactam antibiotics: Ampicillin, cefalotin and cefoxitin. Science of the Total Environment. 2014; 466: 547–55.

[65] da Silva TL, Costa CSD, da Silva MGC, Vieira MGA. Overview of non-steroidal anti-inflammatory drugs degradation by advanced oxidation processes. Journal of Cleaner Production. 2022; 346: 131226.

[66] Addamo M, Augugliaro V, Paola AD, García-López E, Loddo V, Marci G, et al. Removal of drugs in aqueous systems by photoassisted degradation. Journal of Applied Electrochemistry. 2005; 35: 765–74.

[67] Boreen AL, Arnold WA, McNeill K. Photochemical fate of sulfa drugs in the aquatic environment: Sulfa drugs containing five-membered heterocyclic groups. Environmental Science & Technology. 2004; 38(14): 3933–40.

[68] Franquet-Griell H, Medina A, Sans C, Lacorte S. Biological and photochemical degradation of cytostatic drugs under laboratory conditions. Journal of Hazardous Materials. 2017; 323: 319–28.

[69] Chhabra N, Aseri ML, Padmanabhan D. A review of drug isomerism and its significance. International Journal of Applied and Basic Medical Research. 2013; 3(1): 16.

[70] Yoshioka S, Stella VJ. Stability of drugs and dosage forms: Springer Science & Business Media; New York, New York, USA, 2000.

[71] Zamir A, Hussain I, Ur Rehman A, Ashraf W, Imran I, Saeed H, et al. Clinical pharmacokinetics of metoprolol: A systematic review. Clinical Pharmacokinetics. 2022; 61(8): 1095–114.

[72] Hörter D, Dressman JB. Influence of physicochemical properties on dissolution of drugs in the gastrointestinal tract1PII of original article: S0169-409X(96)00487-5. The article was originally published in advanced drug delivery reviews 25 (1997) 3–14.1. Advanced Drug Delivery Reviews. 2001; 46(1): 75–87.

[73] Oashi T, Ringer AL, Raman EP, MacKerell AD, Jr. Automated selection of compounds with physicochemical properties to maximize bioavailability and druglikeness. Journal of Chemical Information and Modeling. 2011; 51(1): 148–58.

[74] Tariq I, Pinnapireddy SR, Duse L, Ali MY, Ali S, Amin MU, et al. Lipodendriplexes: A promising nanocarrier for enhanced gene delivery with minimal cytotoxicity. European Journal of Pharmaceutics and Biopharmaceutics. 2019; 135: 72–82.

[75] Lopes CM, Bettencourt C, Rossi A, Buttini F, Barata P. Overview on gastroretentive drug delivery systems for improving drug bioavailability. International Journal of Pharmaceutics. 2016; 510(1): 144–58.

[76] Karthika C, Sureshkumar R, Zehravi M, Akter R, Ali F, Ramproshad S, et al. Multidrug resistance of cancer cells and the vital role of P-glycoprotein. Life. 2022; 12(6): 897.

[77] Barthe L, Woodley J, Houin G. Gastrointestinal absorption of drugs: Methods and studies. Fundamental and Clinical Pharmacology. 1999; 13(2): 154–68.

[78] Dong X. Current strategies for brain drug delivery. Theranostics. 2018; 8(6): 1481–93.

[79] Feghali M, Venkataramanan R, Caritis S. Pharmacokinetics of drugs in pregnancy. Seminars in Perinatology. 2015; 39(7): 512–19.

[80] Hunter J, Hirst BH. Intestinal secretion of drugs. The role of P-glycoprotein and related drug efflux systems in limiting oral drug absorption. Advanced Drug Delivery Reviews. 1997; 25(2): 129–57.

[81] Weiner IM. Mechanisms of drug absorption and excretion the renal excretion of drugs and related compounds. Annual Review of Pharmacology and Toxicology. 1967; 7: 39–56. Volume 7, 1967.

[82] Lista AD, Sirimaturos M. Pharmacokinetic and pharmacodynamic principles for toxicology. Critical Care Clinics. 2021; 37(3): 475–86.

[83] Rollins DE, Klaassen CD. Biliary excretion of drugs in man. Clin Pharmacokinet. 1979; 4(5): 368–79.

[84] Borowy CS, Ashurst JV. Physiology, zero and first order kinetics: StatPearls. Treasure Island (FL): StatPearls Publishing; Treasure Island, Florida, USA.

Malik Saadullah, Muhammad Yasir Ali, Umair Amin, Abid Mahmood, Nisar ur Rahman, Daulat Haleem Khan, Udo Bakowsky

3 Extraction and drying

3.1 Introduction

It can be defined as the withdrawal or removal of desired constituents from crude drugs through the use of selected solvents in which the desired substances are soluble. This process allows for the isolation of desired substances from the raw material [1]. Extraction is crucial in obtaining purified compounds for various applications. It harnesses the solubility properties of substances to separate them effectively. Since ancient times, medicines have been derived from various parts of plants infused or boiled in water [2]. Through experience, it became evident that liquids like alcohol (such as wine or brandy) and vinegar enhanced ingredient extraction efficiency and acted as preservatives. Before utilizing these liquids for extraction, plants underwent preparation to optimize extraction ease. This involved cutting plants into small pieces, bruising to soften woody parts, or drying and powdering them. These processes were typically conducted in pharmacies. However, with industrialization gaining momentum from the mid-1800s, larger-scale production of medicines became feasible. This transition facilitated the advancement of extraction techniques, enabling the extraction of specific compounds on a much larger scale and with greater efficiency. As a result, extraction played a pivotal role in the advancement of pharmaceutical and chemical industries, allowing for the production of a wide range of medicines and specialty compounds essential for human health and technological progress.

3.2 Components of extraction

3.2.1 Crude drug

A crude drug is a vegetable or animal drug that has not undergone processes except collection, cleaning, and drying [3]. It remains in its raw form without undergoing further refinement or alteration beyond these basic steps. Crude drugs are foundational materials for both pharmaceutical preparations and traditional remedies, embodying the essence and composition of their natural source. They provide the raw framework from which medicines are formulated, maintaining the innate properties and chemical makeup of the original botanical or biological source. These raw materials are crucial in the initial stages of drug development, preserving the therapeutic potential inherent in nature. Through careful processing and refinement, crude drugs transform into potent remedies that address various health concerns. Their utilization under-

https://doi.org/10.1515/9783111438108-003

scores the importance of harnessing natural resources in healthcare. From ancient apothecaries to modern pharmaceutical facilities, crude drugs remain indispensable in the creation of healing formulations.

3.2.2 Extractive

The product of extraction is called extractive. The remaining crude drug, having no active constituent, is known as marc [4]. Crude drugs typically contain a variety of components that dissolve in solvents, making it rare for extracts to consist of just one compound. However, there can be exceptions to this rule. As mentioned before, crude drugs contain both active and inactive ingredients, showcasing the complexity of natural sources. This complexity underscores the importance of understanding the diverse constituents of crude drugs for pharmaceutical and traditional medicine purposes. Researchers strive to extract and utilize the therapeutic potential of these natural resources while navigating their intricate chemical composition.

3.2.3 Menstruum

The solvent used for the extraction is known as menstruum [5]. The menstruum utilized should ideally maximize the dissolution of active ingredients while minimizing the initial dissolution of inactive materials. This balance ensures an efficient extraction process, where the desired therapeutic components are extracted to their fullest potential without the unnecessary inclusion of inert substances. Achieving this optimal menstruum composition is crucial in pharmaceutical preparations and traditional remedies, where the potency and efficacy of the final product depend on extraction efficiency. Careful selection of solvents and extraction methods is essential to strike the right balance between dissolving active constituents and minimizing the dissolution of inactive materials. This approach ensures the production of potent and effective medicines while minimizing unnecessary waste and undesirable side effects [6]. The solvent employed for extraction should be easily removed after extraction, be inert, nontoxic, not inflammable, and noninteractive chemically or physically [7].

3.2.3.1 Solvents

Solvents used in extraction are alcohol (ethanol and isopropyl alcohol), glycerin, water, propylene glycol, acetone, isopropyl alcohol, monoethanolamine, methyl alcohol (methanol), chloroform, and polyethylene glycol (PEG). Extraction requires a solvent system to penetrate cellular material and then dissolve the active constituents. As a result, several physical and chemical processes occur, and the rate at which these processes occur will largely depend upon the homogeneity of structure between the solvent and solute. Since the rate of solvent penetration is enhanced by the increased surface area, the combination of drug particles to reduce the size is often desirable. The movement of liquid through the crude drug primarily occurs via diffusion. This process involves the movement of molecules from an area of higher concentration to one of lower concentration, allowing for the gradual penetration of the liquid through the drug material. Additionally, the movement of liquid through the pores and capillaries present in natural systems is influenced by factors such as the surface tension of the liquid and its wetting properties. These characteristics dictate how effectively the liquid spreads and penetrates the drug material, ultimately impacting the extraction process. Understanding these dynamics is crucial for optimizing extraction efficiency in pharmaceutical and traditional medicine practices.

Selection of solvent: The solvent used is determined by the type of plant, the section of the plant to be extracted, the nature of the bioactive chemicals, and the availability of solvent. In general, polar solvents such as water, methanol, and ethanol are used for polar compound extraction, whereas nonpolar solvents such as hexane and dichloromethane are used in nonpolar compound extraction (Tab. 3.1). The traditional liquid-liquid extraction method uses two miscible solvents, such as water-dichloromethane, water-ether, and water-hexane. Water is present in all of the combinations due to its high polarity and miscibility with organic solvents. To facilitate separation, the substance to be extracted via liquid-liquid extraction should be soluble in an organic solvent but not in water [8]. Furthermore, the extraction solvents are categorized according to their polarity, with n-hexane being the least polar and water being the least polar and water being the most polar [9, 10].

Tab. 3.1: Different extraction solvents grouped in increasing polarity order [11, 12].

S. no.	Solvent	Polarity
1.	n-Hexane	0.009
2.	Petroleum ether	0.117
3.	Diethyl ether	0.117
4.	Ethyl acetate	0.228
5.	Chloroform	0.259
6.	Dichloromethane	0.309
7.	Acetone	0.355
8.	n-Butanol	0.586
9.	Ethanol	0.654
10.	Methanol	0.762
11.	Water	1.000

3.2.3.1.1 Properties of commonly used solvents

Alcohol is the most useful solvent in pharmaceutics, next to water. It is mostly used in a 95.6% concentration. But if concentration is 99.5% then it may be used and is known as absolute and dehydrated alcohol [13, 14]. We have a solution called dilute alcohol, which is 50% each water and alcohol. This dilute alcohol is called a hydroalcoholic mixture. This mixture has both the dissolving capacity of alcohol as well as water-soluble substances. The insoluble drugs must be made soluble in alcohol by dissolving in alcohol, e.g., antimicrobial agents, flavoring agents, and preservatives. Alcohol may alone be used as an antimicrobial agent or may be mixed with other agents like sorbates, benzoates, or parabens. However, it's crucial to note that alcohol is pharmacologically active and may pose health risks. As a result, the FDA imposes restrictions on its use to ensure safety.

Glycerin: Glycerin, also known as glycerol, is a colorless or pale yellow, transparent liquid with a sweet taste and high viscosity. It is miscible with water and alcohol. Due to its thick consistency, glycerin is often used in small quantities to aid in drug dissolution, sometimes requiring heating to facilitate the process. Its viscosity and sweet taste make it suitable for use as a preservative and additional solvent in various applications. Additionally, its antimicrobial properties contribute to its effectiveness in preserving pharmaceutical products. However, its thick consistency can pose challenges in formulation, requiring careful consideration during product development. Despite its versatility in dissolving both hydrophilic and lipophilic compounds, regulatory bodies such as the FDA closely monitor its use in pharmaceuticals to ensure consumer safety. In summary, glycerin's unique characteristics make it a valuable solvent in pharmaceutical applications, though its handling requires caution due to its viscosity and potential formulation challenges [15].

Propylene glycol: It has two hydroxyl groups and is also viscous. It is miscible with water and alcohol. It may dissolve a small amount of volatile oil and is mostly preferred over glycerin. Propylene glycol, characterized by its two hydroxyl groups and thick consistency, serves as a versatile solvent. It effortlessly mixes with water and alcohol, making it a preferred option in many applications. Although it can dissolve small amounts of volatile oils, propylene glycol is often chosen over glycerin due to its unique properties. Its capacity to dissolve diverse substances and its compatibility with other solvents render it invaluable in the pharmaceuticals, cosmetics, and food industries [16]. Furthermore, its stability and low volatility make it ideal for formulations requiring prolonged storage. Despite its benefits, it's crucial to consider potential side effects and regulatory constraints, especially in products intended for topical or oral use. Overall, propylene glycol's diverse traits make it an indispensable solvent across various sectors, offering distinct advantages over alternatives such as glycerin.

Water: It has different grades: a) tap water, which is impure and has dissolved amounts of inorganic salts, including sodium, potassium, calcium, magnesium, iron chlorides, sulfates, and bicarbonates, as well as microorganisms. Due to its impurities, it is deemed unacceptable for use in extraction. Its utilization leads to precipitation and discoloration, compromising the quality and safety of products. Therefore, it is essential to avoid the use of tap water in formulations or processes where purity and consistency are paramount. Proper purification methods should be employed to remove impurities and ensure the integrity of the final product. In industries such as pharmaceuticals, cosmetics, and food processing, adherence to stringent quality standards is essential to safeguard consumer health and satisfaction; b) purified water, which is obtained by distillation, ion exchange, and reverse osmosis.

In distillation, impurities are separated by evaporation of water [17]. Generally, the initial portion of the distillate, approximately one to two drops, is discarded due to the potential presence of dissolved gases and other compounds. It is recommended to discard approximately 10% of the total original water to ensure the removal of impurities. Following this initial discard, the remaining distillate is collected and used for further processing or analysis. The discarded portion helps to eliminate any potential contaminants that may affect the quality or purity of the final product. This precautionary measure is essential in pharmaceutical manufacturing to uphold strict quality standards and ensure product safety. By carefully managing the distillation process and discarding impurities, manufacturers can produce pharmaceutical products of consistent quality and efficacy. Additionally, thorough quality control measures are implemented to verify the purity and integrity of the final product before it is released for distribution or use. Overall, this meticulous approach to distillation contributes to the production of high-quality pharmaceuticals that meet stringent regulatory requirements.

In ion exchange, water passes through cationic and anionic exchangers, where cations from tap water are exchanged for hydrogen ions present in the exchanger materials, and anions from tap water are exchanged for hydroxyl ions from the exchanger [18]. This process effectively removes impurities and ions from the water, enhancing its quality and purity. Cationic exchangers selectively remove positively charged ions, while anionic exchangers target negatively charged ions, resulting in thorough purification of the water. Through this exchange process, undesired ions are replaced by hydrogen and hydroxyl ions, respectively, resulting in cleaner and more purified water. This method finds widespread use in various industries, including water treatment, pharmaceuticals, and electronics manufacturing, to ensure the production of high-quality water for diverse applications.

In reverse osmosis, a pressurized stream of water is passed parallel to the inner side of the filter membrane core. This pressure forces a portion of the feed water to pass through the membrane as filtrate, while the remaining water sweeps tangentially along the membrane and exits the system without undergoing filtration. This process effectively separates contaminants and impurities from the water, producing purified water on one side of the membrane and concentrating impurities on the other [19]. Reverse osmosis is a highly efficient method for water purification, widely used in various industries including desalination, wastewater treatment, and production of ultrapure water for industrial and residential applications.

3.3 Methods of extraction

Methods of extraction play a vital role in obtaining compounds from plants, with maceration and percolation standing out as key techniques. In maceration, plant material is immersed in a solvent, allowing the desired compounds to gradually dissolve. On the other hand, percolation involves passing a solvent through a bed of material, facilitating a controlled extraction process. These techniques are indispensable in industries like pharmaceuticals and food processing, where efficiently extracting valuable ingredients is paramount.

Moreover, extraction processes can be broadly divided into two types based on the scale or quantity of materials being processed:
–   Small-scale extraction
–   Large-scale extraction

3.4 Small-scale extraction and its techniques

This refers to the process of extracting desired substances from raw materials on a smaller, often laboratory scale. Small-scale extraction involves handling smaller quantities of materials and is typically conducted in laboratory settings using equipment suitable for such quantities. Techniques used in small-scale extraction include solvent extraction, solid-phase extraction, liquid-liquid extraction, supercritical fluid extraction (SFE), microwave-assisted extraction (MAE), and ultrasound-assisted extraction (UAE). Small-scale extraction is pivotal for isolating specific active ingredients from plant extracts or complex matrices, aiding in the formulation of herbal remedies and supplements. Additionally, it enables the extraction of potent compounds for experimental studies and preclinical trials in drug development.

The small-scale extraction is carried out by
–   Solid-liquid extraction
–   Liquid-liquid extraction

3.4.1 Solid-liquid extraction

Solid-liquid extraction is a process utilized when a drug exists in solid form, and there's a liquid solvent available. Employing partition chromatography, this method efficiently extracts desired compounds from solid substances. Widely employed in industries such as pharmaceuticals and chemical synthesis, it involves separating target components between a solid matrix and a liquid phase [20]. By immersing the solid material in a suitable solvent, desired compounds dissolve into the liquid phase while undesired substances remain in the solid matrix. This selective dissolution facilitates the extraction of specific compounds, ensuring the purification and concentration of the final product. In pharmaceutical applications, solid-liquid extraction is crucial for isolating active pharmaceutical ingredients (APIs) from plant material or synthetic compounds, maintaining medication potency and purity. Similarly, chemical synthesis plays a vital role in producing fine chemicals and specialty compounds. Solid-liquid extraction's versatility and reliability render it indispensable in various industrial sectors for substance extraction and purification.

Though solid-liquid extraction is a technique that has been known for a long time and is still widely used, there are still many unknown aspects that require further investigation to fully understand the mechanism. In the field of solid-liquid extraction techniques, it is possible to distinguish conventional extraction techniques, including maceration, percolation, squeezing, countercurrent extraction, extraction through Soxhlet, and distillation, from unconventional (or innovative) ones. Conventional extractions have been used for many years, although they have many drawbacks: they require the use of high quantities of expensive and pure solvents since during the process they consume a high amount; they have a low selectivity of extraction; they have a high solvent evaporation rate during the process; and they are generally characterized by long extraction times and by the thermal decomposition of thermolabile compounds [21, 22]. Novel solid-liquid extraction methods, termed nonconventional, like UAE, SFE, MAE, and accelerated solvent extraction, are being introduced, primarily in industry. These techniques offer potential solutions to the limitations of traditional methods, promising improved efficiency, selectivity, and shorter extraction times in pharmaceutical extraction processes.

3.4.2 Steps of solid-liquid extraction

– Reduce the size of crude: The size should be suitable for easy extraction.
– Penetration of solvent: Remove any air bubbles for easy penetration, and maceration time should also be considered.
– Solution of the drug: After penetration, the solvent will form a solution.

- Escape of solution: Agitate the solution. The concentrated solution is removed, and volume is made up.
- Separation of the drug: press the solution to separate the drug.

3.4.3 Types of solid-liquid extraction

3.4.3.1 Maceration

The word "maceration" comes from the Latin macro, meaning to soak or to soften [23]. Maceration refers to the process of softening or breaking down a substance by soaking it in a liquid. This method is commonly used to extract active compounds from plant materials or to prepare solutions for medications. It is often employed in the preparation of herbal extracts, tinctures, and infusions, and it is an important technique in pharmaceutical compounding.

Method of maceration: In this method, the crude drug (only a specific part, not the whole of the plant) in powdered form is generally placed into a solvent system, with or without the application of heat, and the mixture is allowed to stand with occasional agitation for an extended period. The period falls in the range of 2–14 days [24]. Menstruum, when mixed, penetrates the cellular structures of the crude drug and then softens the structures so that the extraction may be possible [25]. After some time, the specific gravity of the menstruum is increased, and it is settled down or localized in the surroundings of the crude drug. So, agitation is occasionally applied. After the appropriate time has elapsed, the system is filtered to remove the undissolved material, and a sufficient quantity of material or menstruum is added to make the required volume. The process of maceration is well suited for the extraction of crude drugs having little or no soft cellular tissue, such as benzoin, aloe, and talc, which are dissolved completely in the menstruum. To address the issue of menstruum settling, one solution is to enclose the crude drug in a permeable membrane or sack, akin to a tea bag. This containment allows the menstruum to extract the desired constituents while facilitating the flow of fresh menstruum to take its place as the first batch settles. By employing this method, continuous extraction is ensured, maximizing the efficiency of the process. This approach minimizes the need for manual agitation or intervention, streamlining the extraction process. Additionally, it helps maintain consistent contact between the crude drug and the menstruum, promoting thorough extraction. Overall, enclosing the crude drug in a permeable membrane or sack offers a practical solution to the challenge of menstruum settling, enhancing the effectiveness of the extraction procedure.

Steps of maceration: For small-scale extraction, maceration generally consists of several steps:

- Firstly, grinding of plant materials into small particles is used to increase the surface area for proper mixing with solvent.
- Secondly, in the maceration process, an appropriate solvent named menstruum is added to a closed vessel.
- Thirdly, the liquid is strained off, but the marc, which is the solid residue of this extraction process, is pressed to recover a large number of occluded solutions.
- The obtained strained and pressed-out liquid is mixed and separated from impurities by filtration.

Occasional shaking in maceration facilitates extraction in two ways: (a) increases diffusion and (b) removes concentrated solution from the sample surface to bring new solvent to the menstruum for more extraction yield.

3.4.3.2 Percolation

Percolation is a method of extracting active compounds from solid substances, such as herbs or drugs, by allowing a solvent to pass through the material. The apparatus used for the percolation process is called a percolator [26].

Percolator: Percolators vary greatly in shape, composition, and capacities. Percolators employed in the large-scale industrial preparation of extractives are made up of stainless steel with greatly varied diameters. Percolators used for the leaf's extraction are 6–8 feet in diameter and 12–18 feet high. But the materials, for example, seeds which are very small in size or have greater densities than the leaves, are percolated in very small-sized percolators.

The cylindrical percolator is highly effective in thoroughly extracting drugs while efficiently conserving menstruum. Its design allows the menstruum to flow over the drug in a tall, narrow column, ensuring each drug particle encounters the solvent multiple times. This repeated exposure enhances extraction efficiency, maximizing the yield of desired constituents. Moreover, the cylindrical shape promotes a consistent flow of menstruum throughout the column, ensuring uniform extraction across the entire batch. This systematic approach minimizes waste and optimizes resource utilization, making it a favored choice in the pharmaceutical and herbal medicine industries. The controlled movement of menstruum within the percolator ensures precise extraction, resulting in extracts of exceptional potency and purity. In summary, the cylindrical percolator serves as a versatile and efficient tool for achieving comprehensive drug extraction while conserving menstruum and maintaining extraction quality (Fig. 3.1). Therefore, maximum time is available for the extraction from each particle [27]. However, if the menstruum is viscous, then wider and shorter columns are used, which pass a shorter distance to travel for the second.

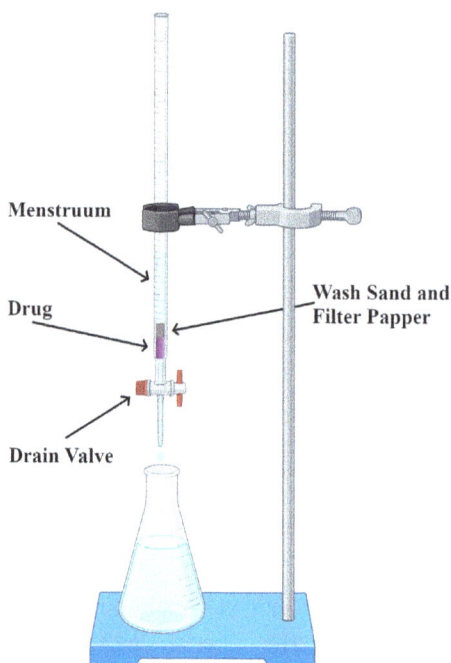

Fig. 3.1: Cylindrical percolator.

Method of percolation: This is the process of extraction in which a comminuted drug is extracted from its soluble constituents by the slow passage of a suitable solvent through the column of the drug. The drug is packed in a special extraction apparatus termed a percolator. Then it is collected, and this extract is called percolate. In this method, menstruum, along with the extractive product, is re-utilized. This is very important because volume can then be adjusted easily only by adding the menstruum rather than removing the solvent from the extract if the volume is increased. The process of percolation is carried out in an apparatus called a percolator, which is just like a funnel. Controlling the rate of liquid movement, often influenced by factors like gravity or external forces, can lead to the extraction of highly concentrated extracts. By allowing the menstruum to move slowly through the crude drug material, more time is afforded for the dissolution and extraction of desired constituents. This deliberate pace enhances the efficiency of the extraction process, enabling the extraction of a greater concentration of active compounds. Additionally, a slow movement of the menstruum helps to minimize the inclusion of unwanted materials, resulting in a purer final extract. This method is particularly advantageous in industries where high concentrations of active ingredients are desired, such as pharmaceuticals and herbal medicine. Through careful control of liquid movement, manufacturers can optimize the extraction process to yield potent and concentrated extracts with enhanced therapeutic potential.

Steps of percolation: a) Preparation of equipment and crude drug: Ensure that the percolator and all related equipment are clean and sterile to prevent contamination of the extract. For the preparation of crude drug initially, the drug material is ground into smaller particles to enhance its surface area. Then, it's mixed with a suitable solvent, resulting in a damp powder. This preparation ensures that the drug and the solvent interact effectively, setting the scene for the extraction process to begin smoothly. By ensuring thorough contact between the two, this step creates the conditions needed for efficient extraction. It marks the crucial starting point where the solvent can penetrate the drug material and dissolve the desired compounds. Through this careful preparation, we establish the groundwork for a successful extraction process, leading to the desired outcomes. b) Positioning of filter paper: Place a filter paper at the base of the percolator to prevent the solid substance from passing through while allowing the solvent to flow freely. c) Packaging of damp material: After dampening the material, it is moistened with a suitable amount of the solvent, left to stand for roughly 4 h, and packed in the percolators. Packages are done in stages. First, take a small amount of crude drug in a percolator and adjust its level. Following that, take another small, damp mass and ensure its level is maintained by gentle pressing, repeating this process until all the material is tightly packed. d) Pouring of solvent: After packing, the percolator top is closed. Pour the additional solvent to produce a shallow layer above the bulk, and the combination is macerated in the closed percolator for 24 h [28]. The solvent percolates through the solid substance, extracting the desired compounds. Even distribution promotes optimal contact between the drug and solvent, enhancing the extraction of a wide range of compounds. Proper saturation enhances the extraction of desired constituents, yielding a potent, concentrated extract. e) Percolation: Allow the solvent to pass through the solid substance slowly, typically through gravity or controlled pressure. This facilitates the extraction of active compounds from the solid. f) Collection of percolates: Finally, the percolate is collected. The time of flow of menstruum should be optimum for the extraction to be maximum. Control over the percolation process can be achieved by reducing pressure at the surface of the menstruum or by lessening the gravitational force. In cases where the percolate is dilute, evaporating the menstruum can yield the desired volume. Adjusting these factors ensures precise control over the extraction process, allowing for optimal extraction efficiency and concentration of desired constituents. This methodical approach ensures that the resulting extract meets the required standards for potency and purity, essential in pharmaceutical and herbal medicine practices [29].

The timing of pouring the solvent and collecting the extract is crucial in percolation. Initially, a filter paper is positioned at the base of the percolator to facilitate the extraction process effectively. This setup aids in separating the extracted material from the solvent. Once this arrangement is in place, the extract is meticulously collected, ensuring a precise and controlled extraction procedure. By monitoring the duration between solvent introduction and extract collection, optimal extraction efficiency is achieved. This systematic approach guarantees that each step is carried out with meticulous attention to detail, ensuring the extraction's accuracy. Through these careful steps, percolation allows for the extraction of desired constituents while maintaining the purity and potency of the resulting extract, meeting the standards expected in pharmaceutical and herbal medicine practices.

3.4.3.3 Decoction

Decoction is a water-based preparation to extract active compounds from medicinal plant materials [30]. In this process, the liquid preparation is made by boiling the plant material with water. Decoction differs from infusion in that the latter is not actively boiled. Decoction is the method of choice when working with tough and fibrous plants, barks, and roots, and with plants that have water-soluble chemicals. The plant material is generally broken into small pieces or powdered. A dried, ground, and powdered plant material is placed into a clean container. Water is then poured and stirred. Heat is then applied throughout the process to hasten the extraction. The process lasts for a short duration, usually about 15 min. The ratio of solvent to crude drug is usually 4:1 or 16:1. It is used for the extraction of water-soluble and heat-stable plant material.

Steps of detection: a) Preparation of crude drug: Obtain the crude drug in the form of small pieces (yanakuna). b) Selection of vessel: Choose earthen pots or tinned copper vessels with clay on the outside. c) Addition of water: Place the crude drug in the selected vessel. Add water according to the recommended ratio. For soft materials: Add four times water per one part of the drug. For moderately hard materials: Add eight times water per one part of the drug. For very hard materials: Add sixteen times the water per one part of the drug. d) Heating process: Heat the vessel containing the crude drug and water mixture over a fire. e) Boiling: boil the mixture on a low flame until it is reduced to one-fourth of the starting volume for soft drugs or one-eighth for moderately or very hard drugs. f) Cooling: Allow the boiled extract to cool down naturally. g) Straining: Strain the cooled extract to remove solid particles. h) Collection: Collect the filtrate in clean vessels for further use.

3.4.3.4 Infusion

Infusions are prepared by soaking a drug in water for a specific period. The process can be either hot or cold, depending upon the type of ingredients present, as decomposition may occur at higher temperatures [7]. The solvent-to-sample ratio is usually 4:1 or 16:1, depending on the intended use. Infusions are generally prepared for immediate use, as preservatives are absent. In some cases, preservatives like alcohol are used, and the infusions are concentrated by boiling.

Steps of infusion: a) Preparation of plant material: The botanical material (e.g., herbs, roots, or flowers) is cleaned, dried, and sometimes ground to increase the surface area and facilitate solvent penetration. b) Immersion in solvent: The prepared plant material is placed in a container and covered with the chosen solvent, ensuring that the material is fully submerged. c) Steeping or soaking: The mixture is left to steep or soak for a specific duration, allowing the solvent to extract the bioactive compounds from the plant material. d) Straining: After the infusion period, the liquid extract is separated from the solid plant material using filtration or straining methods. d) Concentration: In some cases, the extracted liquid may undergo further concentration through methods such as evaporation or distillation to increase the potency of the extract.

3.4.3.5 Hot plate extraction

Hot plate extraction, a variant of solid-liquid extraction, employs the application of heat to accelerate the extraction process. By heating the mixture of solid substance and solvent on a hot plate, this method enhances the solubility of active compounds, resulting in more efficient extraction. The precise control of temperature provided by the hot plate ensures optimal conditions for extraction without compromising the integrity of the desired compounds. Hot plate extraction is widely utilized in pharmaceutical compounding and herbal medicine practices to obtain potent and concentrated extracts for medicinal use. The hot plate serves as a reliable platform for maintaining controlled temperatures during extraction, ensuring optimal conditions for the process. By raising the sample's temperature, molecules gain kinetic energy, aiding in their release into the solvent. This makes hot plate extraction especially effective for capturing delicate compounds prone to degradation at higher temperatures. Researchers often select solvents based on the nature and polarity of the target compounds. Options range from water to various organic solvents like ethanol or methanol, or combinations thereof. The choice of solvent impacts extraction efficiency and selectivity. Enhancements like sonication or stirring can be integrated into hot plate extraction to accelerate mass transfer, thus improving extraction kinetics. Additionally, reflux systems help retain solvents and maintain continuous extraction, preventing evaporation losses. Temperature control plays a critical role in hot plate extraction to prevent degradation of both the sample and the solvent. Consistent monitoring and adjustment of temperature ensure reproducibility and accuracy throughout the process.

Steps of hot plate extraction: a) Preparation of materials: Grind the solid substance to increase surface area and enhance extraction efficiency. Weigh the desired amount of solid substance and place it in a suitable container. b) Selection of solvent: Choose a solvent that is appropriate for extracting the desired compounds from the solid substance. Ensure compatibility with the solid substance and desired extraction outcome. c) Preparation of mixture: Add the solid substance to the solvent in the container, ensuring proper mixing and dispersion. Achieve a suitable ratio of solid to liquid for efficient extraction. d) Heating on a hot plate: Place the container containing the solvent and solid substance on the hot plate. Set the hot plate to the desired temperature, taking into account the solubility of the compounds and the stability of the mixture. e) Extraction: Heat the mixture on the hot plate, allowing the solvent to extract the desired compounds from the solid substance. Stir the mixture periodically to enhance extraction efficiency and promote uniform dissolution. f) Cooling and filtration: After extraction, remove the container from the hot plate and allow the mixture to cool to room temperature. Filter the mixture to separate the liquid extract from any remaining solid particles or debris. g) Storage and analysis: Transfer the filtered extract to a suitable storage container for further analysis or use in pharmaceutical preparations.

3.4.3.6 Soxhlet extraction

It is named after the German agricultural scientist Franz Ritter von Soxhlet and is the best method for continuous extraction of a solid by a hot solvent [31]. It is either a continuous solid/liquid extraction or a continuous heat extraction. The glass equipment is known as a Soxhlet extractor. It has a spherical bottom flask, an extraction chamber, a siphon tube, and a condenser on top (Fig. 3.2).

Steps of extraction: A dried, ground, and finely powdered plant material is snugly packed within a porous bag (thimble) composed of a clean cloth or strong filter paper and tightly closed [32]. The extraction solvent is poured into the bottom flask, and then the thimble is placed in the extraction chamber. The solvent is then heated from the bottom flask, evaporates, and flows down to the extraction chamber, where it condenses and extracts the drug by coming into contact. As a result, when the amount of solvent in the extraction chamber reaches the top of the siphon, the solvent and the extracted plant material flow back to the flask. The entire procedure is repeated until the medication is entirely extracted, at which time a solvent pouring from the extraction chamber leaves no residue behind. This approach is appropriate for plant materials that are partly soluble in the selected solvent, as well as plant materials containing insoluble contaminants. However, it is not appropriate for thermolabile plant materials.

3.4.3.7 Supercritical fluid extraction

People are currently focused on green technologies to build novel extraction procedures that use renewable natural resources or nonhazardous solvents while keeping high-quality and safe extracts, as opposed to standard solvent extraction methods. A

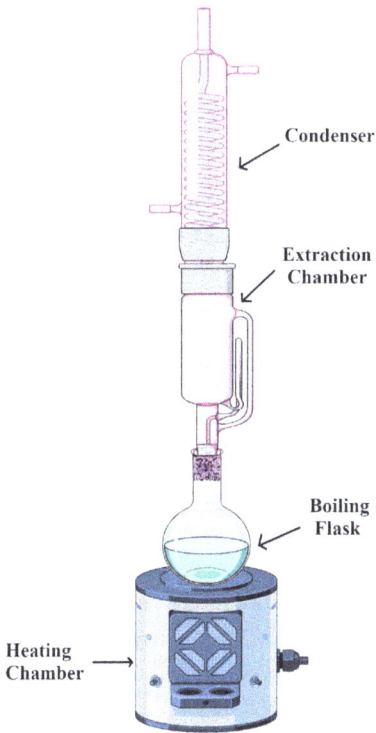

Fig. 3.2: Soxhlet extractor.

novel extraction approach that requires less energy than the previous one is preferable [33, 34]. The introduction of the SFE technique to the extraction field aims to minimize the reliance on traditional organic solvents during extraction processes while enhancing overall throughput. The term "supercritical fluid" refers to any substance that is in an aggregated state at pressures and temperatures higher than the critical temperature and pressure. SFE can be applied to systems on various scales, from the laboratory scale to the pilot plant scale, through to the industrial scale. SFE is often used for solid extraction, but it may also be used for liquid extraction. This sort of extraction is employed in analytical laboratories to prepare samples. The main important applications of extraction using solvents under supercritical conditions are the extraction, refinement, and fractionation of edible oils, fats, and waxes. The extraction process for solid substances primarily consists of two stages: the extraction and the separation of the product from the solvent [35]. In this Extraction process, the separated compounds or chemicals are combined with the supercritical fluids, which give rise to a mobile phase. At temperature and pressures (approximately to the critical temperature and pressure conditions), the mobile phase is enhanced with solvating properties. Supercritical fluids are types of fluids with high density and are noncompressible. Figure 3.3 represents the pressure versus temperature profile for fluids. The pressure and temperature of supercritical fluids are higher than the critical point.

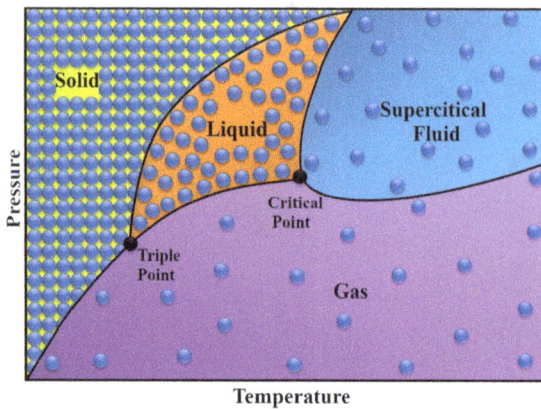

Fig. 3.3: Pressure versus temperature profile for supercritical fluids.

Steps of SFE: a) Formation of supercritical fluid: Under specific conditions of temperature and pressure, CO_2 transitions into a supercritical state, exhibiting properties of both a gas and a liquid. In this state, CO_2 becomes an excellent solvent for extracting various compounds. b) Preparation of sample: The sample to be extracted is typically placed in an extraction vessel or chamber. It can be in the form of solid, liquid, or semi-solid material. c) Extraction process: The supercritical CO_2 is pumped into the extraction vessel, where it comes into contact with the sample. The supercritical CO_2 interacts with the sample, dissolving the target compounds and forming a solution. d) Separation: After extraction, the CO_2 solution containing the extracted compounds is depressurized or cooled, causing the CO_2 to revert to a gas phase. As a result, the dissolved compounds precipitate out of the solution. e) Collection of extract: The extracted compounds are collected, often through a separator or condenser, leaving behind the CO_2, which can be recycled for future extractions (Fig. 3.4).

3.4.3.8 Ultrasound-assisted extraction (UAE) or sonication extraction

Ultrasonication extraction, also known as ultrasound-assisted extraction (UAE), is a technique that utilizes ultrasound waves to enhance the extraction process. This procedure employs ultrasound with frequencies ranging from 20 to 2,000 kHz, which improves cell permeability and causes cavitations. Although this procedure is effective in many circumstances, such as the extraction of anthocyanins and antioxidants, and it is also valuable in the field of nanotechnology, it is restricted due to its high cost. In this method, plant material should be dried first, ground into fine powder, and sieved properly. The prepared sample is then mixed with an appropriate solvent of extrac-

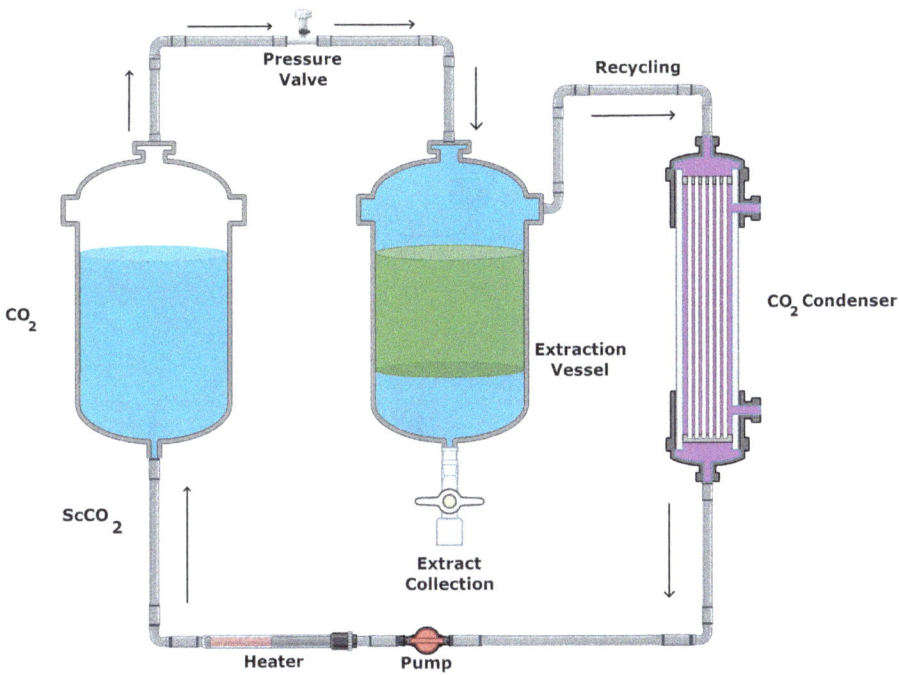

Fig. 3.4: Supercritical fluid extraction.

tion and packed into the ultrasonic extractor. The high sound energy applied hastens the extraction process by reducing the heat requirements [36].

Steps of sonication extraction: a) Preparation of sample: The sample to be extracted is typically prepared by grinding or pulverizing it to increase the surface area and facilitate extraction. b) Solvent selection: A suitable solvent is chosen based on the properties of the target compounds and the sample matrix. The solvent should have good solubility for the target compounds and be compatible with ultrasound. c) Immersion of sample: The prepared sample is placed in a container or vessel containing the selected solvent. d) Application of ultrasound: Ultrasonic waves are then applied to the sample-solvent mixture using an ultrasonic bath or probe. The ultrasonic waves create cavitation bubbles in the solvent, which implode near the surface of the sample. This implosion generates localized hotspots and shockwaves, facilitating the breakdown of cell walls and the release of intracellular compounds. e) Extraction process: The UAE is typically carried out at ambient or slightly elevated temperatures. The duration of the extraction can vary depending on factors such as the type of sample, the solvent used, and the desired extraction efficiency. f) Separation: After the extraction process is complete, the extract containing the target compounds is separated from the residual sample matrix and solvent. This can be achieved through filtration, centrifugation, or decantation (Fig. 3.5).

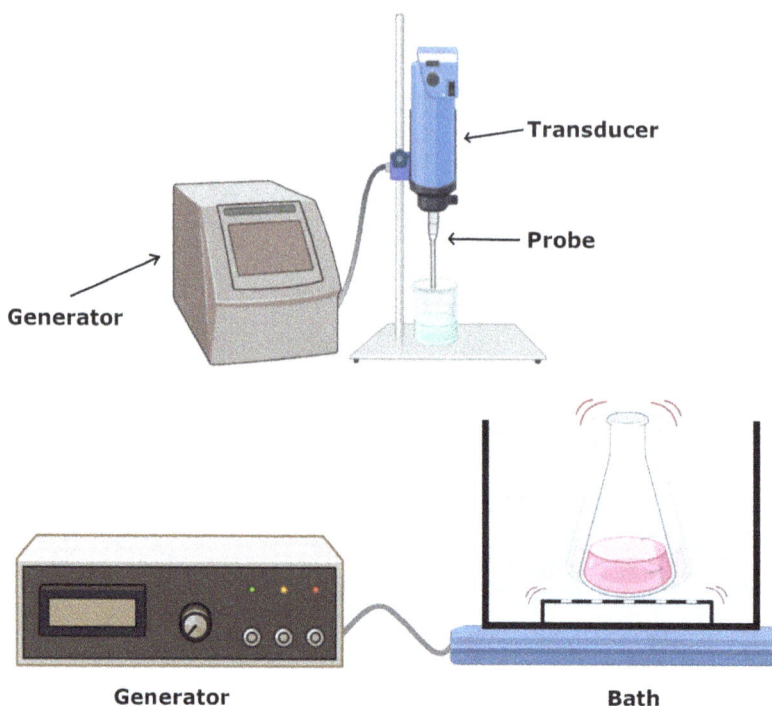

Fig. 3.5: Ultrasound-assisted extraction: a) probe sonicator, b) bath sonicator.

3.4.3.9 Microwave-assisted extraction method

Electromagnetic waves consist of two perpendicularly oscillating fields, namely: the electric field and the magnetic field, which can also be called as microwave. These waves are used as energy vectors or information carriers. Electromagnetic waves are absorbed by the material and converted to heat energy. This is a microwave energy. Microwave energy ranges from 300 MHz to 300 GHz [37]. These waves are nonionizing radiation.

This technique uses the mechanism of dipole rotation and ionic transfer by displacement of charged ions present in the solvent and drug material. This method is suitable for the extraction of flavonoids. It involves the application of electromagnetic radiation in frequencies between 300 MHz and 300 GHz and wavelengths between 1 cm and 1 m. The microwaves applied at a frequency of 2,450 Hz yielded energy between 600 W and 700 W. The technique uses microwave radiation to bombard an object, which can absorb electromagnetic energy and convert it into heat. Subsequently, the heat produced facilitates the movement of solvent into the drug matrix. When a polar solvent is used, dipole rotation and migration of ions occur, increase solvent penetration, and assist the extraction process. However, when a nonpolar solvent is used, the microwave radiation released will produce only small heat; hence, this method does not favor the use of nonpolar solvents.

Steps of MAE: a) Preparation of sample: The sample to be extracted is typically prepared by grinding or pulverizing it to increase the surface area and facilitate extraction. b) Solvent selection: A suitable solvent is chosen based on the properties of the target compounds and the sample matrix. The solvent should have good solubility for the target compounds and be compatible with microwave heating. c) Immersion of sample: The prepared sample is placed in a container or vessel containing the selected solvent. d) Application of microwave radiation: Microwave radiation is applied to the sample-solvent mixture using a microwave oven or dedicated MAE system. The microwaves penetrate the sample and selectively heat the solvent molecules through dipole rotation and ionic conduction. e) Heating of solvent: As the solvent molecules absorb microwave energy, they rapidly heat up, leading to an increase in temperature throughout the extraction vessel. This heating accelerates the extraction process by promoting mass transfer and the dissolution of target compounds into the solvent. f) Evaporation of moisture: In addition to heating the solvent, microwave radiation can also cause the evaporation of moisture present in the sample, further enhancing the extraction efficiency. g) Exertion of pressure: The increase in temperature and pressure within the extraction vessel can lead to enhanced extraction efficiency by promoting the breakdown of cell walls and facilitating the release of intracellular compounds. h) Swelling and rupture: The combination of heat and pressure can cause the sample matrix to swell and rupture, allowing the solvent to penetrate more effectively and extract the target compounds. i) Extraction process: The MAE is typically carried out under controlled conditions of temperature, power, and time. The duration of the extraction can vary depending on factors such as the type of sample, the solvent used, and the desired extraction efficiency.

3.4.3.10 Pressurized liquid extraction (PLE)

Pressurized liquid extraction is done at high temperatures and high pressure. These conditions maintain the solvent in the liquid state. The value of the temperature and pressure of solvents is less than the critical temperature and pressure [38]. This technique is also called accelerated solvent extraction/pressurized hot solvent extraction/ pressurized fluid extraction. PLE is a faster extraction approach than traditional extraction techniques. The amount of solvent used is much less than in a routinely utilized procedure. The characteristics of solvents alter at high temperatures and pressures. There is an increase in mass transfer rates, an increase in analysis solubility, and a decrease in surface tension and viscosity. Such circumstances lead to higher extraction rates. Accelerated solvent extraction (ASE) is a solid-liquid extraction technique that is carried out at extreme temperatures, typically between 500 and 200 °C, and pressures ranging from 10 to 15 MPa. Accelerated solvent extraction is, therefore, a type of pressurized solvent extraction. Increased temperature speeds the extraction kinetics, while increased pressure maintains the solvent liquid, resulting in safe and quick extraction.

Steps of PLE: Solvent is fed to the system with the help of a pump. This pump should withstand the selected pressure. Another function of the pump is to sweep or push away the extract after the process. In the Extraction cell, the extraction process occurs. The cell works at high pressure and is equipped with on and off valves. The oven heats the extraction cell to the desired temperature. The maximum working temperature is about 200 °C. The collection vial collects the sample from the extraction cell.

3.4.4 Liquid-liquid extraction

Liquid-liquid extraction is a method where both the crude drug and the extractant are in liquid form. This process involves transferring specific compounds between two different liquid phases that don't mix, like water and oil. It's a common technique used in different industries, including pharmaceuticals and environmental analysis, to separate and purify compounds based on their solubility properties [39].

This liquid-liquid extraction relies on the principle of differential solubility between two liquids that don't mix, often water and an organic solvent. This method facilitates the transfer of desired components from one liquid phase to another, making it efficient for isolating target compounds [40]. It's widely used in industries like pharmaceuticals and chemical manufacturing due to its versatility and effectiveness in separating and purifying substances. By leveraging differences in solubility, liquid-liquid extraction allows for the selective extraction of specific compounds while leaving unwanted substances behind. In pharmaceuticals, it's used to isolate active ingredients from complex mixtures, ensuring the final product's purity and potency. Similarly, in chemical manufacturing, it's valuable for refining raw materials and creating pure compounds. The versatility and scalability of liquid-liquid extraction make it essential in various stages of product development and manufacturing. Overall, this method is crucial for efficiently extracting and purifying substances in diverse industrial sectors.

In this extraction procedure, we have our active ingredient in the liquid. Now we have to extract the active substance from the liquid. For this purpose, we make use of the distribution law or partition law. According to this law, whenever a drug or any other substance is supplied to a system having two immiscible liquids, then the ratio of the distribution of the drug or substance is always constant [41].

Now, for extraction by this law, we use a menstruum which is immiscible with the liquid from which the extractive product is withdrawn. But remember that it should be efficient enough to extract most of the active substances. For this purpose, we may use two menstruum systems.

In this method of extraction, mix the menstruum and the liquid as these both are immiscible, hence there is no mixing, but the purpose of mixing or agitation is that the maximum area should be in contact. When due to agitation, both the liquids are broken down into globules, and the active ingredient is passed from the liquid to the menstruum. Now, remove the menstruum and pour fresh menstruum because the first has been used to its full capacity, and no further drug is passed from the liquid to the menstruum. When new menstruum is added, the distribution law is operative, but as the amount of drug present in the liquid is now reduced, hence smaller amount is distributed in fresh samples of the menstruum. In this way, the amount of drug decreases in the menstruum. At the industrial level, extracting antibiotics from a solution using centrifugal extractors involves liquid-liquid extraction [42].

3.4.4.1 Types of liquid-liquid extraction

Simple liquid-liquid extraction: This is the basic form of liquid-liquid extraction, involving two immiscible liquids, typically aqueous and organic phases. The solute of interest partitions between these two phases based on its solubility in each. This method is used for a wide range of applications, including the purification and isolation of compounds.

Continuous liquid-liquid extraction: In this technique, the two immiscible liquid phases are continuously contacted and separated in a series of extraction stages. It's particularly useful when dealing with large volumes of feed solution or when precise control over extraction efficiency is required.

Countercurrent liquid-liquid extraction: This is a variation of continuous extraction where the two phases flow in opposite directions. This allows for better utilization of the solvent and more efficient extraction of the solute. Countercurrent extraction is commonly employed in industries such as pharmaceuticals, food processing, and petrochemicals.

Batch liquid-liquid extraction: Unlike continuous extraction, batch extraction involves a single step where the two phases are mixed and allowed to separate. This method is simpler and often used in laboratory settings or small-scale processes.

Micellar liquid-liquid extraction: In this technique, a surfactant is added to one of the phases to form micelles, which can solubilize certain solutes. Micellar extraction is especially useful for extracting polar or charged compounds that are not easily soluble in the organic phase alone.

3.5 Large-scale extraction

Large-scale extraction refers to the process of extracting desired substances from raw materials on a significant industrial or commercial level. It typically takes place in large facilities equipped with specialized machinery capable of handling substantial volumes of raw materials efficiently. Industries such as pharmaceuticals, food processing, chemical manufacturing, and natural resource extraction commonly utilize large-scale extraction methods. These industries employ advanced techniques and processes designed to maximize yield, minimize waste, and uphold product quality and consistency. Large-scale extraction plays a crucial role in meeting the demand for essential products and materials on a mass scale, contributing to economic growth and sustainability across various sectors [43].

For extraction to be carried out on a large scale, we may adopt any one of the following methods.

3.5.1 Recirculation method

In this method, the crude drug is taken in a large tank or vessel and is pressed. At the bottom, we have a perforated layer that prevents the fall of the drug but allows the solvent mixture to pass. Once the tank reaches its maximum capacity, we pour the menstruum into it. After letting it sit for a set time for proper maceration, we run the solvent mixture through pumps, sending it back to the tank to start the extraction process again. This way, we can perform extraction by using the menstruum and drug mixture repeatedly. This method ensures we make the most of our resources and get the highest extraction yield possible. By continuously cycling the solvent mixture through the tank, we make the extraction process more efficient. This approach helps us extract desired compounds from the raw material effectively, making it suitable for large-scale industrial use. Overall, this method allows us to extract valuable substances from raw materials in a practical and resource-conscious manner.

3.5.2 Multistage extraction

In this method, we treat the drug with menstruum and saturate it fully by recirculating it with the help of a pump. Then collect it in tank 1. Repeat this process with the same crude drug but a new sample of menstruum and collect it in tank 2. For the third time, repeat and collect in tank 3. First time, a very concentrated solution is obtained when equilibrium is reached, but no further drug is extracted by the same menstruum. More drugs can be extracted if we use the new sample of menstruum, which is collected in tank 2. Similarly, third tank is filled.

Now, change the new drug by replacing the old one with a new one. Now we use a solvent mixture from tank 2. When this is fully saturated, it is collected in tank 1. Then solvent mixture of tank 3 is collected for extraction, and after extraction, it is collected in tank 2. Then, fresh menstruum is used for extraction, and after extraction is collected in tank 2. Use a sample of tank 2 and collect it in tank 1, and then use a sample of tank 3 and collect it in tank 2. Finally, use the new sample of menstruum. In this way process is completed.

3.5.3 Continuous extraction

Here again, the crude drug is treated with the menstruum and allowed to stand for some time (2–14 days). The drug solution is collected in the receiver when the menstruum is boiled, leaving behind the drug. The vapors of the menstruum are taken to the tank where they are condensed, cooled, and collected again on the surface of the crude drug. In this way, repeated use of the same solvent extraction is carried out [44].

3.6 Drying

Drying is defined as the removal of a liquid from a material by the application of heat and is accomplished by the transfer of a liquid from a surface into an unsaturated vapor phase [45, 46]. There are however many nonthermal methods of drying, for example, the expression of a solid to remove liquid (the squeezing of a wetted sponge), the extraction of liquid from a solvent by use of solvent, the adsorption of water from a solvent by use of desiccants (such as anhydrous calcium chloride), the absorption of moisture from gases by passage through a sulfuric acid column, and desiccation of moisture from a solid by placing it in a sealed container with a moisture-removing material (silica gel in a bottle).

3.6.1 Purpose

Drying is most commonly used in pharmaceutical manufacturing as a unit process in the preparation of granules, which can be dispensed in bulk or converted into tablets and capsules [47]. Another application is found in the processing of materials, e.g., the preparation of dried aluminum hydroxide [48], the spray drying of lactose [49], and the preparation of powdered extracts. Drying can also be used to reduce bulk and weight, and therefore lower the cost of transportation and storage [50]. Other uses are aiding in the preservation of animal and vegetable drugs by minimizing mold and bacterial growth in moisture-laden materials and facilitating comminuting by making substances far more friable than the original water-containing drug [51].

Dried products are often more stable than moist ones [52], as is the case in such diverse substances as effervescent tablets, aspirin, hygroscopic powders, ascorbic acid, and penicillin. The drying reduces the chemical reactivity of the remaining water, which is expressed as a reduction in the water activity of the product [53]. Various methods are used for the removal of moisture in the production of these materials. After moisture is removed, the product is maintained at low water levels by the use of desiccants and low moisture transmission packaging materials.

3.6.2 Classification of dryers

Dryers may be classified into several types depending on the criteria used. Two useful classifications are based on either the method of heat transfer or the methods of solid handling [54]. Classification according to the type of heat transfer is important in demonstrating gross differences in dryer design, operation, and energy requirement [55]. Classification based on the method of solid handling is important and suitable when special attention is given to the nature of the material to be dried.

When dryers are classified according to their methods of solids handling, the major criterion is the presence or absence of agitation of materials to be dried (Tab. 3.2). A dryer that produces excessive agitation is contraindicated when dried material is friable and subject to agitation. On the other hand, if dried material is intended to be pulverized, then drying time can be reduced, and the process made more efficient, by the use of a dryer that produces intensive agitation during the drying cycle [56].

Tab. 3.2: Classification of dryers.

Criterion	Types of dryers
1. Material not agitated – Static bed dryer	– Tray and truck dryers (batch type) – Tunnel and conveyor dryers (continuous type)
2. Material agitated – Moving bed dryers	– Pan dryer (batch type) – Turbo-tray dryers (continuous type)
– Fluidized bed dryers	– Vertical fluidized bed dryers (batch type) – Horizontal vibrating conveyor dryers (continuous type)
– Pneumatic dryers	– Spray dryers (continuous type) – Flash dryer (continuous type)

3.6.2.1 Static bed dryers

The systems in which there is no relative movement among the solids dried, although there may be a bulk motion of the entire drying mass [57]. Only a fraction of the total number of particles is directly exposed to the heat source. The exposed surface can be increased by decreasing the thickness of the bed and allowing drying air to flow through it. In the drying process, only a portion of the total number of particles is directly subjected to the heat source. To enhance drying efficiency, increasing the exposed surface area is crucial. This can be achieved by reducing the thickness of the drying bed, allowing the drying air to flow more effectively through the material. By decreasing the thickness of the bed, more particles are brought into direct contact with the heat source, facilitating faster evaporation. Additionally, improved airflow aids in the removal of moisture from the material, expediting the drying process. This approach is commonly employed in various drying operations, including food processing, pharmaceutical manufacturing, and agricultural applications. By optimizing the exposed surface area, industries can achieve faster and more efficient drying, leading to increased productivity and reduced energy consumption. Overall, maximizing the exposed surface area is essential for enhancing drying performance and achieving optimal results. Static bed dryers are of two types, i.e., tray/truck dryers and tunnel/conveyor dryers.

Tray/truck dryers: These dryers are most commonly used in pharmaceutical plant operations. Tray dryers are sometimes called shelf, cabinet, or compartment dryers. This dryer consists of a cabinet in which the material to be dried is spread on the tier of trays. The number of trays varies with the size of the dryer. Dryers of laboratory size may contain as few as three trays, whereas larger dryers often hold as many as 20 trays [58].

A truck dryer is one in which trays are loaded on trucks that can be rolled into and out of drying cabinets (Fig. 3.6). In plant operations truck dryer is preferred over a tray dryer because it offers greater convenience in loading and unloading. The trucks usually contain one or two tiers of trays, with about 18 or more trays per tier. Each tray is square or rectangular and about 4 to 8 square feet in area. Trays are usually loaded from 0.5 inches to 4.0 inches deep with at least 1.5 inches clearance between the surface and the bottom of the tray above.

Drying in a tray or truck dryer is a batch procedure, as opposed to continuous drying as performed in a moving belt dryer. Batch drying is used extensively in the manufacture of pharmaceuticals for several reasons:
- Each batch of material can be handled as a separate entity.
- The batch sizes of the pharmaceutical industry are relatively small (500 or fewer pounds per batch) compared with the chemical industry (2,000 or more pounds per batch).
- The same equipment is readily adjusted for use in drying a wide variety of materials.

Fig. 3.6: A typical tray dryer (special thanks to Majestic Pharma, Pakistan).

Tray dryers may be classified as direct or indirect [59]. Most tray dryers used in the pharmaceutical industry are of the direct type, in which heating is accomplished by the forced circulation of large volumes of heated air [60]. Indirect tray dryers utilize heated shelves or radiant heat sources inside the drying chamber to evaporate the moisture, which is then removed by either a vacuum pump or a small amount of circulated gas. The trays used have solid, perforated, or wire mesh bottoms. The circulation of drying air in trays with a solid base is limited to the top and bottom of the pan, whereas in trays with a perforated screen, the circulation can be controlled to pass through each tray and the solids on it. The screen trays used in most pharmaceutical drying operations are lined with paper, and the air thus circulates across rather than through the drying material. The paper is used as a disposable tray liner to reduce cleaning time and prevent product contamination. To achieve uniform drying, there must be a constant temperature and a uniform airflow over the material being dried. This is accomplished in modern dryers by the use of a well-insulated cabinet with strategically placed fans and heating coils as integral parts of the unit [61]. The air circulates through the dryer at 200 to 20,000 feet/min. The use of adjustable louvers helps eliminate nonuniform airflow and stagnant pockets. The preferred energy sources for heating the drying air used on pharmaceutical products are steam or electricity. Units fired with coal, oil, and gas produce higher temperatures at lower cost, but are avoided because of possible product contamination with fuel combustion products, and explosion hazards when flammable solvents are being evaporated. Steam is preferred over electricity because steam energy is usually cheaper. If steam is not readily available, and drying loads are small, electric heat is used. In pharmaceutical product drying, steam or electricity is the preferred energy source for heating the drying air. While units fueled by coal, oil, and gas can achieve higher temperatures at lower costs, they are often avoided due to potential contamination risks from fuel combustion products and explosion hazards when flammable solvents are evaporated. Steam is favored over electricity primarily because steam energy tends to be more cost-effective. However, in cases where steam is not readily available and drying loads are small, electric heat may be utilized. This approach ensures that the drying process is conducted safely and efficiently, prioritizing product integrity and worker safety. Overall, the choice of energy source in pharmaceutical drying operations depends on factors such as cost, availability, and safety considerations.

Tunnel/conveyor dryers: Tunnel dryers are adaptations of the truck dryer for continuous drying. The trucks are moved progressively through the drying tunnel by a moving chain [62]. These trucks are loaded on one side of the dryer, allowed to reside in the heating chamber for a time sufficiently long to effect the desired drying, and then discharged at the exit [63]. The operation may be more accurately described as semi-continuous because each truck requires individual loading and unloading before and after the drying cycle. Heat is usually supplied by direct convection, but radiant energy also may be used.

Conveyor dryers are an improvement over tunnel dryers because they are truly continuous. The individual tracks of the tunnel are replaced with an endless belt or screen that carries the wet material through the drying tunnel [64]. Conveyor dryers provide for uninterrupted loading and unloading and are thus more suitable for handling large volumes of materials.

The drying curve characteristic of the material in batch drying is altered considerably when continuous-type dryers are used. As the mass is moved along its drying path in a continuous operation, this mass is subjected to drying air, the temperature and humidity of which are continually changing. As a consequence, the "constant rate" period is not constant, but decreases as the air temperature decreases, although the surface temperature of the wetted mass remains constant. Thus, drying rate curves for batch drying are not equally applicable to continuous drying procedures. In continuous drying operations, the mass being dried is continually exposed to drying air with changing temperature and humidity along its path. This dynamic environment leads to variations in the drying rate throughout the process. Unlike in batch drying, where the drying rate is relatively consistent, the drying rate in continuous drying fluctuates as the conditions change along the drying path. Therefore, drying rate curves developed for batch drying may not accurately represent the drying behavior in continuous drying processes. To optimize continuous drying operations, it is essential to monitor and control the drying parameters, such as temperature, humidity, and airflow, at different stages of the process. Additionally, advancements in process control technologies, such as real-time monitoring and automation, play a crucial role in ensuring uniform drying and consistent product quality in continuous drying operations. Overall, understanding the unique challenges and characteristics of continuous drying is essential for efficient and reliable pharmaceutical product manufacturing.

3.6.2.2 Moving bed dryer

The systems in which the drying particles are partially separated so that they flow over each other. Motion may be induced by either gravity or mechanical agitation. The resultant separation of particles and continuous exposure of new surfaces allow more rapid heat and mass transfer than can occur in static bed dryers. Moving bed dryers are systems designed to partially separate drying particles, allowing them to flow over each other. This movement can be driven by gravity or mechanical agitation. The separation of particles and continuous exposure of new surfaces enable faster heat and mass transfer compared to static bed dryers [52]. These dryers are particularly useful for materials requiring thorough mixing and exposure to drying air. By constantly moving the bed of particles, they ensure uniform drying throughout the material, leading to consistent product quality. Additionally, their enhanced heat and mass transfer rates result in shorter drying times and increased productivity. Moving bed dryers are employed across industries such as pharmaceuticals, food processing, and chemical manufacturing for efficient drying of granular or particulate materials. Overall, they provide a versatile and effective solution for drying operations, improving both product quality and process efficiency.

Turbo-tray dryers: The turbo-tray dryer is a continuous shelf, moving-bed dryer. It consists of a series of rotating annular trays arranged in a vertical stack, all of which rotate slowly at 0.1–1.0 rpm. Heated air is circulated over the trays by turbo-type fans mounted in the center of the stack [65]. Wet mass fed through the roof of the dryer is leveled by a stationary wiper. After about seven-eighths of a revolution, the material being dried is pushed through radial slots onto the tray below, where it is again spread and leveled. The transfer of mass from one shelf to the next is complete after one revolution [66]. The same procedure continues throughout the height of the dryer until the dried material is discharged at the bottom. Because the turbo-tray dryer continuously exposes new surfaces to the air, drying rates are considerably faster than for tunnel dryers. The continuous exposure of new surfaces to the air in turbo-tray dryers significantly accelerates drying rates compared to tunnel dryers. This rapid drying process enhances productivity and efficiency in pharmaceutical manufacturing. Additionally, turbo-tray dryers offer greater flexibility in handling various materials and drying requirements. Their ability to maintain precise temperature and airflow control ensures consistent drying results and product quality. Furthermore, turbo-tray dryers are designed to minimize energy consumption, making them environmentally friendly and cost-effective solutions for pharmaceutical drying applications. Their compact design and modular construction allow for easy installation and integration into existing production lines. Maintenance requirements are also minimal, reducing downtime and optimizing production output. Overall, turbo-tray dryers represent a reliable and efficient solution for pharmaceutical manufacturers seeking high-performance drying equipment [67].

Pan dryer: Pan dryers are moving-bed dryers of the indirect type that may operate under atmospheric pressure or vacuum and are generally used to dry small batches of pastes or slurries [68, 69]. The dryer consists of a shallow, circular jacketed pan having a diameter of 3–6 feet and a depth of 1–2 feet, with a flat bottom and vertical sides. Heat is supplied by steam or hot water. There is a set of rotating plows in the pan that revolve slowly, scraping the moisture-laden mass from the walls and exposing new surfaces to contact with the heated sides and bottom. Atmospheric pan drying allows moisture to escape, whereas, in vacuum dryers, in which the pan is completely enclosed, solvents are recoverable if the evacuated vapors are passed through a condenser. The dried material is discharged through a door on the bottom of the pan. In atmospheric pan drying, moisture escapes freely from the pan, whereas in vacuum dryers, the pan is fully enclosed, allowing for the recovery of solvents. This is achieved by passing the evacuated vapors through a condenser, where they are condensed back into liquid form. As a result, solvents can be reclaimed and reused, minimizing waste and reducing production costs. Once the drying process is complete, the dried material is discharged through a door located on the bottom of the pan. This ensures easy access to the dried product for further processing or packaging. Pan drying offers several advantages, including faster drying times, improved product quality, and the ability to handle heat-sensitive materials. Overall, vacuum dryers provide a versatile and efficient solution for drying a wide range of products in various industries.

3.6.2.3 Fluidized bed dryer

The systems in which solid particles are partially suspended in an upward-moving gas stream. The particles are lifted and then fall back randomly so that the resultant mixture of solid and gas acts as a boiling liquid. The gas-solid contact is excellent and results in better heat and mass transfer than in starting and moving dryers. If a gas is allowed to flow upward through a bed of particulate solids at a velocity greater than the settling velocity of the particles and less than the velocity for pneumatic conveying, the solids are buoyed up and become partially suspended in the gas stream. The resultant mixture of solids and gas behaves like a liquid, and the solids are said to be fluidized [70]. The solid particles are continually caught up in eddies and fall back in a random boiling motion. The gas-solid mixture has a zero angle of repose, seeks its level, and assumes the shape of the vessel that contains it. The fluidization technique is efficient for the drying of granular solids because each particle is surrounded by the drying gas. In addition, the intense mixing between the solids and gas results in uniform conditions of temperature, composition, and particle size distribution throughout the bed [71]. Fluidized bed drying has been reported to offer distinct advantages over conventional tray drying for tablet granulations. In general, tablet granulations have the proper particle size for good fluidization. The only requirements are that the granules are not so wet that they stick together on drying and that the dried product is not so friable as to produce excessive amounts of fine particles through attrition. It was found that the fluidized bed dryer showed a twofold to sixfold advantage in thermal efficiency over a tray dryer. The fluidized bed dryer was also shown to be significantly faster in both drying and handling time than the tray dryer [72]. To avoid electrostatic charge buildup and the resultant explosion hazards, fluid beds are provided with static charge grounding devices.

A fluidized bed dryer available for use in the pharmaceutical industry is of two types, vertical and horizontal. The fluidizing air stream is induced by a fan mounted in the upper part of the apparatus. The air is heated to the required temperature in an air heater and flows upward through the wet material, which is contained in a drying chamber fitted with wire mesh support at the bottom. The airflow rate is adjusted utilizing a damper, and a bag collector filter is provided at the top of the drying chamber to prevent the carryover of fine particles [73]. The unit described is a type of dryer, and the drying chamber is removed from the unit to permit charging and dumping. Dryer capacities range from 5 kg to 200 kg, and average drying time is about 20–40 min [74]. Because of the short drying time and excellent mixing action of the dryer, no hot spots are produced, and higher drying temperatures can be employed than are used in conventional tray and truck dryers [75].

A continuous dryer is more suitable than a batch type for drying larger volumes of materials. The horizontal vibrating conveyor dryer is also used in the pharmaceutical industry [76]. The heated air is introduced in the chamber below the vibrating conveying deck and passes up through the perforated or louvered conveying surface, through the fluidized bed of solids, and into an exhaust hood. A fluidized bed of uniform density and thickness is maintained in any given drying zone by vibration. Residence time in any drying zone is controlled by the length of the zone, frequency and amplitude of vibrations, and the use of dams. The dryers can be divided into several different zones with independent control of airflow and temperature so that drying can take place at the maximum desirable rate in each stage without sacrificing efficiency or damaging heat-sensitive materials. Dryers vary in width from 12 to 57 inches and in length from 10 to 50 feet, with depths of 3 inches. Dryer capacity is limited only by the retention time produced by conveying speeds, which range from 5 to 25 feet/min. In pharmaceutical operations, capacities range as high as 1 to 2 tons/h. Dryers come in a range of widths, spanning from 12 inches to 57 inches, and lengths ranging from 10 feet to 50 feet, with depths of up to 3 inches. The capacity of the dryer is primarily determined by the retention time, which is influenced by conveying speeds typically ranging from 5 to 25 feet/min. In pharmaceutical operations, dryers can handle capacities as high as 1–2 tons/h. This wide range of sizes and capacities allows for flexibility in accommodating various production needs and requirements. Whether it's a small-scale pharmaceutical operation or a large-scale industrial facility, there are dryers available to suit different applications and production volumes. Additionally, advancements in dryer technology continue to enhance efficiency and performance, further expanding the capabilities of these essential pieces of equipment in pharmaceutical manufacturing.

3.6.2.4 Pneumatic dryers

The systems in which drying particles are entrained and conveyed in a high-velocity gas stream. Pneumatic dryers further improve on Fluidized bed dryers because there is no channeling and short-circuiting of the gas flow path through a bed of particles. Each particle is surrounded by an envelope of drying gas. The resultant heat and mass transfer are extremely rapid, thus drying time is short. Pneumatic dryers operate by entraining and conveying drying particles within a high-velocity gas stream. This method enhances fluidized bed dryers by eliminating issues like channeling and short-circuiting of gas flow through the particle bed. Instead, each particle is fully enveloped by drying gas, ensuring thorough heat and mass transfer. Consequently, drying times are significantly reduced due to the rapid heat exchange. These dryers are commonly utilized in industries requiring fast and efficient drying processes, such as pharmaceuticals, food processing, and chemical manufacturing. The precise control over gas velocity and temperature makes pneumatic dryers suitable for a wide range of materials and applications. Additionally, their compact design and high throughput capacity contribute to their popularity in industrial settings. Despite their effectiveness, proper safety measures should be implemented to mitigate potential hazards associated with high-velocity gas streams. Overall, pneumatic dryers offer a reliable and efficient solution for rapid drying needs in various industries [77].

It is of two types:
1. Spray dryers
2. Flash dryers

Spray dryers: Spray dryers differ from most other dryers in that they can handle only fluid materials such as solutions, slurries, and thin pastes [78]. The fluid is dispersed as fine droplets into a moving stream of hot gas, where they evaporate rapidly before reaching the wall of the drying chamber [79]. The product dries into a fine powder, which is carried by the gas current and gravity flow into a collection system [80].

When the liquid droplets come into contact with hot gas, they quickly reach a temperature slightly above the wet bulb temperature of the gas. The surface liquid quickly evaporated, and a tough shell of solids may form in its place. As drying proceeds, the liquid in the interior of the droplet must diffuse through this shell. The diffusion of liquid occurs at a much slower rate than the transfer of heat through the shell to the interior of the droplet. The resultant build of heat causes the liquid below the shell to evaporate at a far greater rate than it can diffuse to the surface. The internal pressure causes the droplet to swell, and the shell becomes thinner, allowing faster diffusion. If the shell is not elastic or impermeable, it ruptures, producing either fragments or bud-like forms on an original sphere. Thus, spray-dried material consists of intact spheres, spheres with buds, ruptured hollow spheres, or spheres with fragments.

There are many types of spray dryers, each designed to suit the material being dried and the desired product characteristics. All spray dryers are made up of the following components: a feed delivery system, an atomizer, a heated air supply, a drying chamber, a solids gas separator, and a product collection system [81–83].

The feed is delivered in an atomizer by gravity flow or by the use of a suitable pump. The rate of feed is adjusted so that each droplet of sprayed liquid is completely dried before it comes in contact with the walls of the drying chamber, and yet dried powder is not overheated in the drying process. The proper feedback is determined by observation of the outlet air temperature and visual inspection of the walls of the drying chamber. If the inlet air temperature is kept constant, a drop in the liquid feed rate is reflected by a rise in outlet temperature. Excessive feed rates produce a lowering of outlet temperature, and ultimately a buildup of materials on the wall of the drying chamber [84].

Spray dryers are of three basic types: pneumatic atomizer, pressure nozzles, and spinning disc atomizer. Pneumatic atomizer, also called two fluids or gas atomizing nozzle, liquid feed is broken up into droplets by a high velocity of air or other gas [85]. The pneumatic atomizer is used to produce small particles and for spraying more viscous liquids than can be handled by pressure nozzles. Pneumatic atomizers require more power than other types of atomizers to achieve the same fine spray. The liquid feed is delivered by a pressure nozzle under high pressure up to 7,000 pounds per square inch and is broken up on coming into contact with the air or by impact on another jet or fixed plate. In a spinning disc atomizer, liquid is fed onto the center of a rapidly rotating disc from 3,000 rpm to 50,000 rpm, where centrifugal force breaks the fluid up into droplets [86].

Flash dryers: In flash dryers, moistened solid mass is suspended in a finely divided state at a high velocity of 3,000–6,000 feet/min, high-temperature 300–1,300 °F air stream. The dispersed particles may be carried in an air stream to an impact mill, or pneumatic flow itself reduces the particle size of friable material. The resultant attrition exposes new surfaces for rapid drying. The dried fine particulate matter passes through the duct with an opening small enough to maintain the desired air-carrying velocities. The dried solid is collected by a cyclone separator, which may be followed by a bag collector or wet scrubber [87]. This is an example of a parallel air flow drying system.

The drying process is referred to as flash drying because the drying time is extremely short [88]. The drying air temperature can drop from 1,300 to 600 °F in 2 s and to 350 °F in 4 s. The temperature of drying solids is kept at 100 °F or less.

3.6.3 Applications of drying

3.6.3.1 Drug formulation

Many pharmaceutical formulations contain moisture-sensitive components or solvents. Drying is employed to eliminate moisture from these formulations, enhancing stability and preventing degradation. For instance, after wet granulation in tablet manufacturing, drying is essential to ensure proper granule flow and stability.

3.6.3.2 Stability

Moisture can expedite drug degradation, diminishing shelf life and potency. Drying is leveraged to extract moisture from drug substances and formulations, thereby augmenting stability and extending shelf life.

3.6.3.3 Powder production

Drying is commonly used to transform liquid formulations into powders. This is particularly crucial for drugs administered in powder form, such as inhalation powders or reconstituted oral suspensions. Techniques like spray drying and freeze drying are frequently employed for this purpose [89, 90].

3.6.3.4 Sterilization

Drying constitutes a vital step in the sterilization of pharmaceutical products. Heat drying methods, such as hot air or infrared drying, are utilized to expel moisture from products after sterilization processes like autoclaving or gamma irradiation.

3.6.3.5 Granulation

Drying is utilized in granulation processes to eliminate moisture from wet granules, resulting in the formation of dry granules with improved flow characteristics. This is indispensable for subsequent steps in tablet manufacturing, such as compression.

3.6.3.6 Coating

Drying is imperative in the coating of pharmaceutical products, like tablets and capsules, to eliminate solvent from the coating solution and facilitate the creation of a uniform and durable coating layer.

3.6.3.7 API manufacturing

Drying is instrumental in the manufacturing of APIs to eliminate solvents or moisture from intermediates or final products. This is vital for ensuring API purity and quality.

References

[1]   Zhang Q-W, Lin L-G, Ye W-C. Techniques for extraction and isolation of natural products: A comprehensive review. Chinese Medicine. 2018; 13: 1–26.
[2]   Halberstein RA. Medicinal plants: Historical and cross-cultural usage patterns. Annals of Epidemiology. 2005; 15(9): 686–99.
[3]   Greenish HG. An introduction to the study of materia medica: Being a short account of the more important crude drugs of vegetable and animal origin, designed for students of pharmacy and medicine: J. & A. Churchill; London, England, 1899.
[4]   Pandey A, Tripathi S. Concept of standardization, extraction and pre phytochemical screening strategies for herbal drug. Journal of Pharmacognosy and Phytochemistry. 2014; 2(5): 115–19.
[5]   Malik J, Mandal SC. Extraction of herbal biomolecules. In: Herbal biomolecules in healthcare applications: Elsevier; Edinburgh, England, 2022. pp. 21–46.
[6]   Devgun M, Nanda A, Ansari S, Swamy S. Recent techniques for extraction of natural products. Research Journal of Pharmacy and Technology. 2010; 3(3): 644–49.
[7]   Visht S, Chaturvedi S. Isolation of natural products. Journal of Current Pharma Research. 2012; 2(3): 584.
[8]   Hossain MA, Al-Hdhrami SS, Weli AM, Al-Riyami Q, Al-Sabahi JN. Isolation, fractionation and identification of chemical constituents from the leaves crude extracts of Mentha piperita L grown in Sultanate of Oman. Asian Pacific Journal of Tropical Biomedicine. 2014; 4: S368–S72.
[9]   Doughari JH. Phytochemicals: Extraction methods, basic structures and mode of action as potential chemotherapeutic agents: INTECH Open Access Publisher Rijeka, Croatia; 2012.
[10]   Grigonis D, Venskutonis P, Sivik B, Sandahl M, Eskilsson CS. Comparison of different extraction techniques for isolation of antioxidants from sweet grass (Hierochloe odorata). The Journal of Supercritical Fluids. 2005; 33(3): 223–33.

[11]   Azwanida N. A review on the extraction methods use in medicinal plants, principle, strength and limitation. Medicinal and Aromatic Plants. 2015; 4(196): 2167–412.

[12]   Azmir J, Zaidul ISM, Rahman MM, Sharif K, Mohamed A, Sahena F, et al. Techniques for extraction of bioactive compounds from plant materials: A review. Journal of Food Engineering. 2013; 117(4): 426–36.

[13]   Vane LM. Separation technologies for the recovery and dehydration of alcohols from fermentation broths. Biofuels, Bioproducts and Biorefining. 2008; 2(6): 553–88.

[14]   Idrees M, Rahman N, Ahmad S, Ali M, Ahmad I. Enhance transdermal delivery of flurbiprofen via microemulsions: Effects of different types of surfactants and cosurfactants. DARU Journal of Pharmaceutical Sciences. 2011; 19(6): 433.

[15]   García JI, García-Marín H, Pires E. Glycerol based solvents: Synthesis, properties and applications. Green Chemistry. 2014; 16(3): 1007–33.

[16]   Mortensen B. Propylene glycol. Nord. 1993; 29: 181–208.

[17]   Saxena G, Chandwankar RR. Distillation in water and used water purification. In: Handbook of water and used water purification: Springer; London, England, 2024. pp. 251–72.

[18]   Kansara N, Bhati L, Narang M, Vaishnavi R. Wastewater treatment by ion exchange method: A review of past and recent researches. ESAIJ (Environmental Science, An Indian Journal). 2016; 12(4): 143–50.

[19]   Garud R, Kore S, Kore V, Kulkarni G. A short review on process and applications of reverse osmosis. Universal Journal of Environmental Research and Technology. 2011; 1(3): 233–238.

[20]   Aguilera JM. Solid-liquid extraction. In: Extraction optimization in food engineering: CRC Press; Florida, USA, 2003. pp. 51–70.

[21]   Wang L, Weller CL. Recent advances in extraction of nutraceuticals from plants. Trends in Food Science & Technology. 2006; 17(6): 300–12.

[22]   Galanakis CM. Recovery of high added-value components from food wastes: Conventional, emerging technologies and commercialized applications. Trends in Food Science & Technology. 2012; 26(2): 68–87.

[23]   Raubenheimer O. History of maceration & percolation. American Journal of Pharmacy. 1910; 82: 32–42.

[24]   Naviglio D, Scarano P, Ciaravolo M, Gallo M. Rapid Solid-Liquid Dynamic Extraction (RSLDE): A powerful and greener alternative to the latest solid-liquid extraction techniques. Foods. 2019; 8(7): 245.

[25]   Saravanabavan N, Salwe KJ, Codi RS, Kumarappan M. Herbal extraction procedures: Need of the hour. International Journal of Basic and Clinical Pharmacology. 2020; 9(7): 1135.

[26]   Abubakar AR, Haque M. Preparation of medicinal plants: Basic extraction and fractionation procedures for experimental purposes. Journal of Pharmacy and Bioallied Sciences. 2020; 12(1): 1–10.

[27]   Husa WJ, Fehder P. Drug extraction. XII. The effect of variation in proportion of moistening liquid on the percolation of jalap. Journal of the American Pharmaceutical Association. 1937; 26(3): 220–22.

[28]   Jibhkate YJ, Awachat AP, Lohiya R, Umekar MJ, Hemke AT, Gupta KR. Extraction: An important tool in the pharmaceutical field. International Journal of Science and Research Archive. 2023; 10(1): 555–68.

[29]   Shante VK, Kirkpatrick S. An introduction to percolation theory. Advances in Physics. 1971; 20(85): 325–57.

[30]   Rasul MG. Conventional extraction methods use in medicinal plants, their advantages and disadvantages. International Journal of Basic Sciences and Applied Computing. 2018; 2: 10–14.

[31]   Jha AK, Sit N. Extraction of bioactive compounds from plant materials using combination of various novel methods: A review. Trends in Food Science & Technology. 2022; 119: 579–91.

[32]   López-Bascón M, De Castro ML. Soxhlet extraction. In: Liquid-phase extraction: Elsevier; Edinburgh, England, 2020. pp. 327–54.

[33] Da Silva RP, Rocha-Santos TA, Duarte AC. Supercritical fluid extraction of bioactive compounds. TrAC Trends in Analytical Chemistry. 2016; 76: 40–51.

[34] Chemat F, Vian MA, Cravotto G. Green extraction of natural products: Concept and principles. International Journal of Molecular Sciences. 2012; 13(7): 8615–27.

[35] Zougagh M, Valcárcel M, Ríos A. Supercritical fluid extraction: A critical review of its analytical usefulness. TrAC Trends in Analytical Chemistry. 2004; 23(5): 399–405.

[36] Pingret D, Fabiano-Tixier AS, Chemat F Ultrasound-assisted extraction. 2013.

[37] Routray W, Orsat V. Microwave-assisted extraction of flavonoids: A review. Food and Bioprocess Technology. 2012; 5: 409–24.

[38] Alvarez-Rivera G, Bueno M, Ballesteros-Vivas D, Mendiola JA, Ibañez E. Pressurized liquid extraction. In: Liquid-phase extraction: Elsevier; Edinburgh, England, 2020. pp. 375–98.

[39] Mazzola PG, Lopes AM, Hasmann FA, Jozala AF, Penna TC, Magalhaes PO, et al. Liquid–liquid extraction of biomolecules: An overview and update of the main techniques. Journal of Chemical Technology and Biotechnology: International Research in Process, Environmental and Clean Technology. 2008; 83(2): 143–57.

[40] Schneider YK, Jørgensen SM, Andersen JH, Hansen EH. Qualitative and quantitative comparison of liquid–liquid phase extraction using ethyl acetate and liquid–solid phase extraction using poly-benzyl-resin for natural products. Applied Sciences. 2021; 11(21): 10241.

[41] Golumbic C. Liquid-liquid extraction analysis. Analytical Chemistry. 1951; 23(9): 1210–17.

[42] Fedeniuk RW, Shand PJ. Theory and methodology of antibiotic extraction from biomatrices. Journal of Chromatography A. 1998; 812(1–2): 3–15.

[43] Cunha T, Aires-Barros R. Large-scale extraction of proteins. Molecular Biotechnology. 2002; 20: 29–40.

[44] Umrethia B, Kalsariya B, Vaishnav P. Classical and modern drug extraction techniques: Facts and figures. Journal of Ayurveda and Integrated Medical Sciences. 2017; 2(04): 277–183.

[45] Mujumdar AS, Devahastin S. Fundamental principles of drying. Exergex, Brossard, Canada. 2000; 1(1): 1–22.

[46] Amin MU, Ali S, Ali MY, Tariq I, Nasrullah U, Pinnapreddy SR, et al. Enhanced efficacy and drug delivery with lipid coated mesoporous silica nanoparticles in cancer therapy. European Journal of Pharmaceutics and Biopharmaceutics. 2021; 165: 31–40.

[47] Briens L, Bojarra M. Monitoring fluidized bed drying of pharmaceutical granules. American Association of Pharmaceutical Scientists PharmSciTech. 2010; 11: 1612–18.

[48] Nalivaiko AY, Ozherelkov DY, Pak VI, Kirov SS, Arnautov AN, Gromov AA. Preparation of aluminum hydroxide during the synthesis of high purity alumina via aluminum anodic oxidation. Metallurgical and Materials Transactions B. 2020; 51: 1154–61.

[49] Chiou D, Langrish T, Braham R. The effect of temperature on the crystallinity of lactose powders produced by spray drying. Journal of Food Engineering. 2008; 86(2): 288–93.

[50] Kothawade S, Pande V, Wagh V, Autade K, Bole S, Sumbe R, et al. Perspective Chapter: Pharmaceutical Drying [Internet]. Drying Science and Technology. IntechOpen; London, England; 2024. p. 112941.

[51] Kumadoh D, Archer M-A, Kyene MO, Yeboah GN, Adi-Dako O, Osei-Asare C, et al. Approaches for the elimination of microbial contaminants from lippia multiflora mold. Leaves intended for tea bagging and evaluation of formulation. Advances in Pharmacological and Pharmaceutical Sciences. 2022; 2022(1): 7235489.

[52] de Oliveira WP, De Freitas LAP, Freire JT. Drying of pharmaceutical products. Transport Phenomena in Particulate Systems, Freire, JT, Silveira, AM, Ferreira, MC, Eds, Bentham Science. 2012; 1: 148–71.

[53] Ohtake S, Shalaev E. Effect of water on the chemical stability of amorphous pharmaceuticals: I. Small molecules. Journal of Pharmaceutical Sciences. 2013; 102(4): 1139–54.

[54] Mujumdar AS. Handbook of industrial drying: CRC press; Florida, USA, 2006.

[55] Mujumdar AS. Classification and selection of industrial dryers. Mujumdar's Practical Guide to Industrial Drying: Principles, Equipment and New Developments. Brossard, Canada: Exergex Corporation. 2000; 23–36.

[56] Lachman L, Lieberman HA, Kanig JL. The theory and practice of industrial pharmacy: Lea & Febiger Philadelphia; USA, 1976.

[57] Yi J, Li X, He J, Duan X. Drying efficiency and product quality of biomass drying: A review. Drying Technology. 2020; 38(15): 2039–54.

[58] Das S, Das T, Rao PS, Jain R. Development of an air recirculating tray dryer for high moisture biological materials. Journal of Food Engineering. 2001; 50(4): 223–27.

[59] Zolqadri R, Malekjani N, Talemy FP, Jafari SM. Belt dryers and tray dryers. In: Drying technology in food processing: Elsevier; Edinburgh, England, 2023. pp. 33–46.

[60] Parikh DM. Solids drying: Basics and applications. Chemical Engineering. 2014; 121(4): 42–45.

[61] Rathoure AK, Ram BL, Aggarwal SG. Unit operations in chemical industries. International Journal of Environmental Chemistry. 2019; 5(2): 11–29.

[62] Ajala A, Ngoddy P, Olajide J. Regular article design and construction of a tunnel dryer for food crops drying.

[63] Ridley GB. Tunnel dryers. Industrial & Engineering Chemistry. 1921; 13(5): 453–60.

[64] Kiranoudis C. Design and operational performance of conveyor-belt drying structures. Chemical Engineering Journal. 1998; 69(1): 27–38.

[65] Mujumdar AS. Handbook of industrial drying. CRC press; Florida, USA, 2006.

[66] Powders I Solid dosage forms: Powders and granules.

[67] Yuting W Energy-efficient industrial dryers of berries. 2013.

[68] Mohamad BA Mass Transfer.

[69] Kudra T, Mujumdar AS. Advanced drying technologies: CRC press; Florida, USA, 2009.

[70] Kunii D, Levenspiel O. Fluidization engineering: Butterworth-Heinemann; Oxford, England, 1991.

[71] Daud WRW. Fluidized bed dryers – recent advances. Advanced Powder Technology. 2008; 19(5): 403–18.

[72] Villegas-Santiago J, Calderon-Santoyo M, Ragazzo-Sánchez A, Salgado-Cervantes MA, Luna-Solano G. Fluidized bed and tray drying of thinly sliced mango (Mangifera indica) pretreated with ascorbic and citric acid. International Journal of Food Science and Technology. 2011; 46(6): 1296–302.

[73] Grbavcic ZB, Arsenijevic ZL, Garic-Grulovic RV. Drying of slurries in fluidized bed of inert particles. Drying Technology. 2004; 22(8): 1793–812.

[74] Pansare JJ, Pagar UN, Dode RH, Mogal PS, Surawase RK. Fluidized bed processing: Versatile technique in dosage form development:. Research Journal of Pharmaceutical Dosage Forms and Technology. 2022; 14(1): 87–93. 2022.

[75] Taghavivand M, Choi K, Zhang L. Investigation on drying kinetics and tribocharging behaviour of pharmaceutical granules in a fluidized bed dryer. Powder Technology. 2017; 316: 171–80.

[76] Lehmann S, Buchholz M, Jongsma A, Innings F, Heinrich S. Modeling and flowsheet simulation of vibrated fluidized bed dryers. Processes. 2021; 9: 52. s Note: MDPI stays neutral with regard to jurisdictional claims in . . .; 2020.

[77] Celik M, Wendel SC. Spray drying and pharmaceutical applications. Drugs and the Pharmaceutical Sciences. 2005; 154: 129.

[78] Bhandari B. Spray drying and powder properties. Food drying science and technology: Microbiology, chemistry, applications: Lancaster, PA, USA: DEStech Publications, Inc; 2008. pp. 215–49.

[79] Cal K, Sollohub K. Spray drying technique. I: Hardware and process parameters. Journal of Pharmaceutical Sciences. 2010; 99(2): 575–86.

[80] Fogler BB, Kleninschmidt RV. Spray drying. Industrial & Engineering Chemistry. 1938; 30(12): 1372–84.

[81] Furuta T, Neoh TL. Microencapsulation of food bioactive components by spray drying: A review. Drying Technology. 2021; 39(12): 1800–31.
[82] Baghdan E. Spray drying for the preparation of innovative nanocoatings and inhalable nanocarriers. Philipps University Marburg; Germany, 2018.
[83] Baghdan E, Pinnapireddy SR, Vögeling H, Schäfer J, Eckert AW, Bakowsky U. Nano spray drying: A novel technique to prepare well-defined surface coatings for medical implants. Journal of Drug Delivery Science and Technology. 2018; 48: 145–51.
[84] Patel R, Patel M, Suthar A. Spray drying technology: An overview. Indian Journal of Science and Technology. 2009; 2(10): 44–47.
[85] Hede PD, Bach P, Jensen AD. Two-fluid spray atomisation and pneumatic nozzles for fluid bed coating/agglomeration purposes: A review. Chemical Engineering Science. 2008; 63(14): 3821–42.
[86] Sungkhaphaitoon P, Plookphol T, Wisutmethangoon S. Design and development of a centrifugal atomizer for producing zinc metal powder. International Journal of Applied Physics and Mathematics. 2012; 2(2): 77.
[87] Banooni S, Hajidavalloo E, Dorfeshan M. A comprehensive review on modeling of pneumatic and flash drying. Drying Technology. 2018; 36(1): 33–51.
[88] Aslaksen EW. Mathematical Model of a Flash Drying Process. Journal of Industrial Mathematics. 2014; 2014(1): 460857.
[89] Ali MY, Tariq I, Ali S, Amin MU, Engelhardt K, Pinnapireddy SR, et al. Targeted ErbB3 cancer therapy: A synergistic approach to effectively combat cancer. International Journal of Pharmaceutics. 2020; 575: 118961.
[90] Ali MY, Tariq I, Sohail MF, Amin MU, Ali S, Pinnapireddy SR, et al. Selective anti-ErbB3 aptamer modified sorafenib microparticles: In vitro and in vivo toxicity assessment. European Journal of Pharmaceutics and Biopharmaceutics. 2019; 145: 42–53.

Rabia Munir, Maria Manan, Muhammad Yasir Ali, Nisar ur Rahman,
Saeed Ahmad, Udo Bakowsky, Imran Tariq, Aisha Sethi

4 Calculations

4.1 The metric system

The most extensively used system in pharmacy is the metric system. The Federal Food and Drug Administration used this system in the United States Pharmacopeia and National Formulary in the manufacturing and labeling of products. It is also used in writing prescriptions and medication orders. Using this system, the unit of weight is the gram, the unit of volume is the liter, and the unit of length is the meter [1]. Weight is indicated in kilogram, gram, milligram, and microgram.

Volume is indicated in liters or milliliters. Length is indicated in meters, centimeters, and millimeters. The metric system is also called the International System of Units. This system was made in the eighteenth century in France. Advantages of this system are the clarity given by the prefixes and base units, simplicity, and ease of scientific communication via the use of a globally acknowledged system. The main units of the metric system are kilogram (weight), meter (length), and liter (volume). Subdivisions, multiples, relative values, and prefixes of these units are expressed in Tab. 4.1 [1].

Tab. 4.1: Metric system with its symbols, prefixes, and multiplication factor.

Prefix	Terms	Multiplication factor	Symbol
Exa	One quintillion	10^{18}	E
Tera	One quadrillion	10^{15}	T
Giga	One trillion	10^{12}	G
Mega	One billion	10^{9}	M
Kilo	One million	10^{6}	K
Hecto	One thousand	10^{3}	H
Deka	One hundred	10^{2}	Da
Peta	Ten	10	P
Deci	One tenth	10^{-1}	D
Centi	One hundredth	10^{-2}	C
Milli	One thousandth	10^{-3}	M
Micro	One millionth	10^{-6}	μ
Nano	One billionth	10^{-9}	N
Pico	One trillionth	10^{-12}	P
Femto	One quadrillionth	10^{-15}	F
Atto	One quintillionth	10^{-18}	A

https://doi.org/10.1515/9783111438108-004

4.1.1 Measure of length

In SI system, the basic unit of length is meter.
    Metric length units are:
- 1 kilometer = 1,000 meters
- 1 hectometer = 100 meters
- 1 decimeter = 0.1 meter
- 1 centimeter = 0.01 meter
- 1 micrometer = 000,001 meter
- 1 nanometer = 0.000,000,0001 meter

These units could also be expressed as
- 1 meter = 0.001 kilometers
- = 0.01 hectometer
- = 0.1 dekameters
- = 100 centimeters
- = 1,000 millimeters
- = 1,000,000 micrometer
- = 1,000,000,000 nanometers

Most frequently used denominations and their equivalencies [1]:
- 100 centimeters = 1 meter
- 1,000 millimeters = 100 centimeters

4.1.2 Measure of volume

The basic unit of volume is a liter. It shows the volume of the cube of one tenth of a meter.
    Metric volume units are:
- 1 kiloliter = 1,000 liters
- 1 hectoliter = 100 liters
- 1 dekaliter = 10 liters
- 1 liter = 1.0 liter
- 1 deciliter = 0.1 liter
- 1 centiliter = 0.01 liter
- 1 milliliter = 0.001 liter
- 1 microliter = 0.000001 liter

These units could also be described as follows:
- 1 liter = 0.001 kiloliter
- 1 liter = 0.01 hectoliter

- 1 liter = 0.1 dekaliter
- 1 liter = 10 deciliters
- 1 liter = 100 centiliters
- 1 liter = 1,000 millimeters
- 1 liter = 1,000,000 microliters

However, the milliliter is roughly the same volume as a cubic centimeter (cm^3). National Formulary describes: "one milliliter is taken as the equivalent of 1 cubic centimeter." Most frequently used denominations:
- 1,000 milliliter = 1 liter
- 100 milliliter = 1 deciliter

These instruments range from micropipettes to large-sized calibrated vessels. For small volume measurement, a pipette or calibrated syringe is used. Conical graduates are calibrated in both apothecary and metric units, while cylindrical graduates are calibrated in metric units. These graduates, in both plastic and glass, are available in a variety of sizes, 5–1,000 mL and higher [1, 2]. A few examples of volume conversion are mentioned in Tab. 4.2.

Tab. 4.2: A few examples of volume conversion.

S. no.	Examples	Solution
1.	Change 2 mL into μL	2 mL = 1,000 μL
2.	Change 3 L into cc	3 L = 1,000 × 3 cc = 3,000 cc
3.	Change 5 μL into L	5 μL = 5 L/1,000000 = 0.000005 L
4.	Change 7.4 mL into L	74 mL = 7.4 cc
5.	Change 4.2 L into cc	$\left(\dfrac{4.2 \times 1,000 \text{ mL}}{L}\right) \times \dfrac{cc}{mL} = 4,200$ cc
6.	Change 0.9 ml into μL	0.9 mL = 900 μL
7.	Change 0.3 mL into L	0.3 μL = 3 × 10^{-7} L
8.	Change 9 mL into L	9 mL = 9 × 10^{-3} L
9.	Change 4.28 L into mL	4.28 L × 1,000 mL/L = 4,280 mL
10.	Change 0.083 cc into μL	0.083 cc × 1,000 μL/mL = 83 μL

4.1.3 Measurement of weight

In pharmacy, a gram is the basic unit of weight [1, 3]. The most common correlations between imperative units of weight are (Tab. 4.3):

   1/1,000 kg = 10^3 mg = 109 ng = 10^6 μg = 1 g
- 1 kg = 1,000 g
- 1 g = 1,000 mg
- 1 mg = 1,000 μg

- $1\,\mu g = 1{,}000\ ng$

Tab. 4.3: A few examples of weight conversion.

S. no.	Examples	Solution
1.	Convert 1,385 mg into g	$\dfrac{1{,}385\ mg}{1{,}000} = 1.385$
2.	Convert 240 mg into μg	$240\ mg \times \dfrac{1{,}000\ \mu g}{mg} = 240{,}000\ \mu g$
3.	Convert 0.012 into mg	$0.012 \times \dfrac{1{,}000\ mg}{g} = 12\ mg$
4.	Convert 644 mg into kg	$664\ mg \times \dfrac{g}{1{,}000\ mg} \times \dfrac{kg}{1{,}000\ g} = 0.000644\ kg$
5.	Convert 0.0035 g into μg	$0.0035\ g \times \dfrac{1{,}000{,}000}{g} = 3{,}500\ \mu g$
6.	Convert 0.023 mg into g	$0.023\ mg \times \dfrac{1{,}000}{mg} = 0.00023\ g$
7.	Convert 0.045 kg into g	$0.045\ kg \times \dfrac{1{,}000\ g}{kg} = 45\ g$
8.	Convert 31 ng into mg	$31\ g \times 10^{-6}\ \dfrac{mg}{ng} = 3.1 \times 10^{-5}\ mg$

4.2 Density

Mass per unit volume of substance is called density. It is generally written as grams per cubic centimeter (g/cc). Mass of 1 cc of water at a temperature of 4 °C is called a gram, so 1 g/cc is the density of water. The United States Pharmacopeia describes that 1 cc is taken as equal to 1 mL, so the water density could be described as 1 g/mL. To find a substance's density in the laboratory, you weigh it on a balance and measure its volume (Fig. 4.1) by using either the water displacement technique for irregular solids or a ruler/graduated cylinder for liquids and regular solids. Density can be determined by dividing mass by volume [1]:

$$\text{Density} = \frac{\text{mass}}{\text{volume}}$$

So, if 20 mL of hydrochloric acid weighs 38 g, the density is

$$\text{Density} = \frac{38}{20} = 1.98\ \frac{g}{mL}\ \text{or}\ 1.98\ g/cm^3$$

It is possible to change volume to weight or vice versa. For example, if the liquid is mobile. It is convenient to measure by volume instead of weighing the liquid. On the

Fig. 4.1: Laboratory apparatus used to measure the density of a substance: from left to right, weighing balance, measuring cylinders.

contrary, if the liquid is viscous, it is convenient to change volume to weight and weight to volume.

Examples
Change 50 mL of liquid Z to its equal weight. Liquid Z exhibits a weight of 0.98 g/mL.
Let X be the weight of 50 mL of liquid Z:

$$\text{Weight (g)} \quad = \quad 0.98 \quad X$$
$$\text{Volume (mL)} = \quad 1 \quad 50$$
$$X = 0.98 \times 50/1$$

Hence, 50 mL of liquid Z has a 49 g weight.
Change 75 g of liquid X to its equal volume (mL). Liquid X has a 4.2 g weight.
Suppose B is the volume of liquid X

$$\text{Weight (g)} \quad = 4.2 \quad 75$$
$$\text{Volume (mL)} = 1 \quad a$$
$$b = 1 \times 75/4.2$$
$$b = 17.85$$

Hence, 75 g of liquid X has a volume of 17.85 mL.
Find out the density of 20 mL of a liquid's weight is 22 g:

$$\text{Density} = 22\,\text{g}/20\,\text{mL} = 1.1\,\text{g}/\text{mL}$$

If an acid weighs 28 g and occupies 88 mL of volume, find out the density:

$$\text{Density} = 28\,\text{g}/88\,\text{mL} = 0.32\,\text{g}/\text{mL}$$

Example
Find out the weight of 50 mL of liquid showing a density of 5.68 g/mL [4]:

$$\text{Density} = \text{weight}/\text{volume}$$

$$\text{Weight} = 50\,\text{mL} \times 5.68\,\text{g}/\text{mL}$$

$$\text{Weight} = 284\,\text{g}$$

Find out the volume of powder if it has a density of 4.4 g/cc and a weight of 3.2 g [2]:

$$\text{Density} = \text{weight}/\text{volume}$$

$$\text{Volume} = \text{weight}/\text{density}$$

$$\text{Volume} = 3.2\,\text{g}/4.4\,\text{g}/\text{cc} = 0.72\,\text{mL}$$

If 160 mL of ethanol weighs 104 g, calculate the density:

$$\text{Density} = \text{weight}/\text{volume}$$

$$\text{Density} = 104/100\,\text{mL} = 0.65\,\text{g}/\text{mL}$$

The weight of a metal is 42.1 g and shows a volume of 3 mL, determine the density [4]:

$$\text{Density} = \text{weight}/\text{volume}$$

$$\text{Density} = 42.1\,\text{g}/3\,\text{mL} = 14.03\,\text{g}/\text{mL}$$

4.3 Ratio

Ratio is the relative magnitude of two same kind. Ratio is sometimes called the quotient of the same numbers. Ratio is normally expressed with regard to an operation, but not an outcome. In another way, it can also be exhibited as a fraction, and the fraction is taken as describing the actual operation constituting the division of the numerator by the denominator. In this way, a ratio basically is a concept of a common fraction that shows a specific correlation between two numbers. When terms in a ratio are divided or multiplied by a similar number, then the final answer remains unaltered. The answer obtained remains the quotient of the first term divided by the second term [3].

If the comparison is withdrawn between 3 and 15, the ratio can be shown as 3:15, 3/15, 0.2, or 20%. Ratio is read as 3 to 15 if it is expressed as a fraction (3/15). This form is used frequently for calculating dosages and to show drug concentrations, for instance, 7 mg/mL, 20 drops/min, and 250 mg/tablet. A ratio is a quotient, governed by the same regulations utilized for decimal and common fractions. Some common rules are:

If the denominator and numerator are divided or multiplied by the same number, the ratio remains the same. For instance, the ratio 3:15 (or 3/15 = 0.2) would not be modified if both values are multiplied by 2. The ratio would become 6:30 or 6/30 = 3/15 = 0.2.

To change a decimal to a ratio, express the decimal in fraction form. Cut the fraction to its lowest form. And express the fraction as a ratio. For instance,

$$0.6 = \frac{6}{10} = \frac{3}{5} = 3.5$$

To modify a ratio to a percent, change the ratio to a percent, change the ratio to a decimal, then multiply by hundred, and write the symbol percent. For instance, 1:20 = 1/20 = 0.05 × 100 = 5%.

Ratios with the same multipliers and values are equal:

$$\frac{3}{7} = \frac{6}{14} \text{ and } 3 \times 14 = 7 \times 6$$

$$2/4 = 48$$

$$\text{i.e., } 2 \times 8\,(16) = 2 \times 8\,(\text{or }16)$$

If ratios are equivalent, the reciprocals are also equivalent:

$$\text{So, } 2/4 = 4/8, \text{ then } 4/2 = 8/4$$

The ratio of 30 and 15 is not shown as 2, i.e., quotient of 30 divided by 15, but generally as a fraction 30/15. So, the ratio would be written as 30:15 and therefore always read thirty to fifteen. Similarly, when fraction 1/3 is expressed as a ratio, it is written as 1:3 and is therefore not read as one third rather one to third. Ratios are very important for setting up any problem in equation form, giving a mathematical way to solve a problem.

The numerator of a fraction is equivalent to the product of another fraction and its denominator:

$$\text{If } 4/12 = 1/3$$

$$4 = 12 \times 1/3 \ [\text{or } 12 \times 1/3] = 4$$

$$1 = 4 \times 3/12 \ [\text{or } 4 \times 3/12] = 1$$

The product of the denominator of a fraction and the numerator of another is equivalent to the product of the numerator of one and the denominator of the other. The above-mentioned statement proposes that cross-products are generally equivalent.

4.4 Proportion

A description of the equivalence of two ratios is called a proportion. It could be shown in any forms mentioned below:

$$e{:}f = g{:}I, \ e{:}f{::} \ g{:}I, \ e/f = g/i$$

It is read as e is to f and g is to i. e and i are named as extremes, which means outer members, while f and g are named as means, which means middle members. Product of the means is equivalent to the product of the extremes. This rule helps us to determine the missing value of a proportion when other values are given. If the missing value belongs to the mean, it will be found by dividing the product of extremes by the provided mean. On the contrary, if the missing value belongs to an extreme, it will be determined by dividing the product of the means by the provided extreme. Utilizing this information, we can derive the below-mentioned equation [1]:

$$\text{If } e/f = g/i$$

$$E = fg/I, \ f = ec/g, \ g = ei/f, \ I = gf/e$$

In a properly set proportion, the location of the unknown value does not make any difference. Although some people put an unknown value at the fourth number, which is the denominator of the other ratio. Label the units in every place, e.g., mg and mL, to ensure correlation between the ratios of a proportion. The use of proportion helps to find solutions to many pharmaceutical problems. Proportions need to have whole numbers. If decimal and common fractions are given, they can be incorporated in proportion without modification of the method. For ease, common fractions should be converted to decimal fractions before setting up the proportion [5].

If 20 milliliters (mL) show 1/8 of the volume of a prescription, how many milliliters would show 1/6 of the volume?

$$1/8 = 0.125 \text{ and } 1/6 = 0.167$$

$$0.125 \, (\text{volume})/0.167 \, (\text{volume}) = 20 \, (\text{mL})/x \, (\text{mL})$$

$$x = 26.72 \, \text{mL}$$

If 2 tablets contain 812 mg of mefenamic acid, then 8 tablets contain how many milligrams?

$$2 \, (\text{tablets})/8 \, \text{tablets} = 812 \, (\text{mg})/x \, (\text{mg})$$
$$x \, (\text{mg}) = 8 \, (\text{tablets}) \times 812 \, (\text{mg})/2 \, (\text{tablets})$$
$$x \, (\text{mg}) = 3,248 \, \text{mg}$$

If 2 tablets contain 812 mg of mefenamic acid, then 3,248 mg are contained in how many tablets?

$$2 \, (\text{tablets})/x \, \text{tablets} = 812 \, (\text{mg})/3,248 \, (\text{mg})$$
$$x \, (\text{tablets}) = 2 \, (\text{tablets}) \times 3,248 \, (\text{mg})/812 \, (\text{mg})$$
$$x \, (\text{tablets}) = 8(\text{tablets})$$

If 8 tablets contain 3,248 mg of mefenamic acid, then how many milligrams should 2 tablets constitute?

$$8 \, (\text{tablets})/2 \, (\text{tablets}) = 3,248 \, (\text{mg})/x \, (\text{mg})$$
$$x \, (\text{mg}) = 3,248 \, (\text{mg}) \times 2 \, (\text{tablets})/8 \, (\text{tablets})$$
$$x \, (\text{mg}) = 812 \, \text{mg}$$

If 8 tablets contain 3,248 mg of mefenamic acid, then 812 mg will be contained in how many tablets?

$$8 \, (\text{tablets})/x \, (\text{tablets}) = 3,248 \, (\text{mg})/812 \, (\text{mg})$$
$$x \, (\text{tablets}) = 8 \, (\text{tablets}) \times 812 \, (\text{mg})/3,248 \, (\text{mg})$$
$$x \, (\text{tablets}) = 2 \, \text{tablets}$$

Proportion does not require containing a whole number. If a decimal fraction is given, this may be written in the proportion without modification of the method. For ease, common fractions should be converted to decimal fractions before setting up the proportion.

If 20 mL demonstrate 1/8 of the volume of a medication order, then ten 1/6 of the volume will be demonstrated by how many milliliters?

$$1/8 = 0.125, \quad 1/6 = 0.166$$

$$x = 0.125 \, (\text{volume})/0.166 \, (\text{volume}) = 20 \, (\text{mL})/x \, (\text{mL})$$

$$x = 26.56 \, \text{mL}$$

If 2 capsules contain 634 mg of acetyl salicylic acid, how many milligrams will be present in 9

$$2\,(\text{capsules})/9\,(\text{capsules}) = 634\,(\text{mg})/\text{x}\,(\text{mg})$$

$$\text{x} = 634\,\text{x}\,9/2 = 2,853\,\text{mg}$$

If 2 capsules contain 634 mg of acetyl salicylic acid, how many capsules constitute 2,853 mg?

$$2\,(\text{capsules})/\,\text{x}\,(\text{capsules}) = 634\,(\text{mg})/2,853\,(\text{mg})$$

$$\text{x} = 2 \times 2,853/634\,\text{capsules} = 9\,\text{capsules}$$

If 9 capsules constitute 2,853 mg of salicylic acid, then how many milligrams will be present in 2 capsules?

$$9\,(\text{capsules})/2\,(\text{capsules}) = 2,853\,(\text{mg})/\,\text{x}\,(\text{mg})$$

$$\text{x} = 9 \times \frac{2,853}{9} = 634\,\text{mg}$$

If 9 capsules constitute 2,853 mg of acetyl salicylic acid, how many capsules contain 634 mg?

$$9\,(\text{capsules})/\text{x}\,(\text{capsules}) = 2,853\,(\text{mg})/634\,(\text{mg})$$

$$\text{x} = \frac{9 \times 634}{2,853} = 2\,\text{capsules}$$

4.5 Specific gravity

The ratio of the weight of a substance to the weight of the same volume of water at keeping same temperature is specific gravity, shown as

Specific gravity = weight of substance/weight of the same volume of water

The pycnometer (specific gravity bottle) method is used to determine the specific gravity. In this method, weigh, clean, and dry the pycnometer (W_1). Fill with distilled water at 25 °C, wipe, and weigh (W_2). Empty, dry, and fill with test liquid at the same temperature, and weigh again (W_3). Calculate the specific gravity by using above-mentioned formula.

According to the United States Pharmacopeia, specific gravity measurements are taken at a standard temperature, which is 25 °C. It is without any unit and a dimensionless entity. Water has a specific gravity of 1; that's why the weight of water in grams is normally equal to the volume of water in milliliters. If the weight of a specific volume of a compound is given, the specific gravity could be determined by dividing its weight by its volume, shown as [3]

$$\text{Specific gravity} = \text{weight of substance (g)/volume of substance (mL)}$$

The equation is the same as density. So, the easiest way to calculate specific gravity is by calculating the density and expressing the answer without units. If specific gravity is given, the inter-conversions between weight and volume are probable utilizing the following expression:

$$\text{Weight of substance (g)} = \text{Volume of substance (mL)}* \text{specific gravity}$$

$$\text{Volume of substance (mL)} = \text{Weight of substance (g)}/\text{specific gravity}$$

If 15 mL of a liquid weighs 20 g, then the specific gravity can be determined as mentioned below:

$$\text{specific gravity of water} = 20 \text{ g/15 g} = 20 \text{ g/15 mL } 1.33.$$

Specific gravity has no units. The values of specific gravity and density in the SI system are the same (i.e., the values of specific gravity and density are equal when shown as g/mL) [4].

Examples

What is the final weight of 130 mL of liquid having a specific gravity of 224?

$$\text{Weight of liquid} = 130 \text{ mL} \times 2.24 = 291.2$$

If 33.34 mL of an essential oil weighs 42.75 g, determine the specific gravity of the oil:

$$33.34 \text{ mL of water weighs} = 33.34 \text{ g}$$

$$\text{Specific gravity} = 33.34 \text{ g}/42.75 \text{ g} = 0.77$$

The weight of a pint of liquid is 586 g. Determine the specific gravity:

$$1 \text{ pint} = 16 \text{ fl. oz}$$

$$\text{The weight of 16 fl. oz of water is 473 g}$$

$$\text{Specific gravity} = 586(\text{g})/473 \text{ (g)} = 1.24$$

If 5 mL of HCl weighs 8 g and 5 mL of water under the same conditions weighs 5 g, determine the specific gravity:

Specific gravity = weight of 5 mL of hydrochloric acid/ weight of 5 mL of water

$$\text{Specific gravity} = 8\,(g)/5(g) = 1.6$$

Specific gravity is the weight of a substance in comparison with the weight of the same volume of another substance chosen as a reference; both substances are at the same temperature. Most frequently used reference is water (the specific gravity of water is one).

4.6 Scaling-up/large formula/enlarging and reducing formula

In professional practice and pharmaceutical production, it is sometimes imperative to enlarge or reduce a formula for preparing the required amount of product. A reference formulation constitutes the specified quantities of every ingredient required to form a defined amount of product. When formulating other quantities smaller or larger, the quantitative correlation of every ingredient to other ingredients in formulae should be retained. For instance, if there are 3 g of component and 5 mL of component D in a formulation of 1,000 mL, one should take 0.3 g of component C and 10.5 mL of component D to form 100 mL. If a formula needs to be enlarged, for instance, from 1,000 mL to 3,785 mL (1 gallon) of a product, the quantity of every component needed is 3.78 times which required to form 1,000 mL of product. Amount of the product formulation is enlarged or reduced, but the quantitative correlation between every component and the product quantity remains the same [1, 6].

Factor method, ratio and proportional analysis, and dimensional analysis may be utilized for obtaining enlarged or reduced formulas. The easiest method among these is the factor method. It is based on the relative quantity of formula to be formed. For example, if 130 mL of a 1,000 mL reference formula is to be formed. The derived factor is mentioned below:

130 mL (to be formed)/1,000 mL (reference formula) = 0.13 (factor)

After that, by multiplying the quantity of every component in the formula by factors, the right amount of the component to be taken is calculated [6].

Examples

Determine the amount of every component needed to form 160 mL of product:

Calamine   70 g
Glycerin   10 g
Zinc oxide 70 g
Bentonite magma 140 mL
Calcium hydroxide solution up to 1,000 mL
Determining by ratio and proportion

$$70\,g/1,000\,mL = x\,g/160\,mL = 11.2\,g\,calamine$$

$$10\,mL/1,000\,mL = x\,mL/160\,mL; x = 1.6\,mL\,glycerin$$

$$70\,g/1,000\,mL = x\,g/160\,mL; x = 11.2\,g\,zinc\,oxide$$

$$140\,mL/1,000\,mL = x\,mL/160\,mL; x = 22.4\,mL\,bentonite\,magma$$

Determining by dimensional analysis

$$160\,mL \times 70\,g/1,000\,mL = 11.2\,g\,calamine$$

$$160\,mL \times 10\,mL/1,000\,mL = 1.6\,mL\,glycerin$$

$$160\,mL \times 70\,g/1,000\,mL = 11.2\,g\,zinc\,oxide$$

$$160\,mL \times 140\,mL/1,000\,mL = 22.4\,mL\,bentonite\,magma$$

Determining by the factor method

160 mL (to be prepared)/1,000 mL (reference formula) = 0.16 factor

The amount of every component in the reference formula will be multiplied by factor:

$$70\,g \times 0.16 = 11.2\,g\,calamine$$

$$10\,mL \times 0.16 = 1.6\,mL\,glycerin$$

$$70\,g \times 0.16 = 11.2\,g\,zinc\,oxide$$

$$140\,mL \times 0.16 = 22.4\,mL\,bentonite\,magma$$

Calculate the amount of every component needed to form a dozen 20 mL:

Polyvinyl lactate 0.2 g
Chlorobutanol   1.3 g
Povidone   1.5 g

Sterile sodium chloride solution 100 mL

20 mL × 12 = 240 mL

240 mL/100 mL = 2.4 (factor). Utilizing the factor 2.4, the amount of every component is determined as mentioned below:

Polyvinyl lactate = 0.2 g × 2.4 = 0.48 g

Chlorobutanol = 1.3 g × 2.4 = 3.1 g

Povidone = 1.5 g × 2.4 = 3.6 g

Sterile sodium chloride solution 240 mL

Determine the amount of every ingredient needed to form 3 lb of gel:

Estradiol   170 g

Methylcellulose   75 g

Polysorbate 80   2 g

1 lb = 454 g

3 lb = 454 g × 3 = 1,362 g

Formula weight = 170 g + 75 g + 2 g = 247 g

135 g/247 g = 5.47 (factor)

Utilizing this factor 5.47, the amount of every ingredient is determined as mentioned below:

Estradiol = 170 g × 5.47 = 929 g

Methylcellulose = 75 g × 5.47 = 410.25 g

Polysorbate 80 = 2 g × 5.47 = 10.94 g

Calculate the amount of every ingredient required to form 3.6 g of ointment:

Dexamethasone sodium phosphate = 44 mg

Mineral oil = 5 g

Lanolin = 3 g

White petrolatum = 100 g

3.6 g/100 g = 0.036 (factor)

Utilizing the factor 0.036, the amount of every ingredient is determined as described below:

Dexamethasone sodium phosphate = 44 mg × 0.036 = 1.58 g

Mineral oil = 5 g × 0.036 = 0.18 g

Lanolin = 3 g × 0.036 = 0.11 g

White petrolatum = 3.6 g

4.7 Dispensing of medication

Dispensing is apprehensive with the compounding and supply of pharmaceutical dosage forms. The main purpose is to prepare the medicinal agents into different dosage forms, such as in solutions, ear and eye drops, suspensions, and lotions. The consumers will be informed about the suitable storage conditions for the specific substance, and the labeling information is necessary to be applied to the final container [7, 8].

4.7.1 Key steps in the dispensing process

Prescription verification
The pharmacist must validate the rationality of the prescription, confirming it is written by a licensed practitioner, is clear, and encompasses all obligatory information (e.g., patient name, medication name, dosage, instructions).

Patient profile review
A review of the patient's medication history and profile is directed to check for potential drug interactions, contraindications, or allergies.

Medication preparation
The pharmacist formulates the medication, which may comprise measuring, counting, or compounding as needed.

Labeling
The medication must be labeled with the patient's information, dosage instructions, medication name, and other important warnings.

Patient counseling
The pharmacist offers information to the patient about the medication, including its potential side effects, how to take it, and what to do in case of missed doses.

Documentation
Appropriate records must be retained, which include details of the dispensing process, counseling provided, and any other significant information on the patient's reaction to the medication.

Examples

Antibiotics (e.g., amoxicillin)
A general practitioner prescribes 500 mg of amoxicillin to a patient suffering from a bacterial infection; the pharmacist verifies the prescription, looks up the patient's al-

lergy history (e.g., penicillin), prepares the medication, labels it with dosage instructions, and reminds the patient of the significance of finishing the course [9].

Anti-hypertensives (e.g., amlodipine)

A patient comes into the pharmacy with a prescription for amlodipine 5 mg for hypertension. The pharmacist evaluates the patient's prescriptions for potential interactions, dispenses the drug and guides the patient on blood pressure monitoring and any side effects such as dizziness.

Chronic disease management (e.g., insulin)

Insulin is prescribed to a patient with diabetes. The pharmacist makes sure the patient knows how to inject the insulin, how important it is to shift up injection locations, and how to spot hypoglycemic symptoms.

4.8 Dose calculation

Dose calculation is done to determine the required volume/number of drug(s) to be given to the patient. The medication is administered in two ways: fluid form and tablet form. To resolve these problems, pharmacists need good knowledge of fractions, decimals, unit conversions, and long division, intravenous dose, and formula oral dose [10, 11].

Stock required (SR) is the amount of drug you need to administer to the patient. In other words, this is what you want.

Stock strength (SS) is the dosage strength available in the current stock. In other words, this is what you have got.

The units for SR and SS need to be the same. If the units are different, one unit needs to be converted before doing any further calculations.

Example

The doctor has prescribed 50 mg of amitriptyline to a patient. There are 10 mg tablets available in stock. How many tablets will you administer?

Stock required: 50 mg, stock strength: 10 mg.

This is an oral dose. So, tablet dosage = stock required (SR)/stock strength (SS)

Tablet dosage = 50 mg/10 mg = 5 tablets

Example

A patient with acute cholecystitis is to receive 90 mg Gentamycin IV. A vial containing 80 mg/ 2 mL. How many mL should be given?

Stock required: 90 mg, stock strength: 80 mg/2 mL

$$\text{Dosage} = \frac{90 \text{ mg}}{80 \text{ mg}} \times 2 \text{ mL} = 2.25 \text{ mL}$$

4.8.1 Unit conversion

Example
A patient is advised to take 0.4 mg of thyroxine. Each tablet contains 200 µg. How many tablets will be administered to him?
Stock required: 0.4 mg, stock strength: 200 µg
Need to convert the SR amount to µg. So, SR = 0.4 mg = $0.4 \times 1,000 = 400 \, \mu g$

$$\text{Dosage} = \frac{400 \, \mu g}{200 \, \mu g} = 2 \text{ tablets}$$

Example
The patient was prescribed oral Phenergan 0.1 g, TDS (three times a day). The pharmacist has dispensed Phenergan 25 mg tablets. How many tablets will he administer? What dose of Phenergan will this patient receive in 24 h?
Stock required (SR): 0.1 g, stock strength (SS): 25 mg/tablet
You need to convert the SR amount to mg. So, SR = 0.1 g = $0.1 \times 1,000 = 100$ mg

$$\text{Dosage} = \frac{100 \, mg}{25 \, mg} = 4 \text{ tablets}$$

One dosage is 4 tablets. So, the dose that the client will receive in 24 h; 4 tablets × TDS (three times a day) = 12 tablets of 25 mg Phenergan.

4.8.1.1 Dose calculation in children

Dose calculation in children is a perilous practice in pediatrics, particularly in radiology, pharmacotherapy, and other medical interventions. Dosing in children often diverges from adults due to variances in body surface area (BSA), metabolic rates, and weight. Here are some references and key points that sketch the approaches for dose calculation in children [12]. There are different formulas used for dose calculation of infants and children as shown in Tab. 4.4.

4.8.1.2 Weight-based dosing

Many pharmaceutical dosage forms are dosed based on the child's weight (mg/kg). This method is a direct method but requires a good understanding of the child's weight [13].

Tab. 4.4: Different formulas are used to calculate the dose of infants and children.

S. no.	Dose proportionate to age	
1	Young's formula	Dose for the child $= \dfrac{\text{Age in years}}{\text{Age} + 12} \times$ Adult dose
2	Dilling's formula	Dose for the child $= \dfrac{\text{Age in years}}{20} \times$ Adult dose
3	Fried's formula	Dose for the child $= \dfrac{\text{Age in months}}{150} \times$ Adult dose
	Dose proportionate to body weight	
4	Clark's formula	Dose for the child $= \dfrac{\text{Weight of the child (kg)}}{70} \times$ Adult dose
	Dose proportionate to surface area	
5	Square area surface area method	Dose for the child $= \dfrac{\text{Surface area of child}}{\text{Surface area of adult}} \times$ Adult dose

Example

If a medication is prescribed at a dose of 10 mg/kg and the child weighs 21 kg, the total dose would be:

$$\text{Total dose} = \left(\frac{10 \text{ mg}}{\text{kg}}\right) \times 21 \text{ kg} = 210 \text{ mg}$$

4.8.1.3 Body surface area (BSA) dosing

Some pharmaceutical formulations are better administered when based on BSA, especially chemotherapeutics. BSA can be calculated using different formulas, with the Mosteller formula being one of the most commonly used [12]:

$$\text{BSA (m)} = \sqrt{\text{height (cm)} \times \text{weight (kg)}/3,600}$$

Example
For a child who is 121 cm tall and weighs 30 kg:

$$\text{BSA} = \sqrt{121 \text{ (cm)} \times 30 \text{ (kg)}/3,600} = 0.91 \text{ m}^2$$

Dosing also differs by sign and indications; therefore, diagnostic information is helpful when calculating doses. The following examples are typically encountered when dosing medication in children.

Example
Amoxicillin oral suspension will be given to a 1-year-old child weighing 23 lb for otitis media at a dose of 40 mg/kg/day in two divided doses. The suspension is available at a concentration of 400 mg/5 mL. How many milliliters should be administered to the child for each dose [13]?

Convert pounds to kg: $\dfrac{23\ \text{lb} \times 1\ \text{kg}}{2.2\ \text{lb}} = 10.45\ \text{kg}$ or 22.99 lb

Calculate the dose in mg: $(10.45\ \text{kg}) \times (40\ \text{mg/kg/day}) = 418\ \text{mg/day}$

Divide the dose by the frequency: $\dfrac{\frac{418\ \text{mg}}{\text{day}}}{2} = 209\ \text{mg/day}$

Convert the mg dose to mL: $\dfrac{\frac{209\ \text{mg}}{\text{dose}}}{\frac{418\ \text{mg}}{5\ \text{mL}}} = 2.42\ \text{mL}$

Example
Ceftriaxone is being prescribed for a 5-year-old child weighing 20 kg for meningitis at a dose of 100 mg/kg IV once daily. After reconstitution, the concentration of ceftriaxone solution in the vial is 40 mg/mL. How many milliliters of the solution should be administered to this child for each dose [13]?

Calculate the dose in mg: $(20\ \text{kg} \times 100\ \text{mg/kg/day}) = 2{,}000\ \text{mg/day}$

Divide the dose by the frequency: $(2{,}000\ \text{mg/day})/1\ \text{daily} = 2{,}000\ \text{mg/dose}$

Convert the mg dose to mL: $(2{,}000\ \text{mg/dose})/40\ \text{mg/mL} = 50\ \text{mL}$

Example
Vincristine is being administered to a 4-year-old child (height 97 cm; weight 37 lb) with leukemia at a dose of 2 mg/m^2. Vincristine is available in a vial at a concentration of 1 mg/mL. How many milliliters should be administered to this child for each dose [13]?

Convert pounds to kg:	$(37 \text{ lb} \times 1 \text{ kg})/2.2 \text{ lb} = 16.8 \text{ kg}$
Calculate BSA:	$\sqrt{16.8 \text{ kg} \times 97 \text{ cm}/3,600} = 0.67 \text{ m}^2$
Calculate the dose in mg:	$(2 \text{ mg/m}^2) \times (0.67 \text{ m}^2) = 1.34 \text{ mg}$
Calculate the dose in mL:	$1.34 \text{ mg} \div 1 \text{ mg/mL} = 1.34 \text{ mL}$

4.8.1.4 Age-based dosing

Some medications have age-based dosing guidelines, which highlight the specific dosage range for different pediatric age groups (neonates, infants, children, and adolescents) [12, 14].

Adjustment for special populations

Dosage may also need to be adjusted for specific conditions such as renal impairment or liver dysfunction [15].

Important considerations

Always verify calculation methods with institutional protocols or guidelines.
Monitor the child for efficacy and adverse effects following dosing.
Consult a clinical pharmacist or pediatric specialist when in doubt.
This information serves as a fundamental overview of pediatric dosing principles and should always be supplemented with clinical judgment and professional guidelines [16].

Examples

Acetaminophen (Tylenol): Common dosing: 10–15 mg/kg/dose every 4–6 h as needed, not to exceed 5 doses in 24 h.
Amoxicillin: Common dosing: 20–40 mg/kg/day, divided into two or three doses.

4.8.2 Dose calculation in adults

Dose calculation in adults is a critical aspect of pharmacotherapy, particularly in ensuring that medications are administered safely and effectively. The determination of appropriate dosages can be influenced by various factors, including body weight, height, age, and specific pharmacokinetic properties of the drugs involved [17].

4.8.2.1 Basic formula for dose calculation

The basic formula for calculating the dose is:

$$\text{Desired dose} = \frac{\text{Available dose}}{\text{Amount to administer}}$$

Example
A physician orders 500 mg of a medication. The medicine is available in 250 mg tablets.
    Desired dose = 500 mg
    Available dose = 250 mg/tablet
    Quantity = 1 tablet

$$\text{Amount to administer} = \frac{\text{Desired dose}}{\text{Available dose}}$$

$$\text{Amount to administer} = \frac{500 \text{ mg}}{250} \text{mg} = 2 \text{ tablets}$$

The patient should take 2 tablets of 250 mg each.

Example
A patient needs a dose of 0.75 g of a medication. The injection is available in vials containing 1 g/mL:
    Desired dose = 0.75 g = 750 mg
    Available dose = 1 g/mL = 1,000 mg/mL

$$\text{Amount to administer} = \frac{\text{Desired dose}}{\text{Available dose}}$$

$$\text{Amount to administer} = \frac{750 \text{ mg}}{1000 \text{ mg/mL}} = 0.75 \text{ mL}$$

The patient should receive 0.75 mL of the injectable medication.

Example

A physician orders 1.5 g of an antibiotic, which is available in a concentration of 500 mg/250 mL:

Desired dose = 1.5 g = 1,500 mg
Available dose = 500 mg/250 mL
First, determine how much volume 1,500 mg includes:

$$\text{Volume required} = \frac{\text{Desired dose}}{\text{Available dose}}$$

$$\text{Volume required} = \frac{500 \text{ mg} \times 250 \text{ mL}}{500 \text{ mg}}$$

Thus, to administer 1.5 g of the antibiotic, 750 mL will be required.

4.9 Isotonicity calculation

Isotonicity refers to the property of a solution that has the same osmotic pressure as another solution, typically a physiological solution, such as blood plasma or intracellular fluid. Isotonic solutions have the same concentration of solutes as the body fluids, which means they do not cause cells to swell (in a hypotonic solution) or shrink (in a hypertonic solution) [18].

To calculate isotonicity, you need to consider the osmolarity of the solution in relation to a standard isotonic solution, which is typically 0.9% saline (sodium chloride solution) for human physiology [19].

4.9.1 Steps for Isotonicity Calculation

Identify the solute

Determine which solute is being used and its concentration in the solution.

Convert concentration

If the solute is a solid, convert the concentration to molarity (moles per liter). The formula is: **molarity = moles of solute/liters of solution**

Calculate osmolarity

Osmolarity accounts for the total number of particles in the solution:

Osmolarity (OsM) = molarity (M) × number of particles per formula unit

For example, NaCl dissociates into two particles (Na^+ and Cl^-), so for NaCl, the osmolarity would be twice the molarity.

Comparison with physiological osmolarity

Human plasma is approximately 280–300 mOsm/L. An isotonic solution should have an osmolarity close to this range.

Adjustment (if needed)

If the calculated osmolarity is significantly different from 300 mOsm/L, you may need to adjust the concentration of your solute.

Example

You have a 0.9% NaCl solution.
    Convert % to grams
    0.9% NaCl means 0.9 g of NaCl in 100 mL of water, which is 9 g/L.
    Calculate molarity
    Molar mass of NaCl = 58.44 g/mol

$$\text{Molarity (M)} = \frac{9}{58.44} = 0.154 \text{ M}$$

Calculate osmolarity

NaCl dissociates into two ions, so

$$\text{Osmolarity} = \text{Molarity} \times \text{no. of dissociation ions}$$
$$\text{Osmolarity} = 0.154 \times 2 = 0.308 \text{ OsM or 308 mOsM/L}$$

References

[1]   Ansel HC. Pharmaceutical calculations: Lippincott Williams & Wilkins; Philadelphia, USA, 2012.

[2]   Allen L, Ansel HC. Ansel's pharmaceutical dosage forms and drug delivery systems: Lippincott Williams & Wilkins; Philadelphia, USA, 2013.

[3]   Teixeira MG, Zatz JL. Pharmaceutical calculations: John Wiley & Sons; New Jersey, USA, 2017.

[4]   Agarwal P. Pharmaceutical calculations: Jones & Bartlett Publishers; USA, 2014.

[5]   Ballington DA, Green TW. Pharmacy calculations for technicians: Paradigm Education Solutions; Dubuque, USA, 2007.

[6]   Ansel HC, Stockton S. Pharmaceutical calculations: 2016.

[7]   Drysdale TA. APhA-ASP. Journal of the American Pharmacists Association. 2006; 46(5): 543–45.

[8]   Li JK, Galvez CC. Board of pharmacy. California Regulatory Notice Register. 2000; 17: 89.

[9]   Cullen DJ, Bates DW, Leape LL, Group ADEPS. Prevention of adverse drug events: A decade of progress in patient safety. Journal of Clinical Anesthesia. 2000; 12(8): 600–14.

[10] McCollough CH, Schueler BA. Calculation of effective dose. Medical Physics. 2000; 27(5): 828–37.

[11] Toney-Butler TJ, Nicolas S, Wilcox L Dose calculation desired over have formula method. 2018.

[12] Diseases CoI. From the American Academy of Pediatrics: Policy statements–Modified recommendations for use of palivizumab for prevention of respiratory syncytial virus infections. Pediatrics. 2009; 124(6): 1694–701.

[13] Ciccone CD. Davis's drug guide for rehabilitation professionals: FA Davis; Philadelphia, USA, 2013.

[14] Zimmerman JJ, Fuhrman BP. Pediatric critical care e-book: Elsevier health sciences; Philadelphia, USA, 2011.

[15] Pediatrics AAo. Pediatric hospital medicine: A case-based educational guide: American Academy of Pediatrics; Itasca, USA, 2022.

[16] Bereda G. Pediatrics pharmacokinetics and dose calculation. Journal of Pediatrics & Neonatal Cares. 2022; 12: 96–102.

[17] Marek KD, Antle L. Medication management of the community-dwelling older adult. Patient Safety and Quality: An Evidence-based Handbook for Nurses. 2008. pp. 499–536.

[18] Bellini F. Isotonicity properties of generalized quantiles. Statistics & Probability Letters. 2012; 82(11): 2017–24.

[19] Savva M, Savva M. Isotonic solutions. Pharmaceutical Calculations: A Conceptual Approach. 2019; 369: 157–80.

Asia Naz Awan, Asiya Farheen, Muhammad Yasir Ali,
Hafiza Arwa Nadeem, Eisha Mashkoor, Muhammad Hashim, Ali Abbas,
Muhammad Fayyaz

5 Buffers

5.1 Introduction

A buffer solution, also known as a buffer, is a combination of a weak acid and its conjugate base (or a mixture of a weak base and its conjugate acid). Buffer solutions maintain their pH when small volumes of strong acids or bases are introduced. The carbonic acid/bicarbonate buffer system maintains the pH of the plasma at 7.4 [1]. The properties of a good buffer are described in Fig. 5.1.

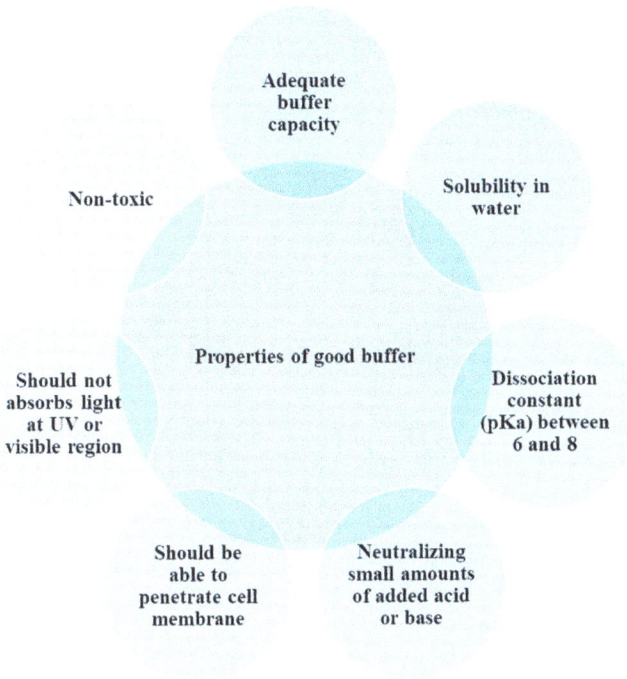

Fig. 5.1: Properties of a good buffer.

https://doi.org/10.1515/9783111438108-005

5.2 Buffer solution examples

- Acetic acid and conjugate base: CH_3COOH and CH_3COO^-
- Formic acid and conjugate base: $HCHO_2$ and CHO_2^-
- Pyridine and conjugate acid: C_5H_5N and $C_5H_5H^+$
- Ammonia and conjugate acid: NH_3 and NH_4^+
- Methylamine and conjugate acid: CH_3NH_2 and $CH_3NH_3^+$
- Many biological systems require buffers for balance

5.3 Types of buffers

5.3.1 Acidic buffer solutions

Acidic buffer is composed of a weak acid and its salt (strong base + weak acid), e.g., acetate buffer ($CH_3COONa + CH_3COOH$). Salt used in an acidic buffer is composed of a strong base and a weak acid because the salt of a weak acid is a strong conjugate base. Strong bases consist of oxides and hydroxides of group I and group II.

Let's consider an example of ethanoic acid, which is a weak acid and partially ionizes, setting up an equilibrium:

$$CH_3COOH_{(aq)} \rightleftharpoons CH_3COO^-_{(aq)} + H^+_{(aq)}$$

The salt fully ionizes as:

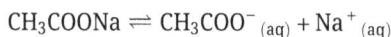

$$CH_3COONa \rightleftharpoons CH_3COO^-_{(aq)} + Na^+_{(aq)}$$

The presence of sufficient H^+ ions and a large number of undissociated acid molecules from the salt solution makes the buffer solution pH acidic.

Adding an acid to this buffer solution: When H^+ ions are added to the acidic buffer, they combine with a few ethanoate ions to make more undissociated acid molecules.

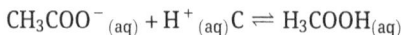

$$CH_3COO^-_{(aq)} + H^+_{(aq)}C \rightleftharpoons H_3COOH_{(aq)}$$

Adding an alkali to this buffer solution: When OH_- ions are added to an acidic buffer, they react with the H^+ ions produced from the dissociated acid molecules. As the H^+ ions are used up, the equilibrium shifts to the right (Fig. 5.2):

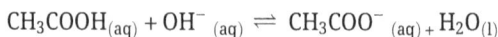

$$CH_3COOH_{(aq)} + OH^-_{(aq)} \rightleftharpoons CH_3COO^-_{(aq)} + H_2O_{(l)}$$

H₃O⁺ added, equilibrium position shifts to the left

$$CH_3COOH_{(aq)} \longleftarrow CH_3COO^-_{(aq)} + H_3O^+$$

OH⁻ added, equilibrium position shifts to the right

$$OH^- + CH_3COOH_{(aq)} \longrightarrow H2O_{(l)} + CH3COO^-_{(aq)}$$

Buffer solution after addition of strong acid

Buffer solution Equimolar in Acid and base

Buffer solution after addition of strong base

Add H₃O⁺

Add OH⁻

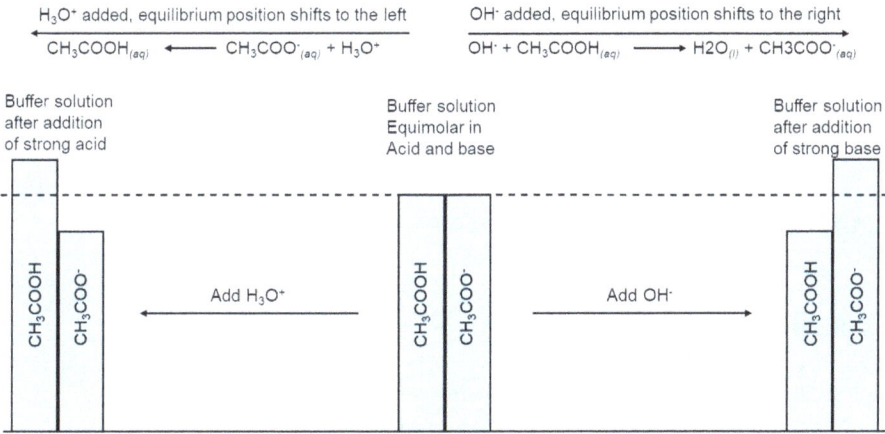

Fig. 5.2: Buffer action after addition of strong acid/base.

5.3.2 Alkaline buffer solutions

Basic buffer is composed of a weak base and its salt (strong acid + weak base), having a pH greater than 7, e.g., ammonia buffer (NH₄OH + NH₄Cl, NH₄OH + NH₄NO₃). In order to increase the ionization of a weak acid or base, strong salts are added.

The salt fully dissociates into

$$NH_4Cl(aq) \rightarrow NH_4{}^+(aq) + Cl^-(aq)$$

The ammonia solution partially dissociates into

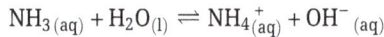

$$NH_{3\,(aq)} + H_2O_{(l)} \rightleftharpoons NH_4{}^+_{(aq)} + OH^-{}_{(aq)}$$

As a result of the presence of sufficient OH₋ ions and a large number of undissociated base molecules from the salt solution, the buffer solution has an acidic pH.

Adding an acid to this buffer solution: When H⁺ ions are added to the basic buffer, they combine with OH⁻ions released when the ammonia molecules dissociate. The equilibrium shifts to the right to replace the OH⁻ ions:

$$H^+{}_{(aq)}\,OH_{(aq)} \rightleftharpoons H_2O_{(l)}$$

Adding an alkali to this buffer solution: When OH₋ ions are added to the basic buffer, they combine with NH₄⁺ ions from the salt to make more undissociated ammonia molecules:

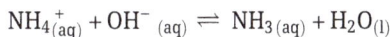

$$NH_4{}^+_{(aq)} + OH^-{}_{(aq)} \rightleftharpoons NH_{3\,(aq)} + H_2O_{(l)}$$

5.3.3 Effective buffers

The most effective buffers contain equal or nearly equal concentrations of an acid and its conjugate base. This type of buffer is effective at neutralizing either added acid or base.

For example, 0.8 M HNO_2 and 1 M $NaNO_2$ is considered as effective buffer, whereas 0.8 M HNO_2 and 0.002 M $NaNO_2$ is not effective buffer.

5.3.4 Biological buffers

Biological buffers should meet the following general criteria [2]:
- They should have a pK_a between 6.0 and 8.0.
- They should not penetrate cell membranes and have a high solubility in water and a low solubility in organic solvents.
- They must not be harmful to cells in any way.
- They must not obstruct any biological function.
- Although salts can be supplied as needed, the salt effect should be negligible.
- Temperature and the medium's ionic makeup should have no impact on buffering capacity.
- Buffers should be stable and resistant to enzymatic degradation.
- The buffer shouldn't absorb in the UV or visible spectrum. These unique properties must be present in the majority of buffers used in cell cultures, cell isolation, enzyme tests, and other biological applications. These requirements are satisfied by Good's zwitterionic buffers. Their pK_a values are at or close to physiological pH. Because their cationic and anionic sites are present as non-interacting carboxylate or sulfonate and cationic ammonium groups, respectively, they show minimal interference with biological processes.

5.3.5 Volatile buffers

In some circumstances, a buffer must be swiftly and totally removed. It is feasible to exclude elements that could cause problems in later processes by using volatile buffers. Ion-exchange chromatography, electrophoresis, and protein digestion followed by peptide or amino acid separation are all applications for volatile buffers. Except for the buffers containing pyridine, the majority of the volatile buffers (Tab. 5.1) are transparent in the lower UV range [3].

Tab. 5.1: Types of system for use as volatile buffers [3].

S. no.	System	pH range
1	87 mL glacial acid + 25 mL 88% HCOOH in 1 L	1.9
2	25 mL 88% HCOOH in 1 L	2.1
3	Pyridine-formic acid	2.3–3.5
4	Trimethylamine-formic acid	3.0–5.0
5	Triethylamine-formic acid (or acetic acid)	3.0–5.0
6	5 mL pyridine + 100 mL glacial acetic acid in 1 L	3.1
7	5 mL pyridine + 50 mL glacial acetic acid in 1 L	3.5
8	Trimethylamine-acetic acid	4.0–6.0
9	25 mL pyridine + 25 mL glacial acetic acid in 1 L	4.7
10	Collidine-acetic acid	5.5–7.0
11	100 mL pyridine + 4 mL glacial acetic acid in 1 L	6.5
12	Triethanolamine HCl	6.8–8.8
13	Ammonia-formic (or acetic) acid	7.0–10.0
14	Trimethylamine CO_2	7–12
15	Trimethylamine CO_2	7–12
16	24 g NH_4HCO_3 in 1 L	7.9
17	Ammonium carbonate ammonia	8.0–10.5
18	Ethanolamine HCl	8.5–10.5
19	20 g $(NH_4)CO_3$ in 1 L	8.9

5.3.6 Broad range buffers

Sometimes, a single buffer system that can cover a broad pH range of maybe five or more pH units is needed. A combination of buffers that adequately spans the desired pH range would be one approach. This could result in nonspecific buffer interactions that need to be adjusted. Using a sequence of structurally related buffers with uniformly spaced pK values, where each pK is separated by roughly 1 pH unit (the buffering capacity limit), is another popular strategy [3].

5.3.7 Pharmaceutical buffers

A buffer solution might consist of weak bases or weak acids. Acetate buffer, citrate buffer, and phosphate buffer are common buffer systems found in pharmaceuticals (Tab. 5.2). Buffers are essential in pharmacies because they shield medicinal preparations from abrupt pH changes. By modifying the pH values of products, buffers are utilized in pharmaceutical systems to guarantee maximum stability [1].

Tab. 5.2: Pharmaceutical buffer systems with their pK_a and pH range.

S. no.	Buffer system	pK_a	pH range
1	Acetic acid/sodium acetate	4.76	3.8–5.6
2	Phosphate acid/sodium phosphate – H_3PO_4/NaH_2PO_4 – NaH_2PO_4/Na_2HPO_4 – NaH_2PO_4/Na_3HPO_4	2.1 (pK_{a1}) 7.2 (pK_{a2}) 12.3 (pK_{a3})	5–8
3	Citric acid/sodium citrate	3.1 (pK_{a1}) 4.8 (pK_{a2}) 9.2 (pK_{a3})	1.2–6.6
4	Boric acid/sodium borate	9.2	7.8–10.6

5.3.8 Drugs as buffers

It is crucial to understand that buffer action is also present in medication solutions that are weak electrolytes. The alkalinity of the glass affects the salicylic acid solution in a soft glass container. The soft glass's sodium ions interact with the salicylate ions to generate sodium salicylate, despite the initial assumption that the reaction would cause a noticeable rise in pH. As a result, a buffer solution consisting of sodium salicylate and salicylic acid forms, which prevents the pH shift. In a similar vein, an ephedrine base solution exhibits a built-in buffer against pH drops. The buffer system of ephedrine plus ephedrine hydrochloride will withstand significant pH variations until the ephedrine is consumed by the acid's reaction. As a result, throughout a specific pH range, a medicine in solution may frequently function as its own buffer. However, this buffer effect is frequently insufficient to reverse pH shifts caused by the air's carbon dioxide and the bottle's alkalinity. Therefore, throughout a specific pH range, a drug in solution may frequently function as its own buffer. However, this buffer effect is frequently insufficient to reverse pH shifts caused by the air's carbon dioxide and the bottle's alkalinity. As a result, extra buffers are usually added to medication solutions in order to keep the system within a specific pH range [1].

5.4 Factors affecting the pH of buffer solutions

5.4.1 Ionic strength

By changing the ionic strength, the addition of neutral salts to buffers modifies the pH of the solution. Alterations in a buffer solution's ionic strength and, consequently, pH can also result from dilution. Although minor water additions do not change pH, they can induce a slight positive or negative variation because they change activity coefficients and because water has the ability to function as a mild acid or base [4, 5].

5.4.2 Temperature

Buffers are also affected by temperature. For many buffers, Kolthoff and Tekelenburg calculated the temperature coefficient of pH, or the pH variation with temperature. It was discovered that whereas the pH of boric acid-sodium borate buffers fell with temperature, the pH of acetate buffers rose. K_w, which appears in the equation of basic buffers and changes significantly with temperature, was discovered to be the reason why the pH of the majority of basic buffers changed more noticeably with temperature, even though the temperature coefficient of acid buffers was very minor [5].

5.4.3 Common ion effect

The impact of an ion on a substance's solubility equilibrium is known as the "common ion effect." The equilibrium point of the solution may vary if a soluble chemical made up of a common ion is added because it may lower the concentration of that ion in the solution. Analyzing the solubility of weak electrolytes, like salts, reveals this.

5.4.4 Common ion effect in buffer solutions

The common ion effect is what causes the pH to alter when a conjugate ion is added to a buffer solution. When a strong electrolyte, such as its conjugate base salt, is added to an acid buffer solution and dissolved, the electrolyte will dissociate and produce both common ions and ions that will partially ionize the acid. The acid's ionization will be offset by the common ions that are produced. The equilibrium will change in favor of the reactants (to the left) as a result of Le Chatelier's principle. As the acid's ionization (and dissociation) diminishes, the pH rises [1, 5].

5.4.5 Dissociation constants (K_a, K_b, and K_w)

5.4.5.1 Ionization constant for water (K_w and pK_w)

K_w is known as the ionization constant for water. The value of K_w is very small, i.e., 1.0×10^{-14} (at 25 °C). It is a neutral solution. K_a and K_b are related to each other through the ion constant for water, K_w [2]:

$$K_w = K_a \times K_b$$

pK_a and pK_b are related by the simple relation:

$$pK_a + pK_b = 14$$

5.4.5.2 Acidic dissociation constant (K_a and pK_a)

The acidic dissociation constant is a measure of how completely an acid dissociates into its component ions in water. The smaller the pK_a value stronger the acid.
pK_a is simply the –log of this constant:

$$pK_a = -\log K_a$$

5.4.5.3 Base dissociation constant (K_b)

It describes how the ions that make up the base split into positive and negative components.
Stronger the base, the greater the value of K_b.

5.5 Relationship between pK_a, pK_b, and pK_w

K_a: dissociation constant of acid

K_b: dissociation constant of base

K_w: dissociation constant of water

whereas
K_w: dissociation constant of water

$$pK_a = -\log K_a$$
$$pK_b = -\log K_b$$
$$pK_w = -\log K_w$$

The acid/base dissociation constant is usually expressed in mol/L.
A simple relationship between the dissociation constant of a weak acid HF and that of its conjugate base F^-.

5.5.1 Inverse logs for pH, pOH, and pK_w

$$pH = -\log H_3O^+ \qquad 10 - pH = [H_3O^+] \text{ or } [H^+]$$
$$pOH = -\log OH^- \qquad 10 - pOH = [OH^-]$$
$$pK_w = -\log(K_w) \qquad K_w = 10 - pK_w$$

K_w is called the ion-product constant of liquid water and is defined as

$$K_w = [H^+][OH^-]$$
$$K_w = [H^+][OH^-] = 1.0 \times 10^{-14}$$
$$pK_w = -\log([H^+][OH^-]) = 14.00$$
$$pK_w = pH + pOH = 14.00$$

Example

$$NH_4 + H_2O \rightleftharpoons NH_4^+ + OH^-$$
$$K_a = 5.8 \times 10^{-10}$$
$$K_b = 1.8 \times 10^{-5}$$
$$pK_a + pK_b = 14$$
$$= -\log 5.8 \times 10^{-1} + -\log 1.8 \times 10^{-5}$$
$$= 9.23 + 4.74 = 14$$
$$HF + H_2O \rightleftharpoons H_3O^+ + F^-$$
$$K_a = 6.8 \times 10^{-10}$$
$$K_b = 1.5 \times 10^{-5}$$
$$pK_a + pK_b = 14$$
$$= -\log 6.8 \times 10^{-1} + -\log 1.5 \times 10^{-11}$$
$$= 3.16 + 10.82 = 14$$

5.5.2 Application of inverse log

pH and pOH calculation

The inverse logarithm is used to calculate pH and pOH values, which are measures of the acidity and basicity of a solution, respectively [6].

Strength of solutions

It is also used to determine the strength of acids, bases, and solutions. By finding the pK_a, pK_b, or pK_w, we can predict the behavior of substances in chemical reactions.

Dose calculation

The inverse logarithm is used in calculating drug doses, particularly in situations where the drug concentration is expressed on a logarithmic scale.

Drug potency

The inverse logarithm is used to express the potency of a drug, which is the amount of drug required to produce a certain effect.

5.5.3 Inverse logs for pK_a, pK_b, and pK_w

pK_a/pK_b is defined as "the negative log of K_a/K_b"

$$pK_a = -\log K_a \qquad\qquad K_a = 10 - pK_a$$

$$pK_b = -\log K_b \qquad\qquad K_b = 10 - pK_b$$

$$pK_w = -\log K_w \qquad\qquad K_w = 10 - pK_w$$

The value of K_w is defined as 1.0×10^{-14} at 25 °C.

Therefore, the pK_w of water is:

$$pK_a + pK_b = pK_w$$

$$pK_w = -\log(1.0 \times 10^{-14})$$

$$pK_w = 14$$

5.6 Henderson-Hasselbalch equation

The Henderson-Hasselbalch equation, named after English scientist E. J. Henderson, is a mathematical equation that estimates the acidity of a solution by relating the concentration of H^+ ions in the solution to its pH (Fig. 5.3):

– This equation shows a relationship between the pH or pOH of the solution, the pK_a or pK_b, and the concentration of the chemical species involved in Fig. 5.3.
– The Hasselbalch equation is used to calculate the pH of the bicarbonate buffer system in blood, as well as the percentage of drugs ionized at that pH.
– Ionization is defined as "the percent dissociation of an acid varies with the concentration of acid [7]."

pH of buffer solution

Acidic Buffer Solution		Basic Buffer Solution	
Weak acid	Salt of its Conjugate base	Weak base	Salt of its Conjugate acid
	acidic buffer		basic buffer
Henderson – Hasselbalch equation		Henderson – Hasselbalch equation	
pH=pK$_a$+ log [conjugate base] / [weak acid]		pH=pK$_a$+ log [conjugate acid] / [weak base]	

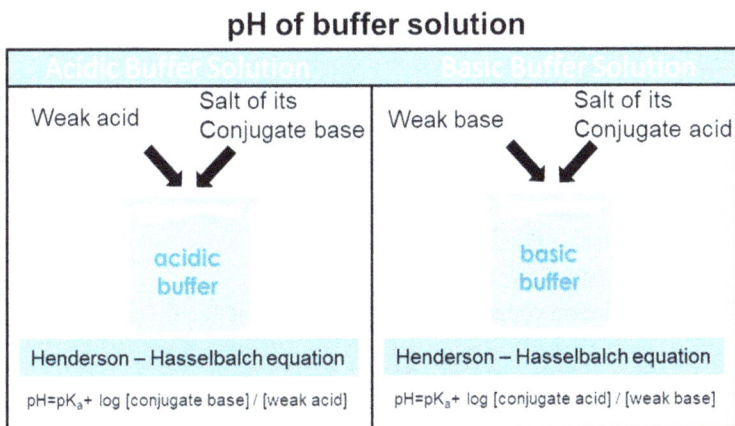

Fig. 5.3: pH of buffer and the Hasselbalch equation.

Let us take an example of the ionization of a weak acid HA:

$$HA \rightleftharpoons H^+ + A^-$$

The dissociation constant of a weak acid HA is as follows:

$$K_a = [H^+][A^-]/[HA]$$

On rearranging the equation,

$$[H^+] = K_a[HA]/[A^-]$$

By taking a log of the above equation,

$$\log [H^+] = \log K_a[HA] / [A^-]$$

$$\text{Or, } \log[H^+] = \log K_a + \log [HA] / [A^-]$$

Multiplying the negative sign on both sides,

$$- \log [H^+] = - \log K_a - \log [HA] / [A^-]$$

Since

$$- \log [H^+] = pH$$

$$- \log K_a = pK_a$$

Therefore,

$$= pK_a - \log [HA] / [A^-]$$

where
– [A⁻] = concentration of salt
– [HA] = concentration of the weak acid

$$pH = pK_a - \log [\text{acid}]/[\text{salt}]$$

The Henderson-Hasselbalch equation can be expressed in two ways:
pH = pK_a + log [conjugate base]/[weak acid](for weak acid)
pH = pK_a + log [conjugate acid]/[weak base](for weak base)
where pK_a is the dissociation constant of an acid and pK_b is the dissociation constant of the base.

5.6.1 Determination of pK_a

pK_a values are generally determined by titration. A carefully calibrated, automated recording titrator is used; the free acid of the material to be measured is titrated with a suitable base, and the titration curve is recorded. The pH of the solution is monitored as increasing quantities of base are added to the solution. Figure 5.4 shows the titration curve for acetic acid. The point of inflection indicates the pK_a value. Frequently, automatic titrators record the first derivative of the titration curve, giving more accurate pK_a values. Polybasic buffer systems can have more than one useful pK_a value.

Example
If [CH₃COOH] is equal to 1.0 mol/dm³ and [CH₃COONa] is 0.1 mol/dm³, then find the pH of the buffer solution:

$$pH = 4.74 + \log 0.1/1$$

$$= 4.74 + \log 1/10$$

$$= 4.74 + \log 10^{-1}$$

$$= 4.74 - 1$$

$$pH = 3.74$$

5.6.2 Applications of the Henderson-Hasselbalch equation

The Henderson-Hasselbalch equation can be applied in the following ways:
– The Henderson-Hasselbalch equation determines which acid or base should be selected to prepare a buffer of the required pH and is routinely used in the treatment of acid-base abnormalities.

- It is used to determine the salt-to-acid ratio to help in pH calculation.
- It is used to determine the amount of acid and base required.
- It is used to determine the change in pH caused by adding a strong base to a weak acid solution.
- It determines the pK_a and pK_b.
- The Henderson-Hasselbalch equation has proved invaluable in aiding the understanding of mammalian acid-base physiology and is routinely used in the treatment of acid-base abnormalities.

5.6.3 Relationship of buffers and the Henderson-Hasselbalch equation

To determine the pH of a buffer system [8], you must know the acid's dissociation constant. This value, K_a (or K_b or a base), determines the strength of an acid (or base). The pK_a of an acid is the negative logarithm of its acid dissociation constant. This is analogous to pH (the negative logarithm of the H^+ concentration).

The Henderson-Hasselbalch equation allows the calculation of a buffer's pH.

For a buffer created from an acid, the equation is:

$$pH = pK_a + \log \frac{[A^-]}{[HA]}$$

For a buffer created from a base, the equation is:

$$pH = pK_b + \log \frac{[B]}{[HB^+]}$$

If the pK_a value is used to indicate an acid's or base's strength, the smaller the pK_a value, the stronger the acid; the greater the pK_a value, the stronger the base.

Example
What is the pH of a buffer that is 0.12 M in lactic acid ($HC_3H_5O_3$) and 0.1 M sodium lactate ($C_3H_5O_3^-$)?
For lactic acid $K_a = 1.4 \times 10^{-4}$

$$HC_2H_5O_3 \rightleftharpoons C_3H_5O^- + H^+$$

$$pK_a = -\log K_a$$

$$pK_a = -\log 1.4 \times 10^{-4}$$

$$pK_a = 3.85$$

Applying the Hasselbalch equation,

$$pH = pK_a + \log [A^-]/[HA]$$
$$pH = 3.85 + \log [0.1]/[0.12]$$
$$pH = 3.77$$

Example

If 0.005 mol of NaOH is added to the above 0.5 L of solution, what will be the resulting pH?

$$M = mol/L$$
$$= 0.005/0.5$$
$$= 0.01 \, M$$

$$NaOH \rightarrow Na^+ + OH$$

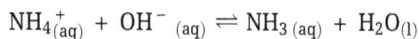
$$NH_{4\,(aq)}^+ + OH^-_{\,(aq)} \rightleftharpoons NH_{3\,(aq)} + H_2O_{(l)}$$

0.2	0.01	0.24
− 0.01	− 0.01	+ 0.01
0.19	0	0.25

$$pH = 9.25 + \log 0.25/0.19$$
$$pH = 9.25 + 0.12$$
$$pH = 9.37$$

Example

If 0.03 mol of HCl is added to 0.5 L of the buffer solution (0.24 M and 0.2 M NH$_4$Cl), what is the resulting pH?

Molarity in 0.5 L = 0.03/0.5 = 0.06 M

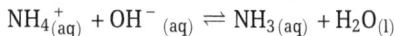
$$NH_{4\,(aq)}^+ + OH^-_{\,(aq)} \rightleftharpoons NH_{3\,(aq)} + H_2O_{(l)}$$

0.24	0.06	0.2
− 0.06	− 0.06	+ 0.06
0.18	0	0.26

$$pH = 9.25 + \log 0.18/0.26$$
$$pH = 9.25 - 0.161$$
$$pH = 9.09$$

5.7 Buffer system in the blood

The normal pH of human blood is about 7.4. The carbonate buffer system in the blood uses the following equilibrium reaction:

$$CO_2(g) + 2H_2O(l) \rightleftharpoons H_2CO_3(aq) \rightleftharpoons HCO_3^-(aq) + H_3O^+(aq)$$

The concentration of carbonic acid, H_2CO_3, is approximately 0.0012 M, and the concentration of the hydrogen carbonate ion, HCO_3^-, is around 0.024 M. Using the Henderson-Hasselbalch equation and the pK_a of carbonic acid at body temperature, we can calculate the pH of blood [9, 10]:

$$pH = pK_a + \log[\text{base}][\text{acid}] = 6.4 + \log 0.0240.0012 = 7.7$$

It may seem strange that the H_2CO_3 concentration is much lower than that of the HCO_3^- ion, but this imbalance results from the fact that the majority of the metabolic byproducts that enter our bloodstream are acidic. Consequently, a higher percentage of base than acid is required to ensure that the buffer's capacity is not exceeded. Our muscles produce lactic acid when we work out. The lactic acid is neutralized by the HCO_3^- ion as it enters the bloodstream, resulting in H_2CO_3. After that, an enzyme speeds up the breakdown of the extra carbonic acid into carbon dioxide and water, which the body can exhale. The body actually employs breathing to control blood pH in addition to the carbonate buffering system's regulating actions. The equilibrium reaction is lowered if the blood's pH drops too much because breathing more causes CO_2 to be expelled from the blood through the lungs. A reduced breath rate raises the blood's CO_2 concentration if it is too alkaline, which pushes the equilibrium reaction in the other direction, raising $[H^+]$ and bringing the pH back to normal.

5.8 Ionization of a drug molecule

The ionization of drug molecules is significant in terms of their absorption into the circulation and distribution to various tissues inside the body. The lesser the ionization, the less will be the absorption as the ionized form of the drug can't cross the lipophilic membrane.

The pK_a value of a drug affects its formulation into a medicine and analytical procedures for its determination.

Alternatively, the percentage of ionization for an acid or a base of a particular pK_a value at a particular pH value is expressed by:

$$\text{Acid: \% ionization} = \frac{10\,pH - pK_a}{1 + 10\,pH - pK_a} \times 100$$

$$\text{Base: \% ionization} = \frac{10 pK_a - pH}{1 + 10 pK_a - pH} \times 100$$

5.8.1 Advantages of ionization

- Ionization of any drug molecule is important to know for that drug to be absorbed in the human body.
- Formulation – Ionized drug is more soluble.
- Absorption – Unionized drugs easily move across the membrane.
- Distribution – Unionized drugs have a high volume of distribution and binding.
- Excretion – Ionized drugs are excreted more readily.

Example

When acetic acid (pK$_a$ 4.76) is in solution at pH 4.76, the Henderson-Hasselbalch equation can be written as follows:

$$pH = 4.76 + \log \frac{[CH_3COO^-]}{[CH_3COOH]}$$

From this relationship for acetic acid, the degree of ionization for acetic acid is determined by:

So, when pH is 4.76, then:

$$4.76 = 4.76 + \log \frac{[CH_3COO^-]}{[CH_3COOH]}$$

$$\log \frac{[CH_3COO^-]}{[CH_3COOH]} = 0$$

$$\frac{[CH_3COO^-]}{[CH_3COOH]} = 10^0 = 1$$

5.8.2 Percent ionization

From the Henderson-Hasselbalch equation

$$pH = pK_a + \log[A^-] / [HA]$$

By arranging this formula:

$$\text{Log}[A^-]/[HA] = pH - pK_a$$

To eliminate this log, we use 10 [10].

    If $x = y$, then it can be written as

$$10^x = 10^y$$

Thus,

$$10^{\log[A^-]/[HA]} = 10^{(pH - pK_a)}$$

So, the equation becomes:

    $[A^-]/[HA] = 10(pH–pK_a)/1$

    Percent ionization is equal to ionized/unionized + ionized×100.

    Thus,

$$\% \text{ ionization} = 10^{(pH - pK_a)} \times 100 \text{ (for acid)}$$

$$1 + 10^{(pH - pK_a)}$$

For base: $BH^+ \longleftrightarrow B + H^+$

$$K_a = [B]/[BH^+]$$

$$pH = pK_a + \log B/BH^+$$

$$\% \text{ ionization} = 10^{(pK_a - pH)}/1 + 10^{(pK_a - pH)} \times 100$$

Example

Calculate the %ionization of acetic acid ($pK_a = 4.76$) at

    (i)   **pH = 3.76 and**

    (ii) **pH = 5.76**

$$\% \text{ ionization} = \frac{10pH - pK_a}{1 + 10pH - pK_a} \times 100$$

At pH = 3.76

$$= 10^{(3.76 - 4.76)}/1 + 10^{(3.76 - 4.76)} \times 100$$

$$= 0.1/1.1 \times 100$$

$$= 9.09\%$$

At pH = 5.76

$$= 10^{(5.76 - 4.76)} / 1 + 10^{(5.76 - 4.76)} \times 100$$

$$= 10/11 \times 100$$

$$= 90.9\%$$

Example

Calculate %ionization, when pH = pK_a (9.25)

$$NH_4^+ \rightarrow NH_3 + H^+$$

$$\% \text{ ionization} = \frac{10pK_a - pH}{1 + 10pK_a - pH} \times 100$$

$$= 10^{(9.25 - 9.25)} / 1 + 10^{(9.25 - 9.25)} \times 100$$

$$= 10^0 / 1 + 10^0 \times 100$$

$$= 1/2 \times 100$$

$$= 50\%$$

Example

This drug (diphenhydramine) contains one basic nitrogen, and, at pH 7.0, its percentage of ionization can be calculated as follows:

$$\% \text{ionization (diphenhydramine)} = 10^{(9.0 - 7.0)} / 1 + 10^{(9.0 - 7.0)} \times 100$$

$$= 10^{(2.0)} / 1 + 10^{(2.0)} \times 100$$

$$= 100 / 101 \times 100$$

$$= 99.0\%$$

Example

This drug (ibuprofen) contains one acidic group, and, at pH 7.0, its percentage of ionization can be calculated as follows:

$$\% \text{ionization (ibuprofen)} = 10^{(7.0 - 4.4)} / 1 + 10^{(7.0 - 4.4)} \times 100$$

$$= 10^{(2.6)} / 1 + 10^{(2.6)} \times 100$$

$$= 398/399 \times 100$$

$$= 99.8\%$$

5.9 Buffer action

Buffer action, in general, is described as the ability of a buffer solution to resist pH shifts when a small amount of an acid or base is added.

To better understand the mechanism of buffer action, consider an acidic buffer composed of a weak acid such as acetic acid and its sodium salt, sodium acetate. This acidic buffer contains equimolar quantities of acetic acid and sodium acetate. Usually, a significant amount of sodium ions (Na^+), acetate ions ($CH3COO^-$), and undissociated acetic acid molecules are present. The salt exists completely as ions as described in Fig. 5.4.

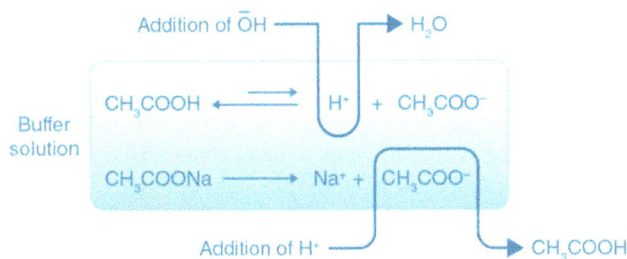

Addition of $\overline{O}H$ ———→ H_2O

CH_3COOH ⇌ H^+ + CH_3COO^-

Buffer solution

CH_3COONa ——→ Na^+ + CH_3COO^-

Addition of H^+ ———→ CH_3COOH

Fig. 5.4: Mechanism of buffer action.

The buffer will contain both acid (CH_3COOH) and the conjugate base (CH_3COO^-). If we add a small quantity of acid, the conjugate base (CH_3COO^-) will remove the hydrogen ions.

The ethanoic acid will be only minimally dissociated in the form of CH_3COOH; therefore, it will not contribute any H^+ ions. As a result, the pH of the final solution will be rather stable. The provided H^+ ions are similarly removed; therefore, there is no significant reduction in pH.

The reaction is completed because CH_3COOH is a weak acid with an increased tendency for its ions to create non-ionized CH_3COOH molecules.

In contrast, when a strong base is added, the OH^- ion is neutralized by the interaction with the acid in the buffer.

It is considered that the OH^- ion can react with the H^+ ion to produce water. The new OH^- ions are eliminated, and the acid equilibrium changes to the right to replace the exhausted H^+ ions. This causes a small shift in pH value. So, when an acid or base is added, the action is practically balanced, and the pH of the solution remains constant.

5.9.1 Buffer action of ammonium acetate solution

Ammonium acetate is almost completely dissociated in aqueous solutions as follows:

$$CH_3COONH_4 \rightarrow CH_3COO^- + NH_4^+$$

In the solution, there is an excess of CH_3COO^- ions and NH_4^+ ions.

On addition of acid: When a few drops of an acid are added to the above solution, the H_3O^+ ions given by the acid combine with the CH_3COO^- ions to form a weakly ionized molecule of CH_3COOH:

$$CH_3COO^- + H_3O^+ \xrightarrow{\hspace{2cm}} CH_3COOH + H_2O$$

The H_3O^+ ion concentration of the solution does not change practically, and hence, the pH of the solution remains almost constant.

On addition of base: When a few drops of base are added to the above solution, the OH^- ions given by the base combine with NH_4^+ ions to form some weakly ionized molecules of NH_4OH:

$$NH_4^+ + OH^- \quad NH_4OH$$

The OH^- ion concentration and hence the H_3O^+ concentration or the pH of the solution remain almost constant.

5.10 Buffer capacity

The ability of a solution to resist change in pH by absorbing or desorbing H^+ or OH^- ions is measured by its buffer capacity. It is also referred to as buffer value, buffer index, and buffer efficiency. The definition of buffer capacity, proposed by Koppel and Spiro, and Van Slyke, is the ratio of the addition of a strong base (or acid) to the slight pH shift that results from it. Buffer capacity depends on the amounts of the weak acid and its conjugate base that are in the buffer mixture, meaning the ratio should be 1:1. Buffer capacity is influenced by both the salt-acid ratio and the total concentration of acid and salt.

Buffer capacity may also be represented as:

$$\beta = \frac{(\Delta A)\text{No. of moles of acid or base added per liter of buffer}}{(\Delta \text{ pH}) \text{ Change in pH}}$$

The maximum buffer capacity occurs when pH is close to pK_a. The maximum buffer capacity where $pH = pK_a$, or, in equivalent terms, and $[H_3O^+] = K_a$. Substitute $[H_3O^+]$ for K_a in both the numerator and the denominator.

Example

Which one has the highest buffer capacity?
a. 0.1 M HF and 0.1 M NaF
b. 1 M HF and 1 M NaF
c. 0.001 M HF and 0.001 M NaF

(b) has the highest buffer capacity due to its high concentration of weak acid and conjugate base available to resist the pH change
- What is the pH of a buffer that is 0.12 M in lactic acid ($HC_3H_5O_3$) and 0.1 M sodium lactate ($C_3H_5O_3^-$)? For lactic acid $K_a = 1.4 \times 10^{-4}$

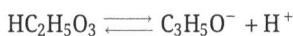

$$HC_2H_5O_3 \rightleftharpoons C_3H_5O^- + H^+$$

$$pK_a = -\log K_a$$

$$pK_a = -\log 1.4 \times 10^{-4}$$

$$pK_a = 3.85$$

Applying Hasselbalch equation,

$$pH = pK_a + \log [A^-]/[HA]$$

$$pH = 3.85 + \log [0.1]/[0.12]$$

$$pH = 3.77$$

- What is the pH of a buffer solution that is 0.24 M NH_3 and 0.2 M NH_4Cl?

$$K_a = 5.6 \times 10^{-10}$$

$$pK_a = -\log K_a$$

$$pK_a = -\log 5.6 \times 10^{-10}$$

$$pK_a = 9.25$$

According to Henderson-Hasselbalch equation:

$$pH = pK_a + \log \text{Base}/\text{Acid}$$

$$NH_3 \rightleftharpoons NH_4^+$$

Base   Acid

0.24 M   0.2 M

$$pH = 9.25 + 0.08$$

$$pH = 9.33$$

- If 0.005 mol of NaOH is added to the above 0.5 L of solution, what will be the resulting pH?

$M = mol/L$
   $M = 0.005/0.5$
   $M = 0.01 M$

$$NaOH \longrightarrow Na^+ + OH^-$$

$$NH_4^+ + OH^- \rightleftharpoons NH_3 + H_2O$$

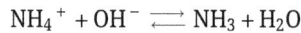

$$pH = 9.25 + \log 0.25/0.19$$

$$pH = 9.25 + 0.12$$

$$pH = 9.37$$

- How many moles of NH_4Cl must be added to 2 L of 0.1 M NH_3 to form a buffer
   whose pH is 9 (assume that the addition of NH_4Cl does not change the volume of
   solution) and $pK_a = 9.25$

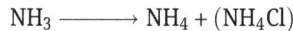

$$NH_3 \longrightarrow NH_4 + (NH_4Cl)$$

0.10 M?
   According to Henderson-Hasselbalch equation:
   $pH = pK_a + \log [Base]/[Acid]$
   Suppose $[NH_4] = x$
   $9 = 9.25 + \log [0.1]/[x]$
   $9 - 9.25 = \log [0.1]/[x]$
   $-0.25 = \log [0.1]/[x]$
   By taking antilog on both sides,
   $1 \times 10^{-25} = 0.1/x$
   $5.62 \times 10^{-1} = 0.1/x$
   $x = 0.1/5.62 \times 10^{-1}$
   $x = 1.778 \times 10^{-1}$ mol for 1 L (1.778 mol/L)
   $x = [NH_4^+]$
   $x = 3.55 \times 10^{-1} M$ for 2 L
   Result: Moles of NH_4Cl added to 2 L is 0.36 mol.

5.10.1 Calculating buffer capacity

Examples

A buffer is made by adding 0.3 mol of CH_3COOH and CH_3COONa to enough water to make 1 L of solution. The pH of the buffer is 4.74.

Calculate the pH of this solution after

(a) –0.02 mol of NaOH

(b) 0.02 mol of HCl

(c) Calculate the pH if 0.02 mol of HCl and NaOH is added to 1 L of water

$$CH_3COOH + H_2O \longrightarrow CH_3COO^- + H_3O^+$$

The pH of the buffer is 4.74.

Now let us consider the effects of adding 0.02 mol of NaOH

Adding 0.02 mol of NaOH,

$$CH_3COOH \longrightarrow CH_3COO^-$$

$$pH = pK_a + \log[salt/acid]$$

$$pH = 4.74 + \log[0.32/0.28]$$

$$pH = 4.74 + 0.056 = 4.796$$

$$CH_3COOH + H_2O \longrightarrow CH_3COO^- + H_3O^+$$

pH = 4.74 + log[0.28/0.32] = 4.606

The pH if 0.02 mol of HCl and NaOH is added to the 1 L water.

M = 0.02/1 L = 0.02 M

In case of acid 0.02 M = [H$^+$]

In case of base 0.02 M = [OH$^-$]

$$pH = -\log(0.02)$$

$$pH = 1.698$$

$$pOH = -\log(0.02) = 1.69$$

$$pOH = 14 - 1.69 = 12.3$$

Example

A 1 L of 0.1 M sodium acetate buffer with a pH of 4.0 is required. The molecular weight of acetic acid is 60; therefore, in a liter of 0.1 M buffer, there will be 6 g of acetic acid. To prepare the 6 g of acetic acid is weighed and made up to ca 500 ml with water. The pH of the acetic acid solution is adjusted to 4.0 by the addition of 2 M sodium hydroxide solution. The solution is then made up to 1 liter with water. Calculate the concentration of acetate and acetic acid in the buffer at pH 4.0. The pK_a is 4.76.

$$pH = pK_a + \log \text{conj. base/weak acid}$$

$$10^{pH - pK_a} = [CH_3COO^-]/[CH_3COOH]$$

$$10^{4 - 4.76} = [CH_3COO^-]/[CH_3COOH]$$

$$10^{-0.76} = 0.173$$

$$[CH_3COO^-]/[CH_3COOH] = 0.173/L$$

The buffer is composed of 1 part of acetic acid and 0.17 part of acetate.
We can assume the total volume of the buffer is $1 + 0.17 = 1.17$ parts.
Therefore,
The mole fraction of acetate ion $= 1/1.17 = 0.854$
The mole fraction of acetate ion $= 0.17/1.17 = 0.145$
Now to calculate the buffer on 0.1 M, multiply the mole fraction by 0.1 mol.

$$CH_3COO^- = 0.8547 \times 0.1 = 0.08547 \text{ M}$$

$$CH_3COOH = 0.1453 \times 0.1 = 0.01453 \text{ M}$$

Therefore, the concentration of acetate ion is 0.08547 M, and the concentration of acetic acid is 0.01453 M in the buffer.

5.11 Buffer range

When the concentration of a weak acid [HA] is 10× greater than the concentration of its conjugate base [A⁻], then the Henderson-Hasselbalch equation is

$$pH = pK_a + \log [A]/[HA]$$

$$pH = pK_a + \log 1/10$$

$$pH = pK_a - 1$$

This is the lower limit of the effective pH range of a buffer solution.
    Now, when the concentration of conjugate base [A⁻] is 10× greater than the concentration of weak acid [HA]:

$$pH = pK_a + \log [A^-]/[HA]$$

$$pH = pK_a + \log 10/1$$

$$pH = pK_a + 1$$

This is the upper limit, and the effective pH range of a buffer is pH = $pK_a \pm 1$.

A buffer works best in a range that is one pH unit on each side of the weak acid or base's pK_a. With a pK_a of 4.76, acetic acid has an effective buffer range of 3.76–5.76. Certain weak acids and bases have more than one buffer range; phosphoric acid, for instance, can be used to make buffers which cover three different pH ranges since it has three ionizable protons with three distinct pK_a values.

The ranges covered by phosphate buffer include the following ionic species:

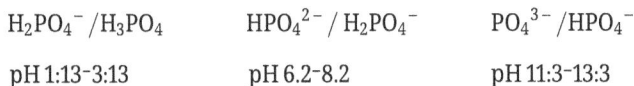

$H_2PO_4^- / H_3PO_4$ $HPO_4^{2-} / H_2PO_4^-$ PO_4^{3-} / HPO_4^-

pH 1:13–3:13   pH 6.2–8.2   pH 11:3–13:3

Some acids have ionizable groups with pK_a values less than 2 pH units apart, resulting in buffers with broad ranges. Succinic acid, with pK_a values ranging from 4.19 to 5.57, has a broad pH buffering range.

Example
Determine the buffer range(s) for the following compounds:
   (i)   Carbonic acid pK_a 6.38, 10.32
  (ii) Boric acid pK_a 9.14, 12.74
 (iii) Glycine 2.34, 9.60
 (iv) Citric acid 3.06, 4.74, 5.4.
Answers: (i) 5.38_7.38, 9.32_11.32. The lower range is not useful because of the ease with which CO_2 is lost from solution; (ii) 8.14–10.14, 11.74–13.74; (iii) 1.34–3.34, 8.6–10.6; (iv) continuous buffering range 2.06–6.4

Example
10 mL of 0.1 M HCl is added to 20 mL of a 0.5 M sodium acetate buffer with a pH of 4.3. Calculate: the pH of the buffer after addition of the HCl, the molarity of the buffer after addition of the HCl, and the resultant pH if the HCl had been added to 20 mL of water.
Answer: pH = 4.04, new molarity¼0.33. Addition of 10 mL of 0.1 M HCl to 20 mL of water would give a pH of 1.48.

5.12 Strong acids and bases as buffers

The ability to resist change in pH on adding acid or alkali is also possessed by relatively concentrated solutions of strong acids and strong bases [10]. If 1 L of pure water having a pH of 7 is added to 1 mL of 0.01 M HCl, the pH is reduced to 5. If the same volume of the acid is added to 1 L of 0.001 M HCl, which has a pH of about 3, the hydronium ion concentration is increased only about 1% and the pH is hardly reduced at all. The nature of this buffer solution is quite different from that when 1 mL of 0.01 M HCl, which represents 0.00001 g-eq of hydronium ions, is added to the 0.0000001 g-eq of hydronium ions in 1 L of pure water, the hydronium ion concentration is increased 100-fold (equivalent to 2 pH units) but when the same amount is added to the 0.001 g-eq of hydronium ions in 1 L of 0.001 M HCl, the increase is only 1/100 the concentration already present. Similarly, if 1 mL of 0.01 M NaOH is added to 1 L of pure water, the pH is increased to 9, while if the same volume is added to 1 L of 0.001 M NaOH, the pH is increased almost immeasurably.

In general, solutions of strong acids of pH 3 or less, and solutions of strong bases of pH 11 or more, exhibit this kind of buffer action by virtue of the relatively high concentration of hydronium or hydroxyl ions present. The USP includes among its standard buffer solution a series of hydrochloric acid buffers, covering a pH range of 1.2–2.2, which also contain potassium chloride. The salt does not participate in buffer action as in the case with the sale of a weak acid; instead, it serves as a non-reactive constituent required to maintain the proper electrolyte environment of the solution.

5.12.1 Selection of buffer

When selecting a buffer, a number of things need to be taken into account. An enzyme's ideal pH, the effects of nonspecific buffers on the enzyme, and interactions with metals or substrates must all be taken into account when studying it. Cost and the buffer's compatibility with various purification methods become crucial factors when purifying a protein.

The first step in choosing the ideal buffer is figuring out a protein's pH optimum. Buffers should be utilized in the vicinity of the pK since this number represents the buffering capacity's maximum. A set of related buffers with a broad pH range can be helpful for figuring out an enzyme's ideal pH [11, 12]. After estimating the ideal pH, several buffers within this range can be investigated for particular buffer effects. Relatively few adverse effects have been reported for the good buffers. Inorganic buffers do, however, have a significant chance of producing particular buffer effects. Several enzymes, such as urease, carboxypeptidase, and numerous kinases and dehydrogenases, are inhibited by phosphate buffer.

5.13 Preparation of buffer solution

When the pK_a (acid dissociation constant) of the acid and pK_b (base dissociation constant) of the base are known, a buffer of known pH can be prepared by altering the ratio of salt and acid or salt and base. Buffers can also be prepared by combining a weak acid and its conjugate base or a weak base with its conjugate acid.

5.13.1 Preparation of some common buffers for use in biological systems

The following information should only be used as a general guide. The use of a sensitive pH meter with the proper temperature setting for the final pH adjustment is strongly encouraged. The final pH value may vary somewhat if additional chemicals are added after the pH has been adjusted. The buffer concentrations in the following tables serve as suggestions. Your experimental requirements will determine whether you choose higher or lower concentrations.

Hydrochloric acid-potassium chloride buffer (HCl-KCl), pH range 1.0–2.2
(a) 0.1 M potassium chloride: 7.45 g/L (M.W.: 74.5)
(b) 0.1 M hydrochloric acid

Mix 50 mL of potassium chloride and the indicated volume of hydrochloric acid. Mix and make up the volume to 100 mL with deionized water. Adjust the final pH using a sensitive pH meter (Tab. 5.3).

Tab. 5.3: pH range of hydrochloric acid-potassium chloride buffer.

mL of HCl	97	64.5	41.5	26.3	16.6	10.6	6.7
pH	1.0	1.2	1.4	1.6	1.8	2.0	2.2

Glycine-HCl buffer, pH range 2.2–3.6
(a) 0.1 M glycine: 7.5 g/L (M.W.: 75.0)
(b) 0.1 M hydrochloric acid

Mix 50 mL of glycine and the indicated volume of hydrochloric acid. Mix and adjust the final volume to 100 mL with deionized water. Adjust the final pH using a sensitive pH meter (Tab. 5.4).

Tab. 5.4: pH range of glycine-HCl buffer.

mL of HCl	44.0	32.4	24.2	16.8	11.4	8.2	6.4	5.0
pH	2.2	2.4	2.6	2.8	3.0	3.2	3.4	3.6

Citrate buffer, pH range 3.0–6.2
(a) 0.1 M citric acid: 19.21 g/L (M.W.: 192.1)
(b) 0.1 M sodium citrate dihydrate: 29.4 g/L (M.W.: 294.0)

Mix citric acid and sodium citrate solutions in the proportions indicated and adjust the final volume to 100 mL with deionized water. Adjust the final pH using a sensitive pH meter. The use of the pentahydrate salt of sodium citrate is not recommended (Tab. 5.5).

Tab. 5.5: pH range of citrate buffer.

mL of citric acid	46.5	10.0	35.0	31.5	25.5	20.5	16.0	11.8	7.2
mL of sodium citrate	3.5	10.0	15.0	18.5	24.5	29.5	34.0	38.2	42.8
pH	3.0	3.4	3.8	4.2	4.6	5.0	5.4	5.8	6.2

Acetate buffer, pH range 3.6–5.6

(a) 0.1 M acetic acid (5.8 mL made to 1,000 mL)
(b) 0.1 M sodium acetate; 8.2 g/L (anhydrous; M.W. 82.0) or 13.6 g/L (trihydrate; M.W. 136.0).

Mix acetic acid and sodium acetate solutions in the proportions indicated and adjust the final volume to 100 mL with deionized water. Adjust the final pH using a sensitive pH meter (Tab. 5.6).

Tab. 5.6: pH range of acetate buffer.

mL of acetic acid	46.3	41.0	30.5	20.0	14.8	10.5	4.8
mL of sodium acetate	3.7	9.0	19.5	30.0	35.2	39.5	45.2
pH	3.6	4.0	4.4	4.8	5.0	5.2	5.6

Citrate-phosphate buffer, pH range 2.6–7.0

(a) 0.1 M citric acid; 19.21 g/L (M.W. 192.1)
(b) 0.2 M dibasic sodium phosphate; 35.6 g/L (dihydrate; M.W. 178.0) or 53.6 g/L (heptahydrate; M.W. 268.0)

Mix citric acid and sodium phosphate solutions in the proportions indicated and adjust the final volume to 100 mL with deionized water. Adjust the final pH using a sensitive pH meter (Tab. 5.7).

Tab. 5.7: pH range of citrate phosphate buffer.

mL of citric acid	44.5	39.8	35.9	32.3	29.4	26.7	24.3	22.2	19.7	16.9	13.6	6.5
mL of sodium phosphate	5.4	10.2	14.1	17.7	20.6	23.3	25.7	27.8	30.3	33.1	36.4	43.6
pH	2.6	3.0	3.4	3.8	4.2	4.6	5.0	5.4	5.8	6.2	6.6	7.0

Phosphate buffer, pH range 5.8–8.0
(a) 0.1 M sodium phosphate monobasic; 13.8 g/L (monohydrate, M.W. 138.0)
(b) 0.1 M sodium phosphate dibasic; 26.8 g/L (heptahydrate, M.W. 268.0)

Mix sodium phosphate monobasic and dibasic solutions in the proportions indicated and adjust the final volume to 200 mL with deionized water. Adjust the final pH using a sensitive pH meter (Tab. 5.8).

Tab. 5.8: pH range of phosphate buffer.

mL of sodium phosphate, Monobasic	92.0	81.5	73.5	62.5	51.0	39.0	28.0	19.0	13.0	8.5	5.3
mL of sodium phosphate, dibasic		8.0	18.5	26.5	37.5	49.0	61.0	72.0	81.0	87.0	91.5	94.7
pH		5.8	6.2	6.4	6.6	6.8	7.0	7.2	7.4	7.6	7.8	8.0

Tris-HCl buffer, pH range 7.2–9.0
(a) 0.1 M Tris(hydroxymethyl)aminomethane; 12.1 g/L (M.W.: 121.0)
(b) 0.1 M hydrochloric acid

Mix 50 mL of Tris(hydroxymethyl)aminomethane and the indicated volume of hydrochloric acid, and adjust the final volume to 200 mL with deionized water. Adjust the final pH using a sensitive pH meter (Tab. 5.9).

Tab. 5.9: pH of Tris-HCl buffer.

mL of HCl	44.2	41.5	38.4	32.5	21.9	12.2	5.0
pH	7.2	7.4	7.6	7.8	8.2	8.6	9.0

Glycine-sodium hydroxide, pH 8.6–10.6
(a) 0.1 M glycine; 7.5 g/L (M.W.: 75.0)
(b) 0.1 M sodium hydroxide; 4.0 g/L (M.W.: 40.0)

Mix 50 mL of glycine and the indicated volume of sodium hydroxide solution, and adjust the final volume to 200 mL with deionized water. Adjust the final pH using a sensitive pH meter (Tab. 5.10).

Tab. 5.10: pH of glycine-sodium hydroxide buffer.

mL of sodium hydroxide	4.0	8.8	16.8	27.2	32.0	38.6	45.5
pH	8.6	9.0	9.4	9.8	10.0	10.4	10.6

Carbonate-bicarbonate buffer, pH range 9.2–10.6
(a) 0.1 M sodium carbonate (anhydrous), 10.6 g/L (M.W.: 106.0)
(b) 0.1 M sodium bicarbonate, 8.4 g/L (M.W.: 84.0)

Mix sodium carbonate and sodium bicarbonate solutions in the proportions indicated and adjust the final volume to 200 mL with deionized water. Adjust the final pH using a sensitive pH meter (Tab. 5.11).

Tab. 5.11: pH range of bicarbonate buffer.

mL of sodium carbonate	4.0	9.5	16.0	22.0	27.5	33.0	38.5	42.5
mL of sodium bicarbonate	46.0	40.5	34.0	28.0	22.5	17.0	11.5	7.5
pH	9.2	9.4	9.6	9.8	10.0	10.2	10.4	10.6

5.14 Applications of buffers

5.14.1 Buffer in protein formulations

Proteins can become more conformationally stable by adding buffers, either by ligand binding or an excluded solute mechanism. Furthermore, they have the ability to modify interfacial damage and change the colloidal stability of proteins. Proteins may become destabilized by buffers as well, and the stability of buffers is demonstrated.

Ensure medication component stability
Prevent the pH of important drug components, such as aspirin, from varying or degrading due to the gastrointestinal environment.

Increase the purity of certain components
Separate and purify certain components, like insulin.

Ensure the drug's solubility
Certain chemicals will only dissolve at a certain pH. In pharmaceutical practice, formulation of ophthalmic solutions contains a buffer solution to increase the solubility of drugs by causing drugs to be in their ionized state.

Ensure biological activity
Some chemicals, such as pepsin, can only sustain their effectiveness at specific pH levels.

Allowing drugs to be used in the human body
Injections are close to the blood pH, whereas eye drops are close to the eye environment.

As an active ingredient
The gastric acid inhibitor contains a buffer that reduces stomach acid.

Maintaining a suitable pH
Biologically, buffer solutions are necessary to keep the correct pH for enzymes in many organisms to work. Many enzymes work only under very precise conditions; if the pH moves outside of a narrow range, the enzymes slow or stop working and can be denatured. In many cases, denaturation can permanently disable their catalytic activity. A buffer of carbonic acid (H_2CO_3) and bicarbonate (HCO_3^-) is present in blood plasma to maintain a pH between 7.35 and 7.45.

Research use
The majority of biological samples that are used in research are made in buffers, specially PBS (phosphate buffer saline) at pH 7.4.

Laboratory techniques
Enzyme assays: Buffers stabilize enzyme activity by maintaining the pH within a specific range.
Polymerase chain reaction (PCR): Buffers in PCR reactions stabilize the pH, facilitating the amplification of DNA.
Chemical reactions
Catalysis: Buffers are essential in catalytic processes, protecting catalysts from pH extremes.
Hydrolysis reactions: Buffers control pH during hydrolysis reactions, preventing rapid and undesirable changes.
Environmental monitoring
Water treatment: Buffers help maintain the pH of water, preventing damage to aquatic ecosystems and infrastructure.
Soil health: Buffers in soils resist changes in pH, influencing nutrient availability and plant growth.

Biological systems
Blood buffering: The bicarbonate ion (HCO_3^-) system in the blood maintains the pH of blood around 7.4, which is crucial for enzyme activity and overall physiological function depending on the biological buffer systems.
Cellular pH regulation: Intracellular buffers regulate the pH within cells, ensuring optimal conditions for biochemical processes.

References

[1]   Beynon R, Easterby J. Buffer solutions: Taylor & Francis; Florida, USA, 2004.

[2]   Van Slyke DD. On the measurement of buffer values and on the relationship of buffer value to the dissociation constant of the buffer and the concentration and reaction of the buffer solution. Journal of Biological Chemistry. 1922; 52(2): 525–70.

[3]   Stoll VS, Blanchard JS. Buffers: Principles and practice. In: Methods in enzymology (Vol. 463): Elsevier; Edinburg, Englandm, 2009. pp. 43–56.

[4]   Rodríguez-Laguna N, Rojas-Hernández A, Ramírez-Silva MT, Moya-Hernández R, Gómez-Balderas R, Romero-Romo MA. The conditions needed for a buffer to set the pH in a system. In: Advances in titration techniques: IntechOpen; Londong, UK, 2017.

[5]   Samuelsen L, Holm R, Lathuile A, Schönbeck C. Buffer solutions in drug formulation and processing: How pKa values depend on temperature, pressure and ionic strength. International Journal of Pharmaceutics. 2019; 560: 357–64.

[6]   Persat A, Chambers RD, Santiago JG. Basic principles of electrolyte chemistry for microfluidic electrokinetics. Part I: Acid–base equilibria and pH buffers. Lab on a Chip. 2009; 9(17): 2437–53.

[7]   Radić N, Prkić A. Historical remarks on the Henderson-Hasselbalch equation: Its advantages and limitations and a novel approach for exact pH calculation in buffer region. Reviews in Analytical Chemistry. 2012; 31(2): 93–98.

[8]   Po HN, Senozan N. The Henderson-Hasselbalch equation: Its history and limitations. Journal of Chemical Education. 2001; 78(11): 1499.

[9]   Gilbert DL. Buffering of blood plasma. The Yale Journal of Biology and Medicine. 1960; 32(5): 378.

[10]   Perrin D. Buffers for pH and metal ion control: Springer Science & Business Media; London, England, 2012.

[11]   Ali S, Amin MU, Ali MY, Tariq I, Pinnapireddy SR, Duse L, et al. Wavelength dependent photo-cytotoxicity to ovarian carcinoma cells using temoporfin loaded tetraether liposomes as efficient drug delivery system. European Journal of Pharmaceutics and Biopharmaceutics. 2020; 150: 50–65.

[12]   Tariq I, Pinnapireddy SR, Duse L, Ali MY, Ali S, Amin MU, et al. Lipodendriplexes: A promising nanocarrier for enhanced gene delivery with minimal cytotoxicity. European Journal of Pharmaceutics and Biopharmaceutics. 2019; 135: 72–82.

Part 2: **Liquid dosage forms**

Ghazala Ambreen, Muhammad Yasir Ali, Khadeja Arshad,
Umaira Maqbool, Mehma Meraal, Syeda Hijab Zahra, Hania Ahmad,
Asia Naz Awan

6 Galenical preparations

6.1 Introduction

Substances that are prepared by Galen's method are known as galenicals. Galenicals are unrefined extracts made from either a vegetable or animal source by using a suitable solvent. Lozenges, tinctures, aromatic water, and spirits are examples of galenicals prepared through Galen's method. In the early days, extemporaneous preparations were made by pharmacists by extracting the active constituents from any natural origin. Galenical preparations were specially devised by Galen, a Greek physician, to organize medicine as a whole. He made major improvements in this field. His works were translated into Latin from Arabic in the twelfth century and gained extreme prominence worldwide. His name is basically the source of the term "galenicals," which refers to medication derived from natural origins.

Over the decades, numerous approaches have been used for producing herbal medicines. However, it has become widely known that particular combinations of natural compounds are needed to get the desired therapeutic outcomes. Identifying the most appropriate method of extraction to yield the intended results can only be made possible by comprehension of the physical and chemical characteristics of the substance being extracted. Galenical preparations would enable more precise dosing, efficient disease therapy, and enhanced accuracy [1].

The components that exhibit curative properties are recognized as active constituents. Crude plants are composed of numerous distinct components, some of which have therapeutic benefits. Inert substances contribute as remaining ingredients that have no medicinal significance. The fundamental concept driving galenicals is an extraction process that employs an appropriate solvent to separate out the desired targeted active ingredients.

6.2 Choice of solvent

The solvent employed for the extraction procedure of medicinal plants is additionally referred to as menstruum. The selection of solvent is impacted by multiple variables, particularly the variety of plants, the part of the plant to be extracted, the chemical makeup of the active, and the accessibility of the solvent.

https://doi.org/10.1515/9783111438108-006

The ideal solvent for extraction ought to possess the characteristics as follows: it should have minimal toxicity, preserve activity, promote rapid absorption of the extract, and not be able to cause the extract to break down and dissociate.

Nonpolar substances are extracted using nonpolar solvents, i.e., hexane, dichloromethane, while polar solvents like water, ethanol, and methanol are typically employed for extracting polar compounds. The usual approach to liquid-liquid extraction requires selecting two miscible solvents, like water and ethane or hexane.

Water exists in every composition owing to its strong polarity and compatibility with organic solvents. The material to be extracted through the method of liquid-liquid extraction must dissolve in an organic solvent, instead of water, to aid in isolation. Furthermore, the solvents are ranked according to their polarity, n-hexane being the least polar in nature and water being the most polar.

6.2.1 Water

Water is the most hydrophilic of all the solvents used and can be employed for separating a wide variety of polar compounds. It is affordable, exceptionally polar, and safe, and dissolves extensively variety of compounds. It could contribute to the disintegration of materials, stimulate the growth of microbes and molds, and concentrate the extract with high heat.

6.2.2 Alcohol

It is polar, miscible with water, and has the potential to isolate second-generation polar molecules. It turns self-preserving at concentrations greater than 20%. At minimal concentrations, the extract is safe, as little heat is required to concentrate it. It is combustible, highly volatile, and cannot disintegrate gums, wax, and fats.

6.2.3 Ether

As a nonpolar solvent, it can be utilized for the extraction of compounds, such as alkaloids, coumarins, terpenoids, and fatty acids. It has a bland taste, compatible with water, and has an inadequately low boiling point. Furthermore, it is a highly stable compound with no sensitivity to metals, acids, or basic substances. It is highly flammable and volatile.

6.2.4 Chloroform

It is a nonpolar solvent that is highly efficient for the extraction of compounds like oils, lipids, terpenoids, and flavonoids. It is transparent and has a pleasant odor and is soluble with alcohol. In addition, it is well processed and consumed by the body. It has both carcinogenic and sedative properties.

6.2.5 Ionic liquid

Ionic liquid extraction solvent has an excellent polarity and thermal stability. It stays liquid up to 3,000 °C which makes it handy in circumstances like high temperatures. Due to its strong affinity with water and other solvents, it is ideal for extracting polar compounds. It can be employed for separation using microwaves due to its exceptional solvent characteristics, which both absorb and propagate microwaves. It is highly polar and has incombustible properties and works well for liquid-liquid extraction. It cannot be employed for preparing tinctures.

6.3 Aromatic water

Aromatic waters are typically clear solutions saturated with volatile oils (e.g., rose oil and peppermint oil) or other aromatic or volatile substances (e.g., camphor) in distilled water, unless otherwise specified. They possess a scent reminiscent of the plant or volatile substance they are derived from and are free from solid impurities and foreign odors [2].

These waters can be utilized for various purposes, such as perfuming, flavoring, or for specific applications, e.g., camphor water is often chosen as the base for ophthalmic solutions due to its capacity to impart a refreshing and invigorating sensation to the formulation. Witch hazel, also recognized as hamamelis water, serves multiple purposes, including as a rub, fragrance, and as an astringent in numerous cosmetic formulations, notably in post-shave lotions.

6.3.1 Types of aromatic waters

6.3.1.1 Simple aromatic water

This type of aromatic water is made by infusing a small amount of aromatic substance, such as essential oils or plant extracts, into distilled water. It typically has a mild aroma and is used for various purposes, such as perfuming, flavoring, or therapeutic applications.

6.3.1.2 Concentrated aromatic water

In contrast, concentrated aromatic water is made by infusing a higher concentration of aromatic substances into distilled water, resulting in a more potent solution with a stronger aroma. This type of aromatic water is often used when a more intense fragrance or flavor is desired, or when higher therapeutic benefits are sought. It may require dilution before use, depending on the intended application [3].

6.3.2 Methods of aromatic water preparation

6.3.2.1 Distillation method

The described process outlines steam distillation, a method employed to extract essential oils or aromatic compounds from plant materials. Coarsely ground aromatic plant parts are placed into a still with purified water. As the water boils, steam carries volatile aromatic compounds from the plant material, which are then condensed back into liquid form. Excess oils from the plant rise to the surface of the condensed liquid and are separated. The resulting solution, saturated with aromatic compounds, undergoes clarification through filtration to remove any remaining impurities. This method ensures the purity and clarity of the final product, commonly used in the production of essential oils, perfumes, and aromatherapy products. Adjustments to parameters such as temperature may be necessary based on the specific plant material being processed. Some of the aromatic waters obtained from the distillation method are peppermint water, rose water, camphor water, hamamelis water, and orange flower.

6.3.2.2 Solution method

The process for crafting aromatic water involves mixing either 2 mL of a liquid volatile substance or 2 g of a solid volatile substance with 1,000 mL of purified water in a suitable container. After gentle shaking for 15 min, the mixture is left undisturbed for at least 12 h, allowing excess oil and solid particles to settle. Without further disturbance, the mixture undergoes careful filtration through wetted filter paper. Purified water is added as needed to adjust the volume of the filtrate to the desired quantity. This method ensures the production of aromatic water with reduced oil and solid impurities, suitable for various applications.

6.3.2.3 Alternative solution method

The method described appears to be a basic procedure for preparing a suspension or emulsion of a volatile oil or finely ground aromatic solid. Here's a breakdown of the steps involved: (a) Preparation of the mixture: Start with the volatile oil or finely ground aromatic solid. Mix it thoroughly with 15 g of talc that is often used as a carrier or bulking agent in such formulations. (b) Addition of purified water: Pour 1,000 mL of purified water into the mixture of the oil/solid and talc. This step creates a slurry, a semi-liquid mixture containing solid particles dispersed in a liquid. (c) Agitation: Vigorously agitate the slurry several times over 30 min. This agitation process ensures thorough mixing of the oil/solid, talc, and water. Adequate agitation is crucial for achieving a homogenous mixture and proper dispersion of the active ingredients. (d) Filtration: After the agitation period, filter the mixture to remove any solid particles or impurities that haven't dissolved or dispersed. Filtering ensures that the final product is free from any undissolved solids, which could affect its appearance, stability, or performance.

This method seems suitable for preparing a suspension or emulsion intended for various applications, such as cosmetics, pharmaceuticals, or industrial formulations. However, specific details, such as the purpose of the formulation, the choice of volatile oil or aromatic solid, and the intended use of the final product would determine the effectiveness and appropriateness of this method. Additionally, factors like particle size, temperature, and the nature of the ingredients could influence the process and may require adjustments for optimal results.

6.3.2.4 Preparation of concentrated aromatic water

They are concentrated alcoholic, nonaqueous preparations containing 2% of volatile oil, 40 times more so than ordinary aromatic waters. Numerous volatile oils contain aromatic and nonaromatic products. The aromatic component is very soluble in weak alcohol in comparison with the nonaromatic component.

By a partial mixture with water of a solution in 90% alcohol of the oil, the aromatic constituent of the oil is left in solution, and the nonaromatic in a somewhat gummy layer at the bottom, forming an oily layer. To solve this, a quantity of 50 g of talc is placed into 1,000 mL of preparation, which acts as a disturbing agent that absorbs the nonaromatic portion. The mixture is stirred up and allowed to stabilize for a couple of hours, after which it is filtered [4]. The water of aromatic water is essential to store since it does not last long. It should be made in small batches and stored in airtight containers that are light-resistant and excluded from harsh light and too much heat.

6.3.3 Uses of aromatic waters

Medicated waters, so-called aromatic water, have extensive utility in the pharmacy and therapeutics on account of their agreeable smell, taste, and gentle curative power. They are applied in the following fields:

Flavoring agent
Aromatic waters find application in enhancing the palatability of oral pharmaceutical dosage forms like syrups, elixirs, and mixtures.

Examples: For an unpleasant flavor of saline purgatives to be disguised, peppermint water is commonly employed.
Vehicle for pharmaceutical preparation
Orally, topically, and ocularly, employed as an oral suspension vehicle and topical or eye ointment where a pleasant odor is required.

Chemical composition
Analysis of rose aromatic waters has shown that they are highly rich in phenethyl alcohol, geraniol, and β-citronellol, which are very different from their essential oil counterpart; it is an indication of the chemical constituents of the material used and the therapeutic opportunities [31].

Cosmetic and skin applications
Flower distillates like rose, orange blossom, and peppermint are treasured in cosmetics due to their refreshing, hydrating, and pH-balancing effects, and can be served as a good choice when developing tonic, mist, and other soft cosmetic products.

Food and beverage industry
Aromatic water's moderate flavor makes the herbal distillates suitable for use as a natural flavoring agent in cooking and beverages, thereby creating a delicate scent without the overwhelming strength of essential oils.

6.3.4 Physicochemical properties

Aromatic waters, also known as hydrosols or floral waters, are the aqueous solutions obtained during the steam distillation of aromatic plants for essential oil extraction. These waters contain trace amounts of essential oil and various water-soluble components from the plant material.

Aromatic waters have a delicate aroma reminiscent of the plant from which they are derived. This aroma is less concentrated than that of the corresponding essential oil but still contributes to the therapeutic and sensory properties of the water. The color of aromatic waters varies depending on the plant material used and the concentration of water-soluble pigments present in the plant. They can range from clear to pale yellow, green, or even blue, depending on the botanical source. The pH of aromatic waters typically ranges from around 4 to 6, making them slightly acidic to neutral. The pH can vary depending on factors, such as plant species, geographic location, and processing methods. Aromatic waters are completely miscible with water and can be easily diluted. They may contain water-soluble components, such as phenols, flavonoids, and other phytochemicals extracted from the plant material during distillation. The density of aromatic waters is close to that of water, with minor variations depending on the specific constituents present in the water. Aromatic waters have a refractive index slightly higher than that of water due to the presence of dissolved compounds from the plant material. This property can be used for quality control and authentication purposes. Aromatic waters contain antimicrobial compounds derived from the plant material, which help inhibit microbial growth and contribute to their shelf stability. However, they are still susceptible to microbial contamination over time, especially if not stored properly. Like most aqueous solutions, aromatic waters have surface tension, which affects their wetting and spreading properties. This property can influence their application in cosmetic, culinary, and therapeutic formulations. Overall, the physicochemical properties of aromatic waters are influenced by various factors, including the plant species, growing conditions, harvesting methods, and distillation techniques, making each aromatic water unique in its composition and characteristics.

6.4 Decoction

Decoction is an extraction method that involves boiling plant material, such as stems, roots, bark, and rhizomes, to extract dissolved chemicals, oils, volatile organic compounds, and other substances. This process begins with mashing the plant material and then boiling it in water [5]. Decoction is particularly suited for extracting substances from the hard parts of plants, including roots, twigs, bark, nuts, and hard seeds.

6.4.1 Preparation methods of decoction

6.4.1.1 Hard boil

After mixing the volatile substance with purified water, heat the mixture to the highest temperature until it rapidly reaches a boil. It is crucial to avoid prolonged boiling; instead, bring it just to the point of bubbling. Prolonged boiling may lead to the degradation of volatile compounds or the loss of aroma. Once the mixture reaches the bubbling point, remove it from the heat to prevent overheating. This method is often used to extract volatile compounds efficiently while preserving their aromatic properties.

Additionally, adjusting the heating process according to the specific characteristics of the volatile substance being used can optimize the extraction and ensure the quality of the final product.

6.4.1.2 Simmer

After bringing the mixture to a brief boil, reduce the heat to the lowest setting and allow it to simmer gently for 45–60 min. This low simmering period helps to further extract the aromatic compounds from the volatile substance, enhancing the fragrance and potency of the resulting mixture. Simmering at a low temperature prevents rapid evaporation of the volatile compounds and minimizes the risk of overheating, which could compromise the quality of the aromatic water. This step is essential for maximizing the extraction efficiency and achieving the desired concentration of aromatic essence in the final product.

6.4.1.3 Resting time

Once the simmering time is complete, carefully remove the mixture from the heat source. This step marks the end of the heating process and ensures that the aromatic water has been adequately infused with the volatile compounds from the original substance. Removing the mixture from the heat source promptly helps prevent any further evaporation or alteration of the aromatic compounds due to prolonged exposure to heat. It's important to handle the mixture with caution to avoid any accidental spills or burns. With the heating process completed, the aromatic water is now ready for the next steps in the preparation process, such as cooling, settling, and filtration.

6.4.1.4 Storage

After allowing the mixture to rest for 2 h, transfer both the liquid and the herbs into a quart jar. Seal the jar securely and let it sit for 8–12 h to allow for further infusion of flavors and aromatic compounds. During this time, the liquid will continue to absorb essences from the herbs, enhancing its fragrance and potency. Once the infusion period is complete, strain the herbs from the liquid using a fine mesh strainer or cheesecloth. Ensure to extract any remaining juice from the herbs by gently squeezing them to release their essence. This step helps to maximize the flavor and aroma extracted from the herbs, resulting in a more concentrated and flavorful aromatic water. The strained liquid can then be further filtered or used as desired in various applications, such as culinary recipes, aromatherapy, or skincare formulations [6].

6.4.2 Concentrated decoctions

A concentrated decoction can be achieved by simmering off additional water from a strained decoction. This concentrated form can then be preserved as a stock remedy or incorporated into more complex products for internal or external use. It's essential to note that this method is suitable only for remedies that can withstand heat without losing efficacy.

Formulation largely relies on common sense, with an example being to start with a normal decoction ratio of about 1:20 (e.g., 50 g of herb decocted in 1 L of water). After straining, the decoction is further reduced by simmering to a ratio of 1:4 (e.g., reducing the volume to 200 mL). This process concentrates the active ingredients, making the remedy more potent [7].

6.4.3 Uses of decoction

Birch bark is an awesome skin-friendly herb and can be used for minor skin irritations. Yucca root or soapwort root can be made to used as a mild cleaning solution for delicate washables in your home [8]. Decoction, a method of extracting active compounds from plant materials by boiling them in water, has various traditional and modern uses across cultures and industries. Decoctions have long been used in traditional medicine systems, such as Ayurveda, traditional Chinese medicine, and Native American herbal medicine. They are employed to extract beneficial compounds from medicinal herbs, roots, bark, and other plant parts. Decoctions can be formulated to address specific health concerns, such as colds, digestive issues, inflammation, and more. Decoctions are often used to prepare herbal remedies and tonics. They can be consumed orally as teas or infusions, applied topically as compresses or washes, or used in steam inhalation therapies. Decoctions allow for the extraction of water-soluble compounds, including vitamins, minerals, antioxidants, and phytochemicals, which contribute to the therapeutic properties of herbal remedies. Decoctions are used in culinary practices to infuse flavors and aromas into food and beverages. For example, decoctions of herbs, spices, or aromatic ingredients like cinnamon, ginger, or cardamom are commonly used to flavor soups, stews, sauces, and beverages, such as teas and punches. Decoctions are utilized in the formulation of various cosmetic and personal care products, including skincare preparations, hair care products, and bath products. They can be used as natural alternatives to synthetic ingredients, providing hydration, nourishment, and soothing properties to the skin and hair. Decoctions derived from plants containing natural dyes are used in textile dyeing and coloring processes. These decoctions impart vibrant colors to fabrics, yarns, and fibers, serving as an eco-friendly and sustainable alternative to synthetic dyes. Decoctions are sometimes used in the preservation and fermentation of food and beverages. For instance, decoctions of fruits, herbs, or spices may be added to homemade jams, jellies, syrups, or fermented beverages to enhance flavor and to extend shelf life.

Overall, decoctions are versatile preparations with diverse applications in medicine, culinary arts, cosmetics, and other fields. They offer a natural and holistic approach to health and well-being, harnessing the therapeutic properties of plants to promote vitality and balance.

6.4.4 Physicochemical properties

Decoction is a method of extraction used to prepare herbal teas or medicinal infusions by boiling plant material in water. The process involves simmering the plant parts (such as roots, bark, or seeds) in water for an extended period to extract their active compounds. The color of decoctions can vary widely depending on the plant material used. It can range from clear to various shades of brown, red, or green, depending on the presence of pigments and other compounds in the plant. The pH of decoctions is typically slightly acidic to neutral, similar to that of water. However, it can vary depending on the pH of the plant material and any acidic or alkaline compounds extracted during the boiling process. Decoctions contain water-soluble compounds extracted from the plant material during boiling. These compounds may include phenolic compounds, alkaloids, flavonoids, and other phytochemicals with various medicinal properties. The density of decoctions is similar to that of water, although it may be slightly altered by dissolved solids and extracts from the plant material. The refractive index of decoctions can vary depending on the concentration of dissolved solids and compounds in the solution. This property can be used for quality control and monitoring the extraction efficiency of decoction processes. Decoctions are typically prepared by boiling the plant material in water for an extended period, which can vary depending on the plant's toughness and desired extraction efficiency. The boiling temperature of the water affects the extraction of heat-sensitive compounds from the plant material. Decoctions may contain antimicrobial compounds extracted from the plant material, which can help inhibit microbial growth and contribute to their shelf stability. However, like any aqueous solution, they are susceptible to microbial contamination if not stored properly. The concentration of active compounds in decoctions can vary depending on factors, such as the plant material's quality, the duration of boiling, and the water-to-plant ratio used in preparation. Overall, the physicochemical properties of decoctions are influenced by the characteristics of the plant material used, the extraction process, and the environmental conditions during preparation. Decoctions have been used for centuries in traditional medicine and herbalism due to their ability to extract a wide range of bioactive compounds from plant material.

6.5 Lozenges

Lozenges are solid, flavored medicated confections typically intended for oral consumption. These small, often disk-shaped tablets are designed to dissolve slowly in the mouth, releasing active ingredients, such as medicinal compounds, soothing agents, or flavorings. Lozenges are commonly used to alleviate symptoms associated with throat irritation, coughing, and mild mouth or throat infections. They are formulated to provide temporary relief by soothing the throat and suppressing coughs. Lozenges come in a variety of flavors, ranging from traditional menthol or eucalyptus to fruity or herbal blends. They often contain ingredients like menthol, eucalyptus oil, honey, lemon, or herbal extracts known for their soothing or antiseptic properties. The texture of lozenges is typically smooth and hard, allowing them to dissolve slowly in the mouth without requiring chewing. Some lozenges may also contain additional ingredients like vitamin C or zinc, which are believed to support immune health.

In addition to their medicinal uses, lozenges are sometimes used as a convenient and portable way to freshen breath or provide a burst of flavor. They are available over the counter in pharmacies and convenience stores and are generally safe for use by adults and children under proper supervision. However, individuals with certain medical conditions or allergies should consult with a healthcare professional before using lozenges, especially if they contain active ingredients or herbal extracts. Overall, lozenges offer a convenient and effective way to address mild throat discomfort and provide temporary relief from coughing and irritation.

6.5.1 Composition of lozenges

The composition of a lozenge can vary depending on its intended use and specific formulation. Fructose, sucrose, and dextrose are commonly utilized as pharmaceutical carriers due to their unique properties. However, here's a general overview of the common components found in lozenges, along with descriptions of their roles:

Active ingredients
These are the medicinal compounds that provide the desired therapeutic effects. Common active ingredients found in lozenges include:

Menthol: Provides a cooling sensation and helps to soothe throat irritation. It can also act as a mild analgesic to relieve minor sore throat pain.

Eucalyptus oil: Known for its antibacterial and decongestant properties, eucalyptus oil helps to relieve nasal congestion and throat irritation.

Zinc: Believed to support immune health and reduce the duration of cold symptoms.

Herbal extracts: Extracts from herbs like licorice root, slippery elm, or marshmallow root may be included for their soothing properties.

Flavoring agents

Lozenges often contain flavorings to improve taste and mask the sometimes bitter or medicinal flavor of the active ingredients. Peppermint oil (flavor no. 113.042, Bell Flavors, Northbrook, Illinois) is recommended as it is sweet, highly acceptable to essentially all patients, and has a menthol effect in the nose without menthol bitterness. Peppermint oil is considered a food product and not a drug. Bell peppermint oil also has proven multi-year stability in directly compressed zinc acetate lozenges. Peppermint oil can be directly absorbed into Emdex(r) with high-pressure micro fine mists of peppermint oil, and thorough mixing. Peppermint oil can also be dried with silica gel (Siloid(r) 244FP – Davison Chemical, Baltimore, Maryland). Peppermint oil will become dry at a 1:1 ratio with silica gel, but flavor losses occur. Perhaps losses result from the retention of peppermint oil by silica gel. For example, lozenges containing 5 mg of peppermint oil sprayed directly into Emdex(r) have a flavor equivalent to lozenges containing 17 mg of peppermint oil plated onto 17 mg of silica gel [9].

Most flavorings can be used in zinc acetate lozenges, given a suitable carrier. Eucalyptol, wintergreen, clove, cinnamon, spearmint, cherry, lemon, orange, lime, menthol, and various combinations are all possible flavorings [10]. However, some flavor oils are not stable in long-term storage in zinc acetate lozenges and may require costly protection from contact with zinc acetate, evaporation, degradation, and oxidation. Common flavors include:

Mint: Provides a refreshing and cooling sensation.

Citrus: Adds a tangy and citrusy flavor.

Berry: Offers a sweet and fruity taste.

Sweeteners

To enhance palatability, lozenges typically contain sweeteners. These can include;

Fructose: Fructose is the sweetest natural sugar, approximately 1.73 times sweeter than sucrose. It is a component of sucrose, a disaccharide, and an isomer of dextrose. Surprisingly, fructose does not visibly react, change color, or form bitter compounds with zinc acetate at higher temperatures, despite being a polyhydroxy ketone, which is typically highly reactive.

Sucrose: Sucrose is a common disaccharide composed of glucose and fructose. It serves as a carrier in pharmaceutical formulations due to its stability and relatively neutral taste. Sucrose is a traditional sweetener used in many lozenges. It provides sweetness and can help mask the taste of other ingredients.

Dextrose: Dextrose, almost identical to glucose, is less sweet than sucrose, being approximately 0.74 times as sweet. Unlike fructose, dextrose is typically considered an inert monosaccharide. However, dextrose can react with zinc gluconate and other zinc compounds over time to form bitter complexes. Interestingly, dextrose does not react with zinc acetate to cause bitterness. The differing reactions between dextrose and zinc compounds, despite their similar structure, may be attributed to the unique properties of acetic acid and gluconic acid, which are closely related to monocarboxylic acids [11].

Artificial sweeteners: Used as sugar substitutes for individuals who need to limit their sugar intake. Sweeteners play a crucial role in making lozenges more palatable, especially since they often contain active ingredients that can have a bitter or medicinal taste. These sweeteners provide sweetness without adding calories and are often used in sugar-free or low-sugar lozenges. Common artificial sweeteners include:
- Aspartame: A high-intensity sweetener that is much sweeter than sugar. It is commonly used in sugar-free lozenges.
- Sucralose: Another high-intensity sweetener that is derived from sucrose. It is also used in sugar-free lozenges and provides sweetness without contributing to tooth decay.
- Stevia: A natural sweetener derived from the leaves of the *Stevia rebaudiana* plant. It is often used in combination with other sweeteners and is known for its intense sweetness.

Natural sweeteners: Some lozenges may use natural sweeteners for those who prefer to avoid artificial additives. These can include:
- Honey: A natural sweetener with additional health benefits such as antimicrobial properties and soothing effects on the throat.
- Maple syrup: Another natural sweetener that adds a distinct flavor to lozenges.
- Agave nectar: Derived from the agave plant, this sweetener is often used as a natural alternative to sugar.

The choice of sweetener in a lozenge depends on factors, such as taste preferences, dietary restrictions, and desired health benefits. It is essential to consider individual needs and preferences when selecting lozenges, especially for individuals with specific dietary requirements or health conditions, such as diabetes.

Binder
This component helps to hold the ingredients together and maintain the shape of the lozenge. Common binders include:

Gelatin: A natural protein derived from animal collagen that helps to form a gel-like consistency.

Pectin: A plant-based polysaccharide that forms a gel when mixed with water, providing structure to the lozenge.

Base

The base of a lozenge provides bulk and stability. It may consist of inert fillers, such as cellulose or maltodextrin, which are commonly used to add volume to the lozenge without contributing significantly to its medicinal properties.

Coating or glazing

Some lozenges may be coated or glazed to improve appearance, texture, and mouth-feel. This can help prevent sticking and make them easier to handle.

Overall, the composition of a lozenge is carefully formulated to deliver therapeutic effects while ensuring palatability and ease of use. It's essential to read the product label and consult with a healthcare professional if you have any questions or concerns about the ingredients in a specific lozenge.

6.5.2 Commercial directly compressible lozenges

Mendel's Sugar tab®

It is a white, free-flowing, inert tablet base comprised of agglomerated sugar. It contains 90–93% agglomerated sucrose, with the remainder being invert sugar. It exhibits a sweetness value identical to sucrose. It produces moderately hard, non-friable tablets when compressed with active ingredients and a suitable lubricant. It possesses good flow characteristics, compressibility, flavor-masking, low hygroscopicity, chemical stability, and non-cloying sweetness. It offers a wide range of compatibilities, smooth disintegration, and a pleasant aftertaste [12].

Mendel's Sweetrex®

It is a directly compressible chewable tablet base with a sweetness value equivalent to sucrose. It does not contain sucrose but is a blend of 70% Emdex® and 30% Krysta® 300 crystalline fructose. It demonstrates a binding capacity of up to 50% active ingredients without significant loss of compressibility. It presents as a white granular powder with an average particle size of 210 microns. Due to its fructose content, Sweetrex® is hygroscopic and should be stored at less than 55% humidity [13].

Mendell's Emdex®

It is known as a deathrate, complying with the official monograph in National Formulary XVI as the hydrated form. It exhibits a sweetness value of 0.74 of sucrose. It is composed almost entirely of free-flowing spray-crystallized porous spheres, possesses outstanding fluidity and compressibility, and making it suitable for direct compression techniques to produce lozenges. It eliminates the need for glidants, allows for maximum press speed, and eliminates induced die feeding. It is soluble in water and commonly used in chewable tablets and lozenges. Its features include outstanding flow, compressibility, lubricity, non-hygroscopicity, controlled particle size, cool mouth-feel, negative heat of solution, and stability to heat and moisture, among other pharmaceutical properties [14].

6.5.3 Preparation methods of lozenges

6.5.3.1 Molded lozenge

Molded lozenges represent a specific variation within the realm of lozenge production, characterized by the shaping of the lozenge paste into form through molds rather than the conventional method of cutting. The process initiates with the careful selection and mixing of active ingredients and excipients, blending them seamlessly with a base substance to form a consistent mixture. Gradual addition of water transforms this amalgamation into a pliable paste, primed for molding. This paste is then poured into specialized molds, meticulously designed to yield the desired shape and size of the lozenge. Once filled, excess paste may be trimmed to ensure uniformity. Following molding, the lozenges undergo a drying phase in a chamber with circulating hot air, essential for solidification and moisture removal. Post-drying, the molded lozenges are meticulously inspected for quality before being packaged securely to preserve their integrity. This method offers distinct advantages, including precise control over shape and size, potential efficiency gains, and opportunities for unique branding through customized shapes and embossments. Ultimately, molded lozenges provide a convenient and effective means of delivering medicinal or therapeutic ingredients, all while upholding standards of consistency and quality [15], for example, bismuth lozenge and liquorice lozenge.

6.5.3.2 Compressed lozenge

Compressed lozenges are a specialized type of lozenge formed through the compression of powdered ingredients into solid tablets. This method involves a series of precise steps beginning with the selection and blending of active ingredients and excipients, which are combined with a suitable base substance. The resulting mixture is then compressed under high pressure into tablet form using a tablet press. This compression not only ensures uniformity in size and weight but also enhances the solidity and stability of the lozenges. Following compression, the tablets may undergo additional processing steps, such as polishing or coating to improve taste, appearance, and dissolution characteristics. Compressed lozenges offer several advantages, including ease of handling, convenient dosing, and improved shelf life due to reduced moisture content. Moreover, their compact size and portability make them a preferred choice for consumers seeking on-the-go relief. Overall, compressed lozenges provide a reliable and effective method for delivering medicinal or therapeutic agents while maintaining consistent quality standards [11].

6.5.4 Uses of lozenges

Soothing of a sore throat
Soothing lozenges offer immediate symptomatic relief against sore throat and minor throat irritations. Evaluation in clinical conditions has shown rapid analgesia onset with a maximum of 2 h.

Treatment of oral infections
Antifungal and antimicrobial agent-based lozenges are prescribed to treat mucosal infections and oral candidiasis, and control mucosal healing as well as the reduction in microbial infection.

Management of cough and respiratory conditions
Anti-cough drops/lozenges with menthol relax the throat and inhibit the cough. They especially come in handy in the temporary cessation of a cough caused by upper respiratory tract infections.

Systemic drug delivery through buccal absorption
The lozenges can be used to deliver the drugs systemically through the mouth mucosa, evading the first-pass metabolism. The most notable example is nicotine lozenges for smoking cessation.

Patient convenience/compliance
Lozenges are preferred by pediatric as well as geriatric patients since the dosage form is palatable and convenient to use.

6.5.5 Physicochemical properties

Lozenges, also known as troches or pastilles, are solid dosage forms designed to dissolve slowly in the mouth, releasing medication or active ingredients for local or systemic effects. Physicochemical properties of lozenges encompass a range of factors crucial to their formulation and effectiveness. These properties include composition, size, shape, hardness, texture, dissolution rate, and taste. The composition of lozenges typically includes active pharmaceutical ingredients (APIs) along with excipients, such as binders, fillers, flavorings, and sweeteners. These ingredients not only contribute to the therapeutic effect but also influence the texture, taste, and dissolution characteristics of the lozenge. Size and shape play a role in patient acceptance and ease of administration, with some lozenges designed to be discreet and portable while others may be larger for extended-release formulations. Hardness affects the lozenge's ability to dissolve slowly in the mouth, providing sustained release of the medication and soothing effects for throat irritation. Texture can vary from smooth to granular, depending on the formulation, and can impact the overall mouthfeel and patient experience. Dissolution rate is critical for ensuring that the lozenge releases the medication at an appropriate rate for optimal efficacy and patient comfort. Taste is an essential consideration, as lozenges are often flavored to mask the bitter or unpleasant taste of medications and improve palatability. By carefully controlling these physicochemical properties, pharmaceutical manufacturers can produce lozenges that effectively deliver medication while providing a pleasant experience for the patient.

6.6 Pill

Pills in botanical preparations are solid dosage forms made from powdered or granulated herbal ingredients, often mixed with binding agents. They deliver therapeutic compounds from plants in a convenient, easily dosable format, commonly used for various health conditions. They offer portability and standardized dosing, but should be used cautiously and preferably under healthcare professional guidance.

6.6.1 Composition

Pills contain active ingredients and excipients. On the one hand, active ingredients are the components which are responsible for producing the desired therapeutic effect, typically derived from medicinal herbs or other sources. These active ingredients may include compounds with pharmacological properties, such as alkaloids, flavonoids, or essential oils, depending on the specific medicinal properties sought.

Excipients, on the other hand, are inactive substances added to the pill formulation to serve various purposes, such as facilitating the manufacturing process, improving stability, enhancing palatability, or aiding in the delivery of the active ingredients. Common excipients in pill formulations include binders, fillers, disintegrants, lubricants, and coatings.

Together, active ingredients and excipients work synergistically to create pills that are effective, stable, and easy to administer while ensuring the desired therapeutic outcomes [16].

6.6.2 Preparation methods of pills

A typical formula or prescription for pills would give the weight of each ingredient required to make one pill and state the number of pills required. Multiplying these two figures together gave the quantities to be weighed. The ingredients were then thoroughly mixed using a pill mortar and pestle.

It was now necessary to add an excipient, an inert substance, to bind all the ingredients together to form a stiff, workable mass. The substance of choice, for most purposes, was syrup of liquid glucose – a very thick and viscous syrup. This had to be added a drop at a time while the ingredients were vigorously worked in the mortar. Just enough was added to form a non-crumbling, stiff, pliable mass. The mass now had to be divided accurately into equal parts equivalent to the number of pills ordered. In earlier days, this procedure was carried out on a pill tile. The mass would be rolled into a ball and then gradually rolled into a long, even pipe. By measuring the pipe and dividing by the number of pills, the length of each dose could be calculated, and the pipe could be cut into portions. Later, tiles had graduations so that the pipe could be rolled to a specific length equal to the required number of graduations.

A later development was the pill machine, an apparatus that allowed for the pipe to be measured and cut into accurate pieces. Grooves in a brass plate corresponded to the number of pills, and the pill mass could be rolled to the number required. A handle with complementary grooves was guided to cut the pipe. Careful manipulation could produce quite rounded portions.

The added advantages of the pill machine were an area to roll the pipe and round the pills and a box to collect the cut pieces (which could also be used to roll the pill mass).

A pill rounder was used to make the cut portions spherical. The portions were roughly rounded between finger and thumb, and placed under the rounded which was manipulated with a circular or figure-of-eight movement. Periodic checks were made, and well-rounded pills were removed. When all the pills were rounded, they were set aside and allowed to dry. When dry, the pills could be given to the patient.

However, it was customary to enhance the appearance of the pills, and this could be done in various ways. A pearlized finish was achieved by again using the pill rounder but rolling the pills in a little talcum powder. A coat of varnish could be applied by rolling the pills in a few drops of varnish in a round-bottomed container. The ultimate finish was a coating of gold or silver, which was achieved with the aid of a pill. Silver, gold, or silver leaf was used to line the inside of the pill. Silver, the pills were moistened by rolling them in a few drops of a liquid vegetable gum then placed inside. The top was put on and the apparatus was rotated for a few minutes. Fees for coating were added accordingly. The resulting pills were packed, usually in round boxes and labeled with the doctor's instructions.

6.6.3 Physicochemical properties

Pills, also known as tablets or capsules, are solid dosage forms used for the oral administration of medications, dietary supplements, or other substances. They are typically composed of APIs along with excipients, which are added to facilitate the manufacturing process, enhance stability, or improve the dosage form's performance. Physicochemical properties of pills include factors, such as size, shape, color, hardness, friability, and dissolution characteristics. The size and shape of pills can vary depending on the dosage and the manufacturer's preference, with some designed for easy swallowing and others for controlled release or specific drug delivery purposes. Color is often added to pills for branding or identification purposes, but may also serve functional roles, such as masking the appearance of the API or enhancing stability. Hardness and friability refer to the pill's resistance to crushing or breaking and its tendency to crumble, respectively, which can impact its handling, packaging, and shelf life. Dissolution characteristics are crucial for oral drug delivery, as they determine how quickly and completely the pill disintegrates and releases the API in the gastrointestinal tract. Various factors, including the composition of the pill, the type of excipients used, and the manufacturing process, can influence these physicochemical properties, ultimately affecting the pill's efficacy, safety, and patient adherence.

6.7 Spirits

Spirits, also referred to as essences, are indeed alcoholic or hydro alcoholic solutions of volatile substances. These volatile substances could include various aromatic compounds, such as essential oils, which give spirits their characteristic flavors and aromas. The term "essence" is often used interchangeably with "spirit" in this context. Typically, spirits are prepared using high-proof alcohol, often at least 90% alcohol by volume (ABV). This high alcohol content helps in extracting and preserving the volatile components from the source material, whether they are solid, liquid, or gaseous.

6.7.1 Comparison of spirits and tinctures

Alcohol content
Spirits typically have a higher alcohol content compared to tinctures. While spirits often use alcohol with a minimum of 90% ABV, tinctures usually employ alcohol with a lower ABV, typically around 40–60%.

Purpose
While both spirits and tinctures can be used for flavoring and medicinal purposes, they may differ in their intended applications. Spirits are often more focused on flavoring and aromatic qualities, whereas tinctures may be used more for medicinal purposes, as they can serve as alcohol-based extracts of medicinal herbs or botanicals.

Overall, spirits play a significant role in various industries, including food and beverage, perfumery, aromatherapy, and traditional medicine, offering a versatile and potent vehicle for delivering the volatile essences of natural substances [17].

6.7.2 Preparation methods of spirit

The following methods can be used for the preparation of spirits.

6.7.2.1 Simple solution method

In the simple solution method, volatile oils are dissolved directly in alcohol to create the desired spirit. This method is straightforward and effective for extracting the aromatic compounds from the volatile oils [18].

In example a, peppermint spirit, 100 mL of peppermint oil is measured and then dissolved in a sufficient quantity of alcohol to produce 1,000 mL of peppermint spirit. This would result in a 10% v/v (volume/volume) concentration of peppermint oil in the final spirit. Peppermint spirit can be used as a carminative, meaning it helps to relieve flatulence and abdominal discomfort.

In example b, spirit ammonia aromatic, ingredients including ammonium bicarbonate, strong ammonia solution, lemon oil, and purified water are combined and dissolved in a closed bottle by heating. This allows for the components to mix effectively and form the aromatic spirit. Spirit of ammonia aromatic can be used both as a carminative and as a cardiac stimulant, meaning it can aid in digestion and stimulate the heart.

Overall, the simple solution method is a versatile approach to spirit preparation, offering flexibility in formulation and application. It is commonly employed in industries, such as pharmaceuticals, perfumery, and culinary arts to extract and preserve the beneficial properties of volatile substances. The simple solution method offers a quick and efficient way to prepare spirits by directly dissolving volatile oils in alcohol. This method is suitable for substances that readily dissolve in alcohol, such as essential oils. By controlling the ratio of volatile oil to alcohol, the concentration of the active compounds in the final spirit can be adjusted to meet specific requirements. In the examples provided, peppermint oil and lemon oil are dissolved in alcohol to create spirits with distinctive flavors and aromatic properties. These spirits can then be used for various purposes, such as aiding digestion or providing cardiac stimulation.

6.7.2.2 Maceration

In the maceration procedure, the crude drug is initially soaked in purified water to extract water-soluble materials. Following this, the moistened macerated leaves are then combined with a specific quantity of alcohol [19]. The mixture is left to macerate for 6 h, allowing for the infusion of the desired compounds. Subsequently, the menstruum is strained, and the marc is pressed to extract any remaining liquid. The strained and expressed liquids are combined and left to settle, after which impurities are filtered out. Finally, peppermint oil is added to adjust the volume. This resulting peppermint spirit, made from peppermint leaves, peppermint oil, alcohol, and water, can be utilized as a carminative. It is essential to store this spirit in a well-closed container to prevent the loss of volatile oil and alcohol.

6.7.2.3 Chemical reaction

This method involves the synthesis of spirits through chemical reactions. One example provided is the production of ethyl nitrate spirit. This process typically includes reacting sodium nitrite ($NaNO_2$) with alcohol and sulfuric acid (H_2SO_4) solution. The chemical reaction results in the formation of ethyl nitrate, which is then collected and utilized as a spirit.

The advantage of this method lies in its ability to produce spirits with specific chemical compositions and properties. By controlling the reaction conditions and ingredients, it's possible to tailor the resulting spirit to meet desired characteristics, such as aroma, flavor, and potency. However, this method requires careful handling of potentially hazardous chemicals and precise control of reaction conditions to ensure safety and product quality.

6.7.2.4 Distillation

Distillation is a widely used method for producing high-quality spirits. It involves the separation of volatile compounds from a mixture through evaporation and condensation. In the context of spirit preparation, distillation is typically used to extract volatile oils and other aromatic compounds from botanical materials.

The spirit of ammonia can be prepared by the distillation process. Initially, volatile oils, alcohol, and water are distilled together to obtain a distillate. This distillate contains the desired aromatic compounds. Subsequently, this distillate is further processed and combined with additional ingredients, such as ammonium bicarbonate, strong ammonia solution, nutmeg oil, and purified water to create the final spirit.

Distillation allows for the concentration and purification of volatile compounds, resulting in spirits with intense aromas and flavors. It's a versatile method that can be adapted to different types of botanicals and desired end products. However, it requires specialized equipment and expertise to perform effectively.

6.7.3 Uses of spirits

Spirits can be utilized in several ways.

6.7.3.1 Inhalation

They can be inhaled to provide aromatic effects, which are often used in aromatherapy or as part of certain religious or cultural practices.

6.7.3.2 Oral use

Spirits can be consumed orally, either directly or as part of a beverage or food preparation. However, it's important to note that some spirits may not be safe for ingestion due to the presence of toxic compounds.

6.7.3.3 External use

Spirits are sometimes applied externally to the skin, often in the form of perfumes, colognes, or as part of skincare products.

6.7.3.4 Flavoring agents

One of the most common uses of spirits is as flavoring agents in food and beverages. They impart distinct flavors and aromas to dishes and drinks, enhancing their overall sensory experience.

Spirits are particularly effective at retaining water-insoluble volatile oils in solution. This property makes them valuable for extracting and preserving the aromatic compounds found in botanical materials.

6.7.4 Physicochemical properties

Spirits, also known as distilled liquors or hard alcohol, encompass a diverse range of alcoholic beverages produced through distillation from fermented grains, fruits, or other raw materials. They exhibit several physicochemical properties that contribute to their unique characteristics and sensory profiles. One key property is alcohol content, typically ranging from 40% to 60% ABV or higher, which is a result of the distillation process concentrating ethanol. This high alcohol content influences other properties, such as density, viscosity, and volatility. Spirits often have a lower density than water due to their alcohol content, and their viscosity can vary depending on factors like aging and any additives used. The color of spirits ranges widely from clear (e.g., vodka) to amber or dark brown (e.g., whiskey), influenced by factors, such as aging in wooden barrels and the addition of coloring agents. Flavor and aroma are complex and influenced by the raw materials used, distillation process, aging, and any added flavorings or botanicals. Additionally, spirits may exhibit specific pH levels, refractive indices, and solubilities depending on their composition. These physicochemical properties collectively contribute to the diversity and complexity of spirits, making them sought-after beverages for enjoyment both neat and in cocktails.

6.8 Tinctures

Tinctures are solutions made by dissolving a nonvolatile drug, derived from either vegetable or chemical sources, in alcohol or a hydro-alcoholic solvent. They are commonly used in both topical and internal applications due to their ability to extract and preserve the active ingredients of various substances [20]. Tinctures are typically composed of 40–45% alcohol, which acts as a solvent for extracting the active constituents of the drug.

The composition of a tincture is usually one part by weight of the drug to four parts of the final product. Alcohol with dilutions ranging from 45%, 60%, to 70% is often used in tincture preparation, serving as a source of the drug for other formulations.

6.8.1 Preparation methods of tincture

6.8.1.1 Maceration

This method involves soaking the drug in alcohol to extract its active ingredients. Tinctures prepared through maceration include tincture of orange [21], tincture of benzoin [22], and tincture of opium [23]. This process allows for the dissolution of both water-soluble and alcohol-soluble compounds present in the drug. Tinctures prepared through maceration are widely used in herbal medicine and pharmaceuticals due to their simplicity and effectiveness in extracting medicinal properties.

In the specific example of preparing tincture of orange, fresh orange peel is utilized as the drug. The process begins by thoroughly macerating the orange peel, typically by chopping or grinding it, to increase the surface area for extraction. This macerated peel is then submerged in 90% alcohol, also known as the menstruum, in an airtight container.

During the maceration period, which lasts for about 7 days, the alcohol gradually extracts the active ingredients from the orange peel. This includes various aromatic compounds, flavonoids, and other bioactive molecules present in the peel.

After the maceration period is complete, the mixture is strained to separate the liquid extract from the solid residue, known as the marc. The liquid extract, now enriched with the extracted compounds, is then filtered to remove any remaining impurities or solid particles from the solution [24]. This filtration step ensures the purity and clarity of the final tincture.

The resulting liquid is the tincture of orange, a concentrated solution containing the active constituents extracted from the orange peel. This tincture possesses the characteristic flavor and aroma of oranges and is often used for its medicinal properties, such as its digestive and anti-inflammatory effects.

Overall, maceration is a straightforward and effective method for preparing tinctures, allowing for the extraction of a wide range of bioactive compounds from botanical sources. It is a time-tested technique that has been used for centuries to harness the healing properties of plants and herbs.

6.8.1.2 Percolation

Percolation is another method for preparing tinctures, particularly when a more concentrated extract is desired. It involves passing a solvent through a bed of the powdered drug to extract the active constituents gradually.

Tinctures, such as tincture of ginger [25], tincture of cardamom [26], tincture of digitalis [27], and tincture of belladonna [28], are prepared through percolation. This method is favored when a more potent and uniform tincture is needed, as it allows for precise control over the extraction process. The percolation process begins by moistening the powdered drug with a suitable solvent, ensuring that it is evenly dampened. The moistened powder is then packed into a percolator, a vessel designed for this purpose. The percolator is covered and left to stand for a specified period to allow for thorough saturation of the drug with the solvent.

Once the drug has absorbed the solvent and swelled adequately, the percolation process begins. Additional solvent is gradually added to the percolator, allowing it to percolate through the drug bed. The solvent dissolves the active constituents present in the powdered drug as it passes through, resulting in the formation of a concentrated liquid extract.

During percolation, care must be taken to control the flow rate of the solvent to ensure optimal extraction efficiency. The process continues until a sufficient volume of solvent has passed through the drug bed, resulting in the desired concentration of the tincture.

After percolation is complete, the liquid extract is collected and may undergo further processing, such as filtration or concentration, to obtain the final tincture product. This tincture is characterized by its potency and uniformity, making it suitable for various medicinal or therapeutic applications.

In summary, percolation is a methodical and controlled approach to tincture preparation, allowing for the extraction of highly concentrated and consistent extracts from powdered drugs. It is a preferred technique in the pharmaceutical and herbal medicine industries where precision and potency are paramount. The process requires moistening the powdered drug with a solvent, allowing it to stand, and then gradually adding more solvent to make the final tincture.

6.8.1.3 Simple solution

The simple solution method for preparing tinctures offers two distinct approaches: chemical reaction and dilution. These methods provide flexibility in tincture preparation, allowing for different techniques depending on the specific characteristics of the drug and desired outcome.

Chemical reaction: This approach involves inducing a chemical reaction between the drug and other reagents to form the desired tincture. An example of a tincture prepared through chemical reaction is iodine tincture. This tincture is produced by reacting iodine with a suitable solvent, often alcohol, to form the iodine tincture solution. The chemical reaction alters the properties of the drug and solvent, resulting in the formation of the tincture with specific therapeutic properties.

Dilution: The dilution method involves diluting a concentrated solution or extract of the drug with a suitable solvent, typically alcohol, to achieve the desired potency and volume. In this approach, a solid form of the drug is dissolved in a strong tincture or extract of another substance, such as ginger. The resulting solution is then further diluted with alcohol to adjust the concentration and volume to the desired level. Tincture of hyoscyamus and tincture of hamamelis are examples of tinctures prepared through dilution.

Both the chemical reaction and dilution methods within the simple solution approach offer distinct advantages and applications in tincture preparation. The choice between these methods depends on factors such as the solubility of the drug, desired potency, and specific therapeutic properties required for the final tincture product.

6.8.2 Uses of tinctures

Tinctures offer several advantages in the extraction and preservation of active compounds from botanical and chemical sources. They provide a convenient and efficient means of delivering therapeutic compounds, allowing for easy dosage adjustment and administration. Tinctures are commonly used in herbal medicine, homeopathy, and conventional medicine for their therapeutic properties.

Moreover, tinctures are versatile and can be applied externally as topical treatments or ingested orally for internal use. They are utilized in various formulations, including herbal remedies, dietary supplements, and pharmaceutical preparations.

Overall, tinctures represent an essential aspect of traditional and modern medicine, offering a reliable method for extracting and administering the beneficial properties of natural and synthetic substances.

6.8.3 Physicochemical properties

Tinctures are alcoholic extracts of herbs, spices, or other plant materials commonly used for medicinal, culinary, or flavoring purposes. Their physicochemical properties vary depending on the specific ingredients used and the extraction process. Typically, tinctures have a high alcohol content, ranging from 25% to 90% ABV, which serves as a solvent for extracting the active compounds from the plant material. This high alcohol content contributes to their preservation and extraction efficiency. Tinctures can exhibit a wide range of colors, flavors, and aromas, influenced by the botanical ingredients and any additional flavorings or additives. The pH of tinctures can vary depending on the acidity or alkalinity of the plant material being extracted. Additionally, tinctures may possess specific gravity, refractive index, and viscosity characteristics that reflect their composition and concentration of dissolved solids. While tinctures share some similarities with spirits in terms of alcohol content and solvent properties, their focus on extracting bioactive compounds from botanical sources distinguishes them and contributes to their diverse range of physicochemical properties.

References

[1] Capasso F, Grandolini G, Izzo AA. Therapeutic overview of galenical preparations. In: Phytotherapy: A Quick Reference to Herbal Medicine. Springer; Berlin, 2003. pp. 45–60.

[2] Canbay HS. Effectiveness of liquid-liquid extraction, solid phase extraction, and headspace technique for determination of some volatile water-soluble compounds of rose aromatic water. International Journal of Analytical Chemistry. 2017; 2017(1): 4870671.

[3] Hamedi A, Moheimani SM, Sakhteman A, Etemadfard H, Moein M. An overview on indications and chemical composition of aromatic waters (hydrosols) as functional beverages in Persian nutrition culture and folk medicine for hyperlipidemia and cardiovascular conditions. Journal of Evidence-based Complementary & Alternative Medicine. 2017; 22(4): 544–61.

[4] Cooper BF, Brecht E. The quantitative evaluation of aromatic waters. Journal of the American Pharmaceutical Association. 1952; 41(7): 394–97.

[5] Green J. The herbal medicine-maker's handbook: A home manual: Crossing Press; New York, USA, 2000.

[6] Zhang X-F, Yang J-L, Chen J, Shi Y-P. Optimization of a decoction process for an herbal formula using a response surface methodology. Journal of AOAC International. 2017; 100(6): 1776–84.

[7] Duric K, Kovac-Besovic E, Niksic H, Sofic E. Antibacterial activity of methanolic extracts, decoction and isolated triterpene products from different parts of birch, Betula pendula, Roth. Journal of Plant Studies. 2013; 2(2): 61.

[8] Ticktin T, Dalle SP. Medicinal plant use in the practice of midwifery in rural Honduras. Journal of Ethnopharmacology. 2005; 96(1–2): 233–48.

[9] Alissa M, Hjazi A, Abusalim GS, Aloraini GS, Alghamdi SA, Rizg WY, et al. Development and optimization of a novel lozenge containing a metronidazole-peppermint oil-tranexamic acid self-nanoemulsified delivery system to be used after dental extraction: In vitro evaluation and in vivo appraisal. Pharmaceutics. 2023; 15(9): 2342.

[10] Hemilä H, Chalker E. The effectiveness of high dose zinc acetate lozenges on various common cold symptoms: A meta-analysis. BMC Family Practice. 2015; 16: 1–11.

[11] Pothu R, Yamsani MR. Lozenges formulation and evaluation: A review. Ijapr. 2014; 1: 290–94.

[12] Ondari CO, Kean CE, Rhodes C. Comparative evaluation of several direct compression sugars. Drug Development and Industrial Pharmacy. 1983; 9(8): 1555–72.

[13] Gopale O, Jethawa S, Shelke S. Medicated lozenges: A review: Artificial intelligence in drug discovery. Asian Journal of Pharmaceutical Research and Development. 2022; 10(2): 129–34.

[14] Umashankar MS, Dinesh SR, Rini R, Lakshmi KS, Damodharan N. Chewable lozenge formulation – a review. International Research Journal of Pharmacy. 2012; 3(7): 5–8.

[15] Jellinek G, Stansby ME. Masking undesirable flavors in fish oils'. Fishery Bulletin. 1971; 69(1): 215.

[16] Singh J. Maceration, percolation and infusion techniques for the extraction of medicinal and aromatic plants. Extraction Technologies for Medicinal and Aromatic Plants. 2008; 67: 32–35.

[17] Porter SC. Tablet coating. Drug Cosmetic Industry. 1981; 128(5): 46–93.

[18] Vasey SA. Guide to the Analysis of Potable Spirits. Baillière, Tindall and Cox; London, 1904.

[19] Thompson HL. The Assay of the Spirit of Peppermint. University of Nebraska–Lincoln; Lincoln, NE, 1914.

[20] Heller D, Einfalt D. Reproducibility of fruit spirit distillation processes. Beverages. 2022; 8(2): 20.

[21] Baladraf DMS, Yusuf Y, Yusuf A. Manufacturing of things and characteristics of chilli, orange skin, and cinnamon extract using maceration method. Journal of Health, Technology and Science (JHTS). 2022; 3(2): 53–62.

[22] Sheffield WJ, Thompson HO. The preparation and evaluation of tangerine peel tincture and tangerine syrup. Journal of the American Pharmaceutical Association (Scientific Ed). 1954; 43(3): 181–85.

[23] Moore WE, Abraham D. A new approach to the chemistry of aromatic ammonia spirit. Journal of the American Pharmaceutical Association (Scientific Ed). 1956; 45(4): 257–59.

[24] Burrin P, Bibbins F. Tincture opium. The Journal of the American Pharmaceutical Association (1912). 1935; 24(11): 964–66.

[25] Moghaddasi MS, Kashani HH. Ginger (Zingiber officinale): A review. Journal of Medicinal Plants Research. 2012; 6(26): 4255–58.

[26] Vijayan K, Madhusoodanan K, Radhakrishnan V, Ravindran P. Properties and end-uses of cardamom: Cardamom: CRC Press; 2002. pp. 285–99.

[27] Rowe L, Scoville WL. Tincture of digitalis. The Journal of the American Pharmaceutical Association (1912). 1933; 22(11): 1087–90.

[28] Lancaster H, Davidson A. Commercial pharmaceutical preparations: 3. – belladonna leaves. Canadian Medical Association Journal. 1927; 17(10 Pt 1): 1187.

Asia Naz Awan, Sabahat Abdullah, Nisar ur Rahman,
Muhammad Yasir Ali, Shakib Kazmi, Ghazala Ambreen, Saeed Ahmad,
Udo Bakowsky

7 Colloids

7.1 Introduction

Over the past several decades, there has been a revival in the study and development of colloid science and technology, driven by the growth and contribution in the fields of physical chemistry, advanced materials, and nanotechnology, and its use in various applications. Colloids are commonly employed in food [1, 2], pharmaceuticals [3, 4], nutraceuticals [5], and cosmeceuticals [5, 6]. Colloid science, therefore, has a long history of being a popular topic of study. Thomas Graham was the first to use the word "colloid" in 1861 to describe a unique way that a substance aggregated [7]. Graham thought the variation in colloids' behavior was attributed to the size of the particle. It was later acknowledged that any substance could be transformed into a colloid by subdividing it into colloidal-sized particles, regardless of its nature and composition [8].

Colloids are dispersions of solid particles in a fluid. The word "colloidal" is from the Greek *kola*, "glue," and *eidos*, "like," old-fashioned glues that formed colloidal dispersion in water. The term "colloid" refers to a homogenous blend of two distinct components: a dispersed phase and a continuous phase (solvent). These components can exist in any state of matter. The dispersed phase comprises small particles that are uniformly distributed across the continuous phase, which is known as the "dispersing medium." Typically, particle sizes range from 1 nanometer (nm) to several micrometers (μm) [9].

Colloidal particles fall within this range, having sizes between those found in solutions (<1 nm) and suspensions (>500 nm). Colloid now has a much broader definition, encompassing more than just particles. Colloids include, for example, proteins, polymers, and certain self-assembled items [10].

7.2 Types of colloids

Colloids are classified into three main types (Fig. 7.1).

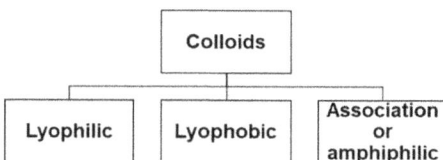

Fig. 7.1: Classification of colloids.

https://doi.org/10.1515/9783111438108-007

Lyophilic colloids are those in which the dispersed phase has an affinity for the dispersing medium. When the affinity of the dispersed phase is for water as the dispersing medium, the colloids are called hydrophilic colloids. Examples include the colloidal dispersion of gum, starch, and protein in water. The affinity of starch and protein for water is due to hydrogen bonding between hydroxyl groups (–OH) of starch and amino groups (–NH_2) of protein with water molecules [11, 12]. When the affinity of the dispersed phase is for a nonaqueous dispersing medium, the colloids are named as lipophilic colloids. An example of this type is polystyrene dispersed in an organic solvent.

Lyophobic colloids have a dispersed phase that lacks affinity for the dispersing medium because there is no solvent sheath around the particles. They typically consist of inorganic particles dispersed in water. Examples include sulfur or gold dispersed in water.

Association or amphiphilic colloids consist of dispersed phases, known as amphiphiles. It contains two distinct regions within a single molecule, each having opposing affinities for solutions. When the concentration of these amphiphilic monomers reaches the critical micelle concentration (CMC), they aggregate to form micelles, which have diameters within the colloidal range, thereby resulting in the formation of amphiphilic colloids. An example of this type is surface active agents in water or nonpolar solvents.

For amphiphiles in water, the hydrocarbon chains orient themselves inward, creating a hydrophobic environment within the micelle, while the polar portions associate with the water molecules of the dispersing medium (Fig. 7.2). Similarly, in the case of amphiphiles in an organic solvent, the orientation of the molecules is reversed, and the hydrocarbon chains are linked with the organic molecules of the dispersing medium [11–13].

In Water   In Non-Polar Solvent

Hydrophilic

Hydrophobic Chain

Fig. 7.2: Orientation of hydrocarbon chains.

7.3 Preparation of colloidal solutions

Lyophilic colloids, which have an affinity for the solvent, are easily prepared by warming the solid in the dispersion medium (Fig. 7.3). For example, dissolving starch, acacia, or gelatin in warm water creates a lyophilic colloidal solution. On the other hand, to prepare lyophobic colloidal solutions, which have no affinity with the solvent, special methods are required. These methods can be categorized into two types: dispersion and condensation methods [13].

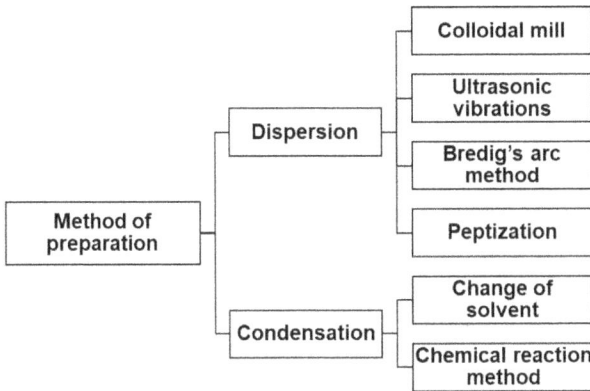

Fig. 7.3: Method of preparation of colloids.

In dispersion methods, coarse particles are reduced to colloidal size using different techniques. One such technique is mechanical dispersion using a colloid mill. In this method, the solid and liquid dispersion medium is fed into a mill with two steel plates rotating in opposite directions at high speed. This process grinds the solid particles to a colloidal size and disperses them into the liquid. Although this approach is not remarkably effective and only decreases a minimal portion of the particles to the colloidal size range, it is still adopted in the preparation of printing ink. Another dispersion method involves high-intensity ultrasonic generators operating at frequencies exceeding 20,000 cycles per second. This method is employed to prepare colloidal particles of mercury, where mercury in the form of a layer disintegrates into small particles in water.

Bredig's arc method is a highly effective technique for producing hydrosols of metals such as gold, silver, and platinum. It involves creating an electric arc in a liquid, which disperses metal electrodes into vapors that condense, resulting in the formation of colloidal particles. This method is invaluable for creating fine metal dispersions. Colloidal particles of nonmetals can be prepared by suspending the material's coarse particles in the dispersion medium and creating an arc between two iron electrodes. In addition to this, peptization offers a reliable way to disperse recently precipitated ionic particles into a water-based colloidal solution. This approach involves adding electrolytes with common ions to the water, allowing the precipitate to adsorb the common ions and change into charged particles. It is instrumental in dispersing substances like ferric hydroxide $Fe(OH)_3$ and silver chloride (AgCl) using appropriate peptizing agents.

$Fe(OH)_3$ (precipitates) + $FeCl_3$ (peptizing agent) → $[Fe(OH)_3Fe]_3$ + Cl_3 split from the other precipitate → $[Fe(OH)_3Fe]_3$ + Cl_3 soluble in water to give colloidal solution.

Condensation methods involve the formation of colloidal-size particles by condensing ions or molecules; this can be made possible by changing the solvent or by chemical reaction. Reduced solubility causes sulfur colloids to form when excess water is introduced to a concentrated ethanol-based sulfur solution. In ethanol, sulfur exists as an atom. Changing the solvent decreases the solubility and causes supersaturation, leading to the condensation of sulfur atoms into colloidal particles in water.

Chemical reaction methods include double decomposition, oxidation, reduction, and hydrolysis. Hydrogen sulfide gas (H_2S) is gradually passed through a cold solution of arsenious oxide (As_2O_3) in the double decomposition process to produce an arsenic sulfide (As_2S_3) colloid until it reaches maximum yellow color intensity:

$$As_2O_3 + 3H_2S \rightarrow As_2S_3(yellow) + 3H_2O$$

Reduction occurs when neutral or slightly alkaline noble metallic salt solutions are added to a reducing agent, such as pyrogallol or formaldehyde, that causes the atoms in the colloidal solution to combine to form colloidal particles. For instance, silver colloid is obtained by using organic reducing agents like formaldehyde (HCHO) in a dilute silver nitrate solution:

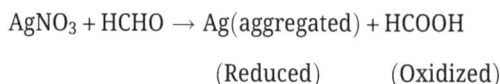

$$AgNO_3 + HCHO \rightarrow Ag(aggregated) + HCOOH$$

$$(Reduced) \qquad (Oxidized)$$

The oxidation of H_2S by sulfur dioxide (SO_2) yields colloidal particles of sulfur via the formation of sulfur atoms:

$$2H_2S + SO_2 \rightarrow 2H_2O + 3S$$

The process of hydrolysis involves the creation of a red color colloidal particles of $Fe(OH)_3$ by the hydrolyzing of a solution of ferric chloride ($FeCl_3$) in water. Chromium and aluminum salts also undergo hydrolysis analogously:

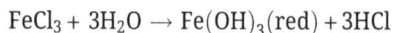

$$FeCl_3 + 3H_2O \rightarrow Fe(OH)_3(red) + 3HCl$$

7.4 Purification

When the methods discussed earlier are employed to generate colloidal solutions, a sizable number of electrolytes are also present in addition to colloidal particles. Eliminating these electrolytes is necessary to produce pure colloids. There are three ways to purify colloids [13].

Dialysis is a crucial process that uses diffusion over a semipermeable membrane to extract small molecules and electrolytes from solution. This membrane, made of collodion or cellophane, has pores small enough to retain colloidal particles but large enough to allow ions and small molecules, such as sodium chloride, glucose, and urea, to pass through. In a typical setup, two compartments, A and B, are separated by the semipermeable membrane. Compartment A contains the solution with electrolytes and small molecules, while compartment B contains pure water. During dialysis, these small molecules and ions diffuse from compartment A to B through the membrane. When the membrane is in equilibrium, the concentration of small molecules is equal on both sides, and the colloidal particles stay in compartment A. One can produce colloidal material in compartment A, devoid of small molecular impurities, by repeatedly extracting the liquid from compartment B.

Ultrafiltration is the procedure of separating colloidal particles from sub-colloidal material by passing them through an ultrafilter (Fig. 7.4). Treating ordinary filter paper with collodion or cellophane (regenerated cellulose) transforms it into ultrafilter paper. Unlike dialysis, ultrafiltration utilizes an ultrafilter instead of a semipermeable membrane and applies pressure to compartment A to accelerate the filtration process. This versatile technique can successfully separate colloidal particles of varying sizes by utilizing ultrafilters of different grades.

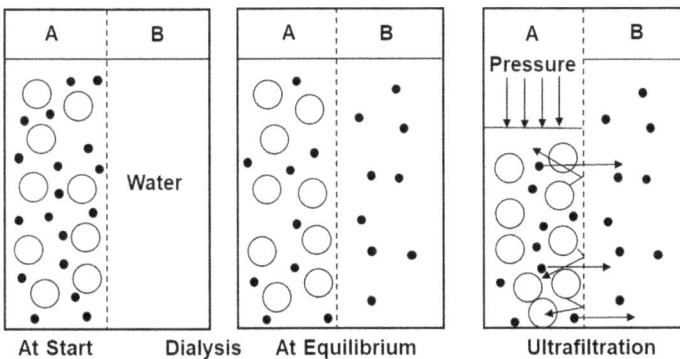

Fig. 7.4: Ultrafiltration of colloidal systems.

A novel technique called electrodialysis uses an electromagnetic field to improve dialysis (Fig. 7.5). The metal screens supporting the membranes are subjected to an electric current to accomplish this. The migration of ions to the opposite electrode is facilitated by applying an electric potential, which significantly speeds up the dialysis procedure. Notably, nonelectrolyte contaminants like sugar and urea cannot be removed by electrodialysis.

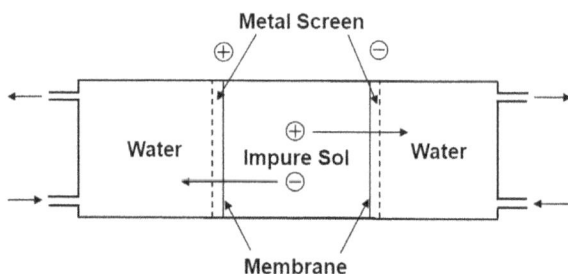

Fig. 7.5: A schematic presentation of electrodialysis.

7.5 Properties of colloids

The physicochemical, morphological, optical, electrical, kinetic, and magnetic characteristics of colloids are crucial for drug delivery procedures. These properties offer insights into the drug kinetics of ADME (absorption, distribution, metabolism, and elimination), in addition to interactions with cell membranes and the behavior of drug molecules in various environments. By defining the size and charge of colloidal particles, they significantly influence the drug loading capacity and transport processes. Furthermore, the osmotic pressure, stability, sedimentation rate, and biocompatibility of colloidal drug carriers are all impacted by colloidal particles. Viscosity, interfacial tension, and aggregation qualities are a few characteristics that influence the choice of excipients and the standards for manufacturing procedures. Together, these elements affect the distribution and bioavailability of colloidal carriers throughout the human body [14].

The properties of particles, such as light scattering, the Tyndall effect, and turbidity, are critical for formulating drugs. These characteristics aid in our comprehension of the structure, size, and form of colloidal particles. Moreover, the kinetic characteristics of colloids, including diffusion, osmosis, Brownian motion, sedimentation, and viscosity, provide vital information about the movement of colloidal carriers through cell membranes and throughout the body. Furthermore, the physicochemical properties, which include physical state, drug polymorphs [15], lyophobicity, lyophilicity [13], and interfacial properties [16], are essential for selecting suitable colloidal carriers for specific drugs. Understanding zeta potential [17] and conductivity [18] also aids in selecting excipients and comprehending how colloids interact with the biological membranes. Lastly, leveraging the magnetic properties of colloids for targeted drug delivery can significantly enhance drug efficacy while minimizing toxicity [19].

7.5.1 Morphological properties

The shape, size, and surface area of colloids significantly influence the characteristics of drug delivery systems, such as micro- and nanoparticles. Changes in these morphological properties can affect the distribution, biocompatibility, and stability of drug carriers [14]. Atomic force microscopy is employed to study various morphologies of colloids, which provides insights into the adsorption characteristics of colloidal particles.

The size range of colloidal particles is around 1 nm to 1 μm, and most colloidal systems exhibit heterodispersity. Thus, particle size is a crucial parameter in determining the drug's route of administration. It directly impacts the surface area for absorption of the colloidal drug delivery system. To distinguish systems with heterodispersity, it is essential to establish the particle size distribution. The diversity in the sizes of colloids has an impact on the distribution of drugs in the body. The size of the colloids can be measured in the following ways [20, 21]:

- Martin's diameter (d_m) measures the length of a line that dissects the colloidal image.
- Feret's diameter (d_f) measures the separation between two tangents on opposing sides of the colloid, which are parallel to a set direction.
- Projected area diameter (d_a) measures the diameter of a circle that shares the same area as the colloids.

Particles in the colloidal range can be successfully administered intravenously and penetrate in tissues with a disrupted capillary endothelium, specifically in solid tumors and in inflamed and infected areas. Small particles can penetrate the tissue because of the approximately 100 nm gap between endothelial cells. When administering drugs parenterally with colloidal emulsions, the size of the particles is a pivotal feature. Emulsions with large particle sizes are not suitable for clinical use due to the risk of embolus formation, while smaller particles enhance physical stability, enabling longer circulation times in the body [22].

One important property that results from colloidal particle size is surface area, which also affects the charge of the particles. By repelling one another, these charged particles maintain dispersion. The surface area has a noteworthy impact on the colloidal qualities of aggregation, viscosity, and interfacial tension and is decisive in selecting colloidal drug delivery systems. For example, smaller emulsion droplet sizes result when lecithin and hexanoyl oil are combined in water to form a microemulsion. This enhances the stability and surface area, thereby enhancing the drug's residence time [22]. An inherent characteristic of colloidal particles is their enormous surface area. For instance, a standard drug delivery micellar system with 0.1 M amphiphile offers 40,000 m^2 of interfacial area per liter of solution. In dispersion, the form of the colloidal particles is significant in many ways. The greater specific surface area provided by an extended particle increases the possibility of the dispersed phase and the dispersion medium developing attractive forces. Any alteration in particle shape can significantly impact the sedimentation rate and osmotic pressure [13]. Moreover, the shape of colloids directly influences their stability, biocompatibility, and distribution as drug carriers. Changes in particle shape can increase the propensity for aggregation, which in turn affects the selection of excipients for drug delivery. Furthermore, drug release kinetics and mechanisms inside the biological system are affected by the shape of the colloidal particles. The cleavage planes of crystals and potential weak points inside the crystals affect the form of the particles, which can be measured using a variety of microscopy techniques. It is worth noting that the surface properties of colloids are instrumental, with the angular or spherical shape directly impacting drug delivery efficacy.

7.5.2 Optical properties

When an intense beam of light passes through a colloidal solution and is observed at a right angle, the path of the light appears as a hazy beam or cone. It happens because the colloidal particles scatter the light in all directions after absorbing it, known as light scattering. This scattering illuminates the path of the light in the colloidal dispersion known as the Tyndall effect or the Faraday-Tyndall effect. The illuminated beam or cone resulting from this scattering is often known as the Tyndall beam or Tyndall cone. The Tyndall effect is a significant optical property of colloids. In contrast, true solutions do not demonstrate the Tyndall effect because the particles in these solutions, such as ions or solute molecules, are too small, allowing light to pass through without scattering, making the light beam invisible. This effect is employed to distinguish between a colloidal solution and a true solution owing to this distinction. Colloids with larger particles may appear turbid in ordinary light due to Tyndall scattering. Interestingly, the Tyndall effect becomes more pronounced as the particle size decreases. This scattering of light by colloidal particles can offer valuable insights into how these particles interact with biological membranes, thus helping to determine the appropriate route of administration for various applications.

The brightness of light passing through a colloidal solution decreases due to scattering of the light, known as turbidity. It expresses the ratio of the intensity of light scattered in all directions to the intensity of the incident light, I_o. This characteristic can be functional in calculating the average molar mass of lyophilic colloids, providing valuable information for the development, processing, and quality assessment of drug delivery systems. Turbidity at a particular dispersed phase concentration is directly proportional to the lyophilic colloid's molecular weight. However, due to the low turbidity of most lyophilic colloids, measuring the scattered light is often more convenient than measuring the transmitted light. The molecular weight of colloids can be calculated using the following equation:

$$\frac{Hc}{\tau} = \frac{1}{M} + 2Bc$$

where τ is turbidity, c is the solute concentration (g/cm^3), M is the weight-average molecular weight, and B is the interaction constant [13]. Utilizing turbidity for determining molecular mass provides a foundation for understanding the absorption, distribution, and interaction of colloidal carriers with biological membranes.

Colloids exhibit dynamic light scattering. This property measures the size of the particles from 0.6 nm to 6.0 µm using the principle of Brownian motion. It determines particle size by calculating the hydrodynamic radius, the diameter of a sphere that disperses at the same pace as the particle under investigation. The hydrodynamic radius depends on the temperature, the suspending fluid's characteristics, the particle's density, and shape [15]. For instance, DLS measurements demonstrate that the mean size of water droplets depends on the water concentration: larger droplet sizes are due to higher water concentrations. At 25 °C, measurements revealed that water with a volume percentage of 3–20% guaranteed that the droplet size was within the nano-emulsion range [23].

The size and properties of the dispersed particles influence the wavelength of light they scatter, which determines the color of a hydrophobic colloid. For example, colloidal silver particles exhibit distinct colors depending on their particle diameters (Tab. 7.1).

Tab. 7.1: Particle size versus size of silver colloids.

S. no.	Color of silver colloids	Particle diameter (mm)
1.	Orange yellow	6×10^{-5}
2.	Orange red	9×10^{-5}
3.	Purple	13×10^{-5}
4.	Violet	15×10^{-5}

As the particle diameter increases, the color of the silver colloids shifts, demonstrating the relationship between particle size and the wavelength of the scattered light.

A conventional microscope cannot view colloidal particles because of their ultra-microscopic size. Zsigmondy developed the ultra-microscope by taking advantage of the Tyndall phenomenon. In this setup, an intense beam of light passes through a glass vessel containing the colloidal solution. A microscope is positioned at the right angle to the beam to observe the light's focus. Bright specks of light against a dark background represent individual colloidal particles, with the dispersion medium as the background. Although the particles cannot be observed, the larger halos of scattered light around them indicate their presence. Counting and observing these bright spots allows for the identification of the particles. However, the ultra-microscope does not offer details about the shape and size of the particles; it only reveals the presence and distribution of the colloidal particles through the scattered light.

It is currently common practice to examine the size, shape, and structure of colloidal particles using an electron microscope, which can create images of particles, even ones that are getting close to molecular dimensions. An electron beam is focused onto a photographic plate in an electron microscope by applying magnetic and electric fields. It is possible to take pictures of the individual particles by passing this concentrated laser through a screen of colloidal particles. The electron microscope offers magnification 10,000 times and much greater resolution than an optical microscope, making it possible to visualize and study the detailed characteristics of colloidal particles [13].

7.5.3 Kinetic property

The kinetic properties of colloid materials, which deal with the movement of particles about the dispersion medium, are the most fascinating subset of these properties. Brownian motion, diffusion, and osmosis are three examples of how thermal energy can cause motion. Gravity can cause sedimentation, while external forces can influence motion through viscosity. The dispersed particles in a colloidal dispersion experience constant buffeting from the moving molecules of the solvent, resulting in an erratic movement called "Brownian motion." This random directional movement of colloidal particles was first noted in 1827 by Robert Brown. The particles exhibit a zigzag movement due to random collisions with molecules in the suspending medium. The particle is moving because an uneven number of molecules are striking it from opposing directions; the particle changes direction when more molecules strike one side than the other. As the particle size decreases, the velocity of colloidal particles increases, making the Brownian motion more pronounced [13, 24, 25].

By adding viscosity enhancers like glycerin or similar agents, one can increase the viscosity of the medium, which in turn decreases and eventually stops Brownian motion. Therefore, grasping the concept of Brownian motion helps in choosing excipients that influence the viscosity of colloidal drug delivery systems. It also aids in understanding the interaction of carriers and the transport mechanism across the cell membrane. In suspensions, the particles are much larger, reducing the likelihood of uneven bombardments. The forces from molecules hitting the particle from different sides cancel each other out, so the particles do not exhibit Brownian motion. It also explains how colloidal particles generally settle due to gravity, but the continual movement of solvent molecules opposes this effect. The constant motion imparted by the solvent molecules has a stirring effect, preventing the particles from settling [26].

Colloidal particles move randomly due to Brownian motion, causing them to disperse from higher-concentration regions to lower-concentration regions until the concentration becomes uniform throughout. Diffusion decreases as the particle radius increases.

Small colloidal particles can readily diffuse through membranes, such as porous plugs. However, their diffusion is slower than that of ions or small molecules. Within the size range of colloids, there is an opposition between diffusion forces, such as Brownian motion, and gravitational forces that induce particles to sediment. In this way, Brownian motion typically outweighs gravitational forces, keeping particles suspended.

According to Sutherland and Einstein's equation:

$$D = \frac{RT}{6\pi\eta rN}$$

where D is the diffusion coefficient, R is the molar gas constant, T is the absolute temperature, η is the viscosity, r is the radius of the particle, and N is Avogadro's number.

The equation states the following principles about diffusion: it increases with higher temperatures and decreases as there is a reduction in the particle size, and increasing the viscosity of the medium further decreases the phenomenon of diffusion.

Osmosis and osmotic pressure are essential for colloidal drug delivery systems. Osmotic pressure occurs during osmosis when molecules move from the regions of lower concentration to those of higher concentration. This pressure develops in colloidal solutions when the solvent is removed from the solution by a membrane that allows solvents to pass through but is impermeable to solute molecules. The colloidal dispersion gets diluted when pure solvent passes through the membrane. Osmotic pressure, or the difference in pressure between the two compartments, arises when the colloidal particles cannot cross the membrane. Osmotic pressure plays a crucial role in the absorption of colloidal carriers during topical administration. For instance, in a transferosomal formulation of diclofenac applied as a gel on the skin, dehydration caused by water evaporation prompts lipid vesicles to sense the osmotic gradient. They respond by migrating along this gradient to mitigate complete drying, thereby enhancing the efficacy of the delivered drug [26, 27].

The osmotic pressure (π) of a dilute colloidal solution containing spherocolloids may be represented by van't Hoff's equation:

$$\pi = cRT \ \text{ or } \ \pi = \frac{c_g RT}{M} \tag{7.1}$$

where c is the molar concentration of solute, R is the gas constant, T is the temperature, c_g is the amount of solute per liter, and M is the molecular mass of the solute.

For linearly lyophilic colloids, the equation is more complex due to deviations from ideal behavior. The solute molecules become solvated, reducing the concentration of "free" solvent. Therefore, we must graphically determine parameter B to estimate the asymmetry of particles and their interactions with the solute. The osmotic pressure of linearly lyophilic colloids is represented as follows:

$$\frac{\pi}{c_g} = RT\left(\frac{1}{M} + Bc_g\right) \tag{7.2}$$

Here, B is a constant specific to a particular solute/solvent system, and it depends on the degree of interaction between the solute and the solvent molecules.

Sedimentation plays a fundamental role in the characteristics of colloids, exerting a substantial impact on the processing, production, and introduction of drug delivery systems. Gravitational force is the primary factor affecting sedimentation. The velocity (V) at which spherical particles with density (ρ) settle in a medium with density (ρ_0) and viscosity (η) is determined by Stokes' law:

$$V = \frac{2r^2 (\rho - \rho_0)g}{9\eta} \tag{7.1}$$

Here, g denotes the acceleration due to gravity. When particles experience only gravitational force, the lower size limit of those conforming to Stoke's equation is approximately 0.5 μm (500 nm). At this scale, Brownian motion becomes significant, opposing sedimentation induced by gravity and facilitating mixing. Therefore, to induce the sedimentation of colloidal particles precisely and measurably, a powerful force must be applied. This is achieved using an "ultracentrifuge," capable of exerting a force one million times greater than gravity. In a centrifuge, $\omega^2 x$ replaces the acceleration due to gravity (g), where ω signifies the angular velocity and "x" denotes the distance of the particle from the center of rotation. Consequently, eq. (7.1) is modified into the following equation:

$$V = \frac{2r^2 (\rho - \rho_0)\omega^2 x}{9\eta} \tag{7.2}$$

Colloidal dispersions, which fall within a specific size range and exhibit Brownian motion, are minimally affected by gravity, so sedimentation is negligible. In contrast, coarse particles tend to settle out of the solution, even if they carry an electric charge, because gravity's influence is stronger than the electrical forces that maintain particle dispersion. When the attractive forces between particles exceed the repulsive forces, the particles aggregate into larger masses and eventually precipitate due to gravity. At lower concentrations, the attraction force is insufficient to bond particles into larger masses, and the groups remain light enough that gravity does not pull them out of the solution.

Particles that develop and aggregate outside of the colloidal size range have the potential to destabilize colloidal systems and cause them to separate from the medium. Therefore, sedimentation offers important information indicating drug loading and entrapment efficiency regarding the concentration of colloidal dispersion in drug delivery systems. Manufacturers, for instance, create topical solutions that are administered externally for protective, cosmetic, and dermatological uses. Parenteral suspensions may contain 0.5–30% solid particles, whereas these suspensions may have dispersed phase concentrations surpassing 20%.

The viscosity of a colloid depends on the shape of the particles in the medium, the type of colloidal system, and the molecular weight of the particles. The shape of the dispersed phase particles primarily influences the viscosity of a colloidal dispersion. Spherical colloids create dispersions with low viscosity, while systems with linear particles are more viscous. The degree of solvation of particles is influenced by the relationship between shape and viscosity. When placed in a low-affinity solvent, a linear colloid tends to "ball up" or take on a spherical shape, which reduces its viscosity. This behavior helps detect changes in the shape of flexible colloidal particles and macromolecules.

Einstein's equation for the viscosity of dilute colloidal dispersions of spherical particles is given by:

$$\eta = \eta_o(1 + 2.5\phi)$$

where η is the viscosity of the dispersion, η_o is the viscosity of the dispersion medium, and ϕ is the volume fraction of colloidal particles.

This equation implies that the viscosity of the dispersion increases linearly with the volume fraction of the colloidal particles. The factor 2.5 is specific to spherical particles and represents intrinsic viscosity [13].

7.5.4 Electrical property

The electrical property of colloids is crucial for applications like drug delivery, as it depends on the charge present on the surface of the particles. Dispersed particles in colloids possess either a positive or negative surface charge. Charges can arise through two main mechanisms: adsorption of ions and ionization of surface groups.

When ions from a surrounding solution are adsorbed onto the surface of colloidal particles, they impart a charge (Fig. 7.6). For instance, if a dilute solution of potassium iodide (KI) is added to a 1 M solution of silver nitrate (AgNO$_3$), a colloidal precipitate of silver iodide (AgI) particles forms. Here, silver ions (Ag$^+$) are in excess and get adsorbed onto the AgI particles, resulting in positively charged sol particles. Conversely, when a dilute solution of AgNO$_3$ is added to a 1 M solution of KI, AgI particles form with iodide ions (I$^-$) in excess. These iodide ions get adsorbed, creating negatively charged colloidal particles. Since the overall colloid remains electrically neutral, the surface charge of the particles is counterbalanced by oppositely charged ions known as counterions.

Colloidal particles can acquire a charge through ionization of their surface groups. For example, protein colloidal particles have both acidic carboxylic (–COOH) and basic amino (–NH$_2$) groups. In an aqueous solution, the NH$_2$ group gains a proton (H) to become NH$_3$, resulting in a positively charged particle. The COOH group loses a

Counter Ions

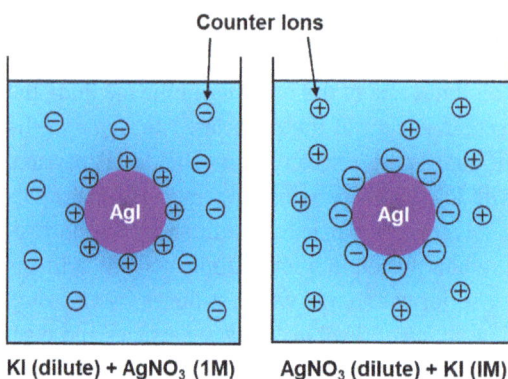

KI (dilute) + AgNO₃ (1M)   AgNO₃ (dilute) + KI (IM)

Fig. 7.6: Ions from the surrounding medium adsorbing on a colloidal particle.

proton (H) to form COO–, resulting in a negatively charged particle. At an intermediate pH, known as the isoelectric point, the particles are electrically neutral as the positive and negative charges balance out. The presence of positive and negative charges on colloidal particles can be experimentally determined by electrophoresis, which measures the movement of particles under an electric field. In short, the electrical properties of colloids, influenced by surface charge, are vital for their stability and applications, such as drug delivery systems.

The distribution of ions around charged particles is explained by the concept of an electrical double layer. When a colloidal particle acquires a positive charge, it will form a layer of positive ions around it through selective adsorption. This layer, in turn, attracts counterions from the surrounding medium, creating a second layer of negative charges. This combination of two layers of positive charges adjacent to the particle and negative charges in the medium was initially described as the Helmholtz double layer. Helmholtz believed that the positive charges near the particle surface were fixed, while the negative charges in the medium were mobile. However, more recent considerations have shown that the double layer is made of a compact layer of positive and negative charges fixed firmly on the particle surface and a diffuse layer of movable counterions (negative ions) diffused into the medium containing positive ions.

While the example given involves positively charged colloids, the same principles apply to negatively charged colloids with the roles of the charges reversed. The diffuse layer, loosely bound to the particle surface, moves in the opposite direction when an electric field is applied. Because of the charge distribution around the particle, a potential difference exists between the compact layer and the bulk solution across the diffuse layer. This potential difference is known as the zeta potential (Fig. 7.7).

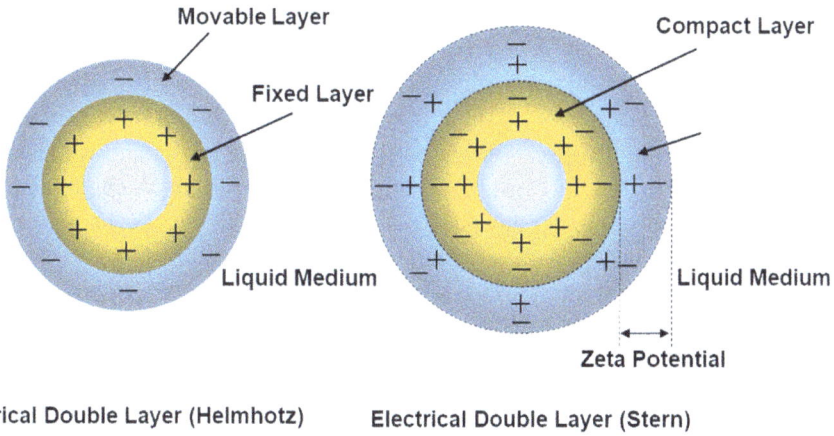

Fig. 7.7: Zeta potential on colloidal particle.

In a dispersion medium, the movement of charged particles depends on the zeta potential. The rate of migration increases along with an increase in zeta potential as the particle velocity increases. Although pH affects the zeta potential, it stabilizes quickly. It can be calculated as follows:

$$\zeta = \frac{v}{E} \times \frac{4\pi\eta}{\varepsilon} \times \left(9 \times 10^4\right)$$

where ζ is the zeta potential (V), v is the velocity of migration (cm/s) in an electrophoresis tube of definite length (cm), η is the viscosity (P), ε is the dielectric constant of the medium, E is the potential gradient (V/cm), and the term "v/E" is known as the mobility [13].

The passage of colloidal particles via an electric potential is called electrophoresis. There is a positive charge on the particles if they migrate toward the negative electrode; a negative charge is present if they migrate toward the positive electrode. The colloidal particles are electrically neutral if they do not migrate to any electrode. For example, consider a sol of $Fe(OH)_3$ prepared by the hydrolysis of $FeCl_3$:

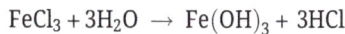

$$FeCl_3 + 3H_2O \rightarrow Fe(OH)_3 + 3HCl$$

The arms of the U-shaped tube are filled with the colloid and injected with deionized water to keep the colloid at a particular level. The level of the colloid drops on the side of the positive electrode and increases on the side of the negative electrode when a potential difference is applied between two platinum electrodes submerged in the deionized water. The finding implies that the colloidal particles have moved along the direction of the negative electrode, suggesting a positive charge:

$$\mathrm{Fe(OH)_3 + FeCl_3 \rightarrow \left[Fe(OH)_3Fe\right]^{3+} + 3Cl^-}$$

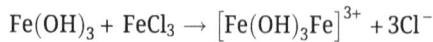

In contrast to electrophoresis, electroosmosis describes the dispersion medium's movement when an electric potential is applied. The dispersion medium exhibits an opposing charge to the dispersed particles in a colloidal dispersion, an electrically neutral substance. The dispersion medium, therefore, flows in the opposite direction to the dispersed phase under an applied electric potential. It moves about the charged surface when the dispersed phase is immobilized, such as by forming a porous plug.

Consider a plug of moist clay, a negative colloid fixed on the bottom of a U-tube. Consider a plug of moist clay, a negatively charged colloid fixed on the bottom of a U-tube. Platinum electrodes are submerged in water and then poured into both limbs of the U-tube at the same level. When a potential difference is applied across the electrodes, the water level rises on the negative side and falls on the positive side. It indicates that the dispersion medium has a positive charge.

The zeta potential between the colloidal particles and the dispersion medium directly results in electroosmosis. It is caused by the diffusion of ions when the applied voltage is greater than the zeta potential.

Sedimentation potential refers to the generation of an electric potential when particles settle or sediment in a fluid as a result of gravity or centrifugal force. It is the reverse of electrophoresis, where an electric field causes particle movement. The formation of an electric potential when a liquid causes itself to pass through a porous material, like a plug or a bed of particles, is known as streaming potential. Unlike electroosmosis, which involves the movement of a liquid through a membrane under an applied electric field, streaming potential arises from the liquid's pressure-driven flow.

The Donnan membrane effect describes the uneven distribution of diffusible charged ions across a semipermeable membrane at equilibrium due to the presence of nondiffusible charged ions. Imagine a semipermeable membrane with a negatively charged colloid containing its counterions (R^-Na^+) on the one side and sodium chloride (NaCl) solution on the other. Anionic particles (R^-) in colloidal form cannot traverse the membrane freely, although sodium (Na^+) and chloride (Cl^-) ions can.

The following steps are involved in the system to attain equilibrium: At first, the concentrations of sodium ions [Na^+] on the outside (o) and inside are equal at 99. However, there is a concentration gradient for chloride ions [Cl^-], with a maximum concentration of 99 outside and zero concentration inside.

Chloride ions [Cl^-] try to diffuse inside due to this gradient, but are partially repelled by the colloidal anionic particles (R^-). As a result, some chloride ions [Cl^-] = 33 diffuse inside, while the majority [Cl^-] = 66 remain outside. To maintain electroneutrality, sodium ions [Na^+] = 33 also diffuse inside.

At equilibrium, the concentrations are as follows: Outside: sodium ions [Na^+] = 66 and chloride ions [Cl^-] = 66. Inside: sodium ions [Na^+] = 132, negative colloidal ions [R^-] = 99, and chloride ions [Cl^-] = 33; this makes a total of 132.

At equilibrium, the principle of electroneutrality is maintained, with equal concentrations of positive and negative ions inside (132) and outside (66). The unequal distribution of ions across the semipermeable membrane is due to nondiffusible colloidal anions (R^-) inside (Fig. 7.8). This results in different osmolalities of the solutions on the opposite side of the membrane. The molecular weight of the colloidal particles can be calculated using the consequent difference in osmotic pressure.

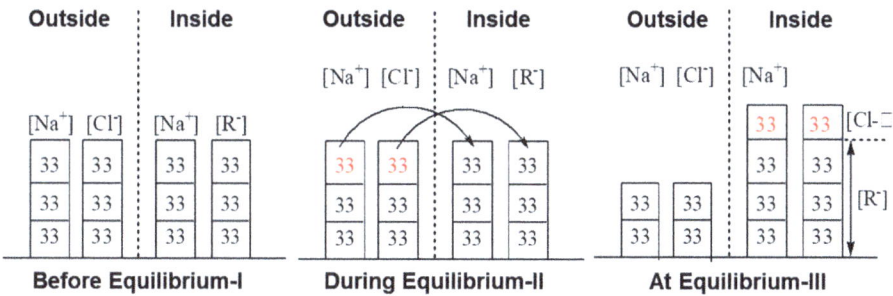

Outside		Inside			Outside		Inside			Outside		Inside	
					$[Na^+]$ $[Cl^-]$		$[Na^+]$ $[R^-]$			$[Na^+]$ $[Cl^-]$		$[Na^+]$	
$[Na^+]$ $[Cl^-]$		$[Na^+]$ $[R^-]$										33	33 $[Cl^-]$
33	33	33	33		33	33	33	33				33	33
33	33	33	33		33	33	33	33		33	33	33	33 $[R^-]$
33	33	33	33		33	33	33	33		33	33	33	33

Before Equilibrium-I	**During Equilibrium-II**	**At Equilibrium-III**

Fig. 7.8: Concentrations of ions inside and outside of the semipermeable membrane during reaction.

According to the principle of escaping tendencies:

$$[Na^+]_0[Cl^-]_0 = [Na^+]_i[Cl^-]_i \tag{7.1}$$

On the outside:

$$[Na^+]_0 = [Cl^-]_0 \tag{7.2}$$

and inside:

$$[Na^+]_i = [R^-]_i + [Cl^-]_i \tag{7.3}$$

By substituting the values of $[Na^+]_0$ and $[Na^+]_i$ from eqs. (7.2) and (7.3) into (7.1), we have:

$$[Cl^-]_0^2 = ([R^-]_i + [Cl^-]_i)[Cl^-]_i$$

$$= [Cl^-]_i^2 + [R^-]_i[Cl^-]_i$$

$$= [Cl^-]_i^2 \left(1 + \frac{[R^-]_i}{[Cl^-]_i}\right)$$

$$\frac{[Cl^-]_0^2}{[Cl^-]_i^2} = \left(1 + \frac{[R^-]_i}{[Cl^-]_i}\right)$$

$$\frac{[Cl^-]_0}{[Cl^-]_i} = \sqrt{\left(1 + \frac{[R^-]_i}{[Cl^-]_i}\right)} \tag{7.4}$$

Equation (7.4), which represents the Donnan membrane equilibrium, demonstrates the ratio of diffusible anion concentrations inside and outside the membrane at equilibrium. This principle shows how the equilibrium concentration ratio of a diffusible anion is influenced by the presence of a colloidal anion (R^-) inside a semipermeable membrane. Ions with similar charges are drawn out of the membrane by this effect.

By applying our concentration values, we can better understand this relationship:

$$\frac{[Cl^-]_0}{[Cl^-]_i} = \sqrt{\left(1 + \frac{[R^-]_i}{[Cl^-]_i}\right)} = \sqrt{\left(1 + \frac{99}{33}\right)} = \sqrt{(1+3)} = \sqrt{4} = 2 \text{ or } 2/1$$

On the other hand, when we take the ratio of concentrations of $[Cl^-]_0$ and $[Cl^-]_i$ we also get the same ratio:

$$\frac{[Cl^-]_0}{[Cl^-]_i} = \frac{66}{33} = \frac{2}{1}$$

Drug delivery is significantly influenced by the Donnan membrane equilibrium, particularly when improving the absorption of medications like potassium, sodium salicylate, and benzylpenicillin. Semipermeable membranes prevent the diffusion of sodium carboxymethyl cellulose, an anionic polyelectrolyte. If it is diffusible, it can nevertheless improve drug absorption when used with another anionic medication. Similarly, anions from the digestive tract tend to be driven into the bloodstream by ion-exchange resins and slowly diffusing sulfate and phosphate ions.

The Donnan membrane equilibrium has been modified for anions of both a diffusible drug and a nondiffusible polyelectrolyte (sodium carboxymethyl cellulose) at equilibrium:

$$\frac{[D^-]_0}{[D^-]_i} = \sqrt{\left(1 + \frac{[R^-]_i}{[D^-]_i}\right)}$$

It will be observed that when $[R^-]_i/[D^-]_i = 8$, the ratio $[D^-]_0/[D^-]_i = 3$ and when $[R^-]_i/[D^-]_i = 99$, the ratio $[D^-]_0/[D^-]_i = 10$. Therefore, the drug should diffuse out of the membrane more readily when anionic polyelectrolyte is added to a solution containing a diffusible drug anion.

7.5.5 Magnetic property

Hematite and yttrium oxide particles are the most common magnetic colloidal particles. The surface characteristics and magnetic behavior of the particles affect their magnetic properties. Especially for drug delivery applications, this idea is pragmatic in designing hybrid systems with an outer shell frequently made of biodegradable polymer. An active pharmacological component in magnetically modified systems, such as magnetic nanoparticles and magnetic microemulsions, can be selectively localized by applying an external magnetic field to a particular location. The drug is released precisely at the intended site, offering significant advantages. This enables targeted drug delivery and a reduction in the required drug dosage to one-tenth of the dose needed for free drug administration [28].

7.6 Stability of colloids

Particles in a colloidal solution do not separate or agglomerate because they are stable. Two main elements affect the stability of colloids: the solvent layer around the particles and their charge. The distributed particles in a hydrophobic colloid have an identical electric charge, either positive or negative. Particles in a hydrophobic fluid push away from one another and resist clumping together because like charges oppose one another. Nevertheless, the charges on the particles are neutralized by adding an electrolyte to the solution, which causes the particles to aggregate and precipitate.

Lyophilic (or hydrophilic) colloids are stable, carry similar charges that repel each other like hydrophobic colloids, and possess a layer of solvent (such as water) surrounding each particle. For instance, a layer of water surrounds a gelatin particle, which carries a negative charge. Adding NaCl to a colloidal solution containing gelatin causes the particles to remain suspended because the aqueous layer around each particle prevents the Na^+ ions from neutralizing the negative charges and causing the particles to precipitate. Thus, lyophilic colloids exhibit considerable stability as compared to lyophobic colloids.

Particle sensitization or coagulation can occur when a small quantity of a hydrophilic or hydrophobic colloid is added to a hydrophobic colloid of the opposite charge. It happens because the addition lowers the zeta potential below the critical value (20–50 mV), reducing the repulsion between the hydrophobic particles and thinning the ionic layer surrounding them. Therefore, appropriate concentrations of electrolytes can easily precipitate hydrophobic colloids.

Conversely, adding a large amount of a hydrophilic colloid stabilizes the hydrophobic system. In this case, the hydrophilic colloid adsorbs onto the hydrophobic particles, causing the hydrophobic colloid to behave like a hydrophilic colloid and resist precipitation when electrolytes are added. This process is called "protection," and the hydrophilic colloid that needs to be added is referred to as a "protective colloid." The minimal weight (in milligrams) of the protective colloid required to stop 10 mL of a gold solution from turning violet when 1 mL of a 10% NaCl solution is added is known as the "gold number." This gold number is used to determine the effectiveness of the protective colloid.

A dispersion of tragacanth (gum) suspended in bismuth subnitrate illustrates the sensitization and protective effect of colloids. Bi^{3+} ions cause the tragacanth in the mixture to coagulate, forming a gel that solidifies at the bottom of the container. Following that, the bismuth subnitrate particles and flocculated gum combine to create a gel or cake. By lowering the bismuth particle's zeta potential and avoiding excessive coagulation, the addition of phosphate, citrate, or tartrate shields the gum. Systems with partial flocculation typically cake less than those with deflocculation.

Lyophobic colloids are thermodynamically unstable; however, surface charges are responsible for their stability. These like charges repel each other, which prevents the particles from coagulating. Despite this repulsion, they remain unstable and readily precipitate when you add electrolytes. In contrast, lyophilic and association colloids exist as a single-phase solution and are thermodynamically stable. The appropriate amount of electrolytes does not result in coagulation since they are stable. However, the particles may settle if a lot of salt is incorporated.

DLVO theory explains how colloidal dispersions are stabilized by a combination of electrostatic forces and van der Waals interactions rather than by the steric repulsion caused by polymeric solubilizing agents. Derjaguin, Landau, Verwey, and Overbeek developed this theory in the 1940s. The attraction and repulsion forces of a colloidal system govern its stability, according to DLVO theory. The attractive force between particles of the same kind arises from van der Waals interactions. The attractive force (V_A), when the particles are larger than the separation distance, can be written as follows:

$$V_A = -\frac{Aa}{12H}$$

where A is the Hamaker constant, a is the radius of the particle, and H is the distance between the particles [29].

The size, shape, and chemical composition of the colloidal particles determine the intensity of the van der Waals interactions, as shown by the Hamaker constant (A). Van der Waals forces are always attractive for particles similar to one another. Repulsive forces occur due to the charges on the surfaces of colloidal particles, which arise either from the ionization or the adsorption of ions. These charges lead to repulsion between similarly charged particles. Generally, the stability of the colloidal dispersion depends on how these repulsive and attractive forces interact. The stability of the colloidal dispersion is affected when salt is added to the solution because it changes the electrostatic interactions [13].

When a particle carries a negative charge on its surface, it attracts positive ions to create the Stern layer. A diffuse layer is formed by the accumulation of positive and negative ions, known as the electrical double layer (EDL). The interplay of these electrical double layers around the dispersed particles gives rise to electrostatic forces. Particles repel one another if they have identical surface charges, whether positive or negative.

The strength of the electrostatic repulsive force decreases as there is a drop in the thickness of the EDL. Attraction (V_A) and repulsion (V_R) are the two potential energies produced between the particles. The sum of the electrostatic repulsive energy (V_R) and the attractive potential energy (V_A) is known as the total potential energy of interaction (V_T):

$$V_T = V_A + V_R$$

The minimum and maximum energy levels are depicted in the accompanying figure by plotting the V_A, V_R, and composite potential energy (V_T) curves against interparticle distance (H) for particles in dispersion.

The attraction curve (V_A) reaches its maximum negative value, indicating strong attraction, when the interparticle distance is zero. This attraction decreases as the interparticle distance increases. Conversely, the repulsion curve (V_R) reaches its maximum positive value, indicating strong repulsion, when the interparticle distance is zero. This repulsion decreases as the distance increases.

Adding V_A and V_R produces a composite curve (V_T). A greater negative value of the primary minimum (indicated by the depth of the curve) signifies that the particles have an irreversible attraction, leading to aggregation.

Particle aggregation is averted if the primary maximum (barrier) is high enough to keep the particles from coming into close contact. The secondary minimum depth is crucial for determining the system's stability because coagulation or flocculation occurs at this point.

The stability of colloids can be influenced by the addition of electrolytes. When there is a low concentration of electrolytes, the double layer is large, causing the repulsive force (V_R) around the particles to be widespread. The total energy curve, obtained by summing V_R and V_A, has a high primary maximum (barrier) and no secondary minimum, indicating a highly stable colloid. Adding more electrolytes causes V_R to decay rapidly, leading to a small primary maximum but, more importantly, a secondary minimum. Flocculation may occur in the secondary minimum of this electrolyte concentration, resulting in a steady dispersion.

The small primary maximum prevents aggregation in the primary minimum, making the colloid kinetically stable. When the primary maximum is small and the secondary minimum is deep, the colloid undergoes slow coagulation. If the primary maximum is zero, coagulation occurs rapidly at the critical coagulation concentration (CCC), making the colloid unstable. Conversely, when there is no primary maximum or secondary minimum, attractive forces alone determine the colloid's stability, resulting in very rapid coagulation and high instability.

The valence of the oppositely charged ions, or counterions, determines how much of an electrolyte is present in V_R. The effect of the additional counterions on V_R will increase with their valence. The Schulze-Hardy rule refers to these principles.

Colloids can be stabilized by repulsive forces resulting from absorbed water molecules, macromolecules, or surfactants on their surfaces. Stabilization occurs due to three factors: (a) entropic effect, (b) osmotic effect, and (c) enthalpic stabilization.

The decrease in mobility for the chains of adsorbed molecules is known as the entropic effect. When particles with these adsorbed stabilizing chains come close to each other, the chains interact, leading to steric interaction. As the distance between the particles decreases, this steric interaction increases. This interaction causes a loss of conformational freedom in the chains, resulting in a negative change in entropy ($\Delta S = S$ (final) $- S$ (initial) = negative). The loss of conformational freedom in each chain increases the system's free energy, leading to repulsion between the particles. $\Delta G = \Delta H - T\Delta S = (+) - (+) (-) = $ positive.

The macromolecular chains on adjacent particles invade the space of the other, causing the osmotic effect, which raises the concentration of chains in the overlap zone. This concentration gradient generates a repulsive force due to the resulting differences in osmotic pressure. Additionally, as particles approach closely, releasing adsorbed water molecules results in a positive change in entropy and an increase in enthalpy, further contributing to the repulsion between the particles.

Upon close approach, the adsorbed water molecules are released ($\Delta S = S$ (final) $- S$ (initial) = positive), resulting in a positive change in entropy and an increase in enthalpy ($\Delta H = $ positive), leading to repulsion ($\Delta G = $ positive) among the particles:

$$\Delta G = \Delta H - T\Delta S = (+) - (+) (+) = \text{positive} \ (\Delta H > T\Delta S).$$

7.7 Pharmaceutical applications of colloids

- Colloidal formulations are used in treating and diagnosing various disease conditions. For example, administering colloidal copper for cancer treatment, colloidal arsenic is employed for treating eye diseases, and iron-dextran complexes are applied to treat anemia. Colloidal mercury serves as a diagnostic agent for syphilis, while colloidal gold is utilized for diagnosing paresis.
- Colloids as a drug delivery system for therapeutics include hydrogel, microparticles, microemulsions, liposomes, micelles, nanoparticles, and nanocrystals [30, 31]. They are extensively employed to modify the properties of pharmaceutical agents, most commonly to enhance their solubility, and as drug delivery systems. Certain medicines exhibit increased therapeutic activity when formulated as colloids. For example, colloidal silver compounds, which include silver iodide, silver chloride, and silver protein, are effective disinfectants.
- Colloidal electrolytes and association colloids are preferred to improve the stability and solubility and mask the taste of active pharmaceutical ingredients in water and oil-based formulations. The effectiveness of these compounds is enhanced when they are in a colloidal state because of their extensive surface area. These include kaolin in adsorbing toxins from the gastrointestinal tract (GIT) and aluminum hydroxide as an antacid.

– Colloidal natural macromolecules, such as starch and cellulose, and synthetic polymers are used as layering materials for solid dosage forms to protect them from degradation in the acidic medium of the gastric mucosa or moisture.
– Colloidal hydroxyethyl starch (HES), dextran, PVP, and gelatin are hydrophilic colloids that aid in restoring or maintaining the blood volume as blood plasma substitutes.

References

[1]   Pirsa S, Hafezi K. Hydrocolloids: Structure, preparation method, and application in food industry. Food Chemistry. 2023; 399: 133967.
[2]   Dickinson E. Colloids in food: Ingredients, structure, and stability. Annual Review of Food Science and Technology. 2015; 6(1): 211–33.
[3]   Trinh TND, Do HDK, Nam NN, Dan TT, Trinh KTL, Lee NY. Droplet-based microfluidics: Applications in pharmaceuticals. Pharmaceuticals. 2023; 16(7): 937.
[4]   Myburgh J, Liebenberg W, Willers C, Dube A, Gerber M. Investigation and evaluation of the transdermal delivery of ibuprofen in various characterized nano-drug delivery systems. Pharmaceutics. 2023; 15(10): 2413.
[5]   Morganti P, Morganti G, Chen H-D, Coltelli M-B, Gagliardini A. Smart tissue carriers for innovative cosmeceuticals and nutraceuticals. Cosmetics. 2024; 11(1): 20.
[6]   Bai Y, Li X, Wang X, Wang X, Yang X, Xin H, et al. Nanoscale composite lignin colloids with tunable visible colors used for anti-UV cosmetics. Langmuir. 2023; 40(1): 554–60.
[7]   Graham TX. Liquid diffusion applied to analysis. Philosophical Transactions of the Royal Society of London. 1861; 151: 183–224.
[8]   Rajagopalan R, Hiemenz PC. Principles of colloid and surface chemistry. Marcel Dekker, New-York. 1997; 8247: 8.
[9]   Hunter R, O'Brien R. The electrokinetic effects. Foundations of Colloid Science. 1989; 2.
[10] Sarangapani PS, Hudson SD, Jones RL, Douglas JF, Pathak JA. Critical examination of the colloidal particle model of globular proteins. Biophysical Journal. 2015; 108(3): 724–37.
[11] Ali MY, Tariq I, Sohail MF, Amin MU, Ali S, Pinnapireddy SR, et al. Selective anti-ErbB3 aptamer modified sorafenib microparticles: In vitro and in vivo toxicity assessment. European Journal of Pharmaceutics and Biopharmaceutics. 2019; 145: 42–53.
[12] Ali MY, Tariq I, Ali S, Amin MU, Engelhardt K, Pinnapireddy SR, et al. Targeted ErbB3 cancer therapy: A synergistic approach to effectively combat cancer. International Journal of Pharmaceutics. 2020; 575: 118961.
[13] Swarbrick Jam A. Colloids. In: Physical pharmacy (Vol. 545, 3rd edition ed.): Philadelphia: PA: Lea & Febiger; 1991. pp. 471–86.
[14] Yang L, Alexandridis P. Physicochemical aspects of drug delivery and release from polymer-based colloids. Current Opinion in Colloid & Interface Science. 2000; 5(1–2): 132–43.
[15] Wong J, Brugger A, Khare A, Chaubal M, Papadopoulos P, Rabinow B, et al. Suspensions for intravenous (IV) injection: A review of development, preclinical and clinical aspects. Advanced Drug Delivery Reviews. 2008; 60(8): 939–54.
[16] Vink R, de Virgiliis A, Wolfsheimer S, Schilling T, Horbach J, Binder K, editors. Interfacial properties of colloidal model systems: NIC Symposium; Jülich, 2006.
[17] Wiącek AE. Electrokinetic properties of n-tetradecane/lecithin solution emulsions. Colloids and Surfaces A: Physicochemical and Engineering Aspects. 2007; 293(1–3): 20–27.

[18] Sripriya R, Raja KM, Santhosh G, Chandrasekaran M, Noel M. The effect of structure of oil phase, surfactant and co-surfactant on the physicochemical and electrochemical properties of bicontinuous microemulsion. Journal of Colloid and Interface Science. 2007; 314(2): 712–17.

[19] Vyas S, Khar R. Nanocrystals and nanosuspensions In: Vijay S, Editor Vyas. Targeted and controlled drug delivery: New Delhi: CBS Publishers and Distributors; 2002.

[20] Allen T. Particle size measurement: Springer; Berlin/Heidelberg, Germany, 2013.

[21] Tariq I, Ali MY, Janga H, Ali S, Amin MU, Ambreen G, et al. Downregulation of MDR 1 gene contributes to tyrosine kinase inhibitor induce apoptosis and reduction in tumor metastasis: A gravity to space investigation. International Journal of Pharmaceutics. 2020; 591: 119993.

[22] Kawaguchi E, Shimokawa K-I, Ishii F. Physicochemical properties of structured phosphatidylcholine in drug carrier lipid emulsions for drug delivery systems. Colloids and Surfaces B: Biointerfaces. 2008; 62(1): 130–35.

[23] Chiesa M, Garg J, Kang Y, Chen G. Thermal conductivity and viscosity of water-in-oil nanoemulsions. Colloids and Surfaces A: Physicochemical and Engineering Aspects. 2008; 326(1–2): 67–72.

[24] Everett DH. Basic principles of colloid science: Royal society of chemistry; London, 2007.

[25] Ali S, Amin MU, Tariq I, Sohail MF, Ali MY, Preis E, et al. Lipoparticles for synergistic chemo-photodynamic therapy to ovarian carcinoma cells: In vitro and in vivo assessments. International Journal of Nanomedicine. 2021; 16: 951–76.

[26] Ogemdi IK. Properties and uses of colloids: A review. Colloid and Surface Science. 2019; 4(2): 24.

[27] Cevc G. Transfersomes, liposomes and other lipid suspensions on the skin: Permeation enhancement, vesicle penetration, and transdermal drug delivery. Critical Reviews™ in Therapeutic Drug Carrier Systems. 1996; 13(3–4): 257–388.

[28] Fanun M. Colloids in drug delivery: CRC Press; Boca Raton, Florida, 2016.

[29] Horinek D. DLVO theory. Encyclopedia of Applied Electrochemistry. 2014; 343–46.

[30] Amin MU, Ali S, Ali MY, Tariq I, Nasrullah U, Pinnapreddy SR, et al. Enhanced efficacy and drug delivery with lipid coated mesoporous silica nanoparticles in cancer therapy. European Journal of Pharmaceutics and Biopharmaceutics. 2021; 165: 31–40.

[31] Arshad S, Asim MH, Mahmood A, Ijaz M, Irfan HM, Anwar F, et al. Calycosin-loaded nanostructured lipid carriers: In-vitro and in-vivo evaluation for enhanced anti-cancer potential. Journal of Drug Delivery Science and Technology. 2022; 67: 102957.

Humaira Gul, Nisar ur Rahman, Muhammad Yasir Ali, Adeeba Ishaq,
Amna Shakeel, Bisma Islam, Javeria Batool

8 Oral solutions, syrups, elixirs, and spirits

8.1 Introduction

This chapter provides a detailed explanation of four liquid dosage forms: oral solutions, syrups, elixirs, and spirits commonly used in pharmaceutical practice. These formulations consist of active drug ingredients combined with various adjunct components designed to deliver therapeutic effects when taken orally. The chapter comprehensively covers the preparation methods for each liquid dosage form. It emphasizes the critical considerations necessary to ensure both efficacy and quality in the final product. Detailed descriptions of the specific techniques and processes involved in creating oral solutions, syrups, elixirs, and spirits highlight the importance of precise formulation to achieve the desired therapeutic outcomes. Additionally, the chapter explores the inherent advantages and drawbacks associated with these liquid dosage forms. For example, syrups are often preferred for their palatability, particularly in pediatric care, while elixirs, which contain alcohol, are valued for their solubilizing properties and stability. These insights help practitioners understand how to best utilize each type of formulation to maximize patient benefit within clinical settings. A key focus of the chapter is the discussion of pivotal physicochemical properties that govern the behavior and performance of liquid dosage forms. Parameters such as solubility, pH, viscosity, and density are meticulously analyzed, as they significantly influence formulation stability, drug release kinetics, and overall therapeutic efficacy. By elucidating these fundamental principles, the chapter equips readers with a deeper understanding of the factors that shape the design and optimization of liquid pharmaceuticals. In summary, liquid dosage forms are essential in modern pharmaceutical administration. They offer versatility, rapid onset of action, and enhanced patient compliance. Through an in-depth exploration of their formulation, properties, and practical considerations, this chapter underscores their significance in contemporary pharmacotherapy. It serves as a vital resource for healthcare practitioners aiming to optimize patient outcomes and therapeutic efficacy, ensuring that patients receive the most effective and appropriate treatments available.

8.2 Oral solution

Oral solutions are liquid preparations in which one or more chemical substances are dissolved in a suitable solvent or mixture of mutually miscible solvents [1].

https://doi.org/10.1515/9783111438108-008

Oral solutions are the drug delivery systems used for the delivery of drugs that are poorly water-soluble drugs is an essential element in the development of formulations for pediatric patients [2].

Oral solutions may contain a sweetening agent, albeit with a lesser degree of sweetness compared to syrups and elixirs, while maintaining consistency in all other components. The reduced sweetness is attributed to the lower sucrose content, specifically less than 60% [3].

8.2.1 Method of preparation

The preparation method for oral solutions mirrors that of syrups or elixirs. Research indicates that esters of p-hydroxy benzoic acid, such as methyl, ethyl, propyl, and butyl paraffins, are commonly employed as preservatives in oral solutions. These preservatives exhibit a partitioning capability between the solutions and flavoring oils, leading to a lower overall availability of these preservatives [4]. It is essential to note that when diluting flavoring agents that are insoluble in water, the initial step involves solubilizing them in a different solvent before incorporating them into the solution.

8.2.2 Advantages of oral solutions

Easier to administer
Oral solutions are generally easier to administer, especially for children or individuals who have difficulty swallowing pills. The liquid form makes it easier to control the dosage and intake.

Faster absorption
Liquids and solutions are typically absorbed more quickly by the body compared to solid dosage forms like tablets or capsules. This can lead to faster onset of action for the active ingredients.

Flexible dosing
Oral solutions allow for more flexible and precise dosing compared to pre-measured spirits or elixirs. The dosage can be easily adjusted based on the individual's needs.

Better bioavailability
The liquid form of oral solutions can enhance the bioavailability of certain active ingredients compared to solid forms, leading to improved therapeutic effects.

Easier handling
Oral solutions are generally easier to handle and transport compared to breakable glass bottles of spirits or elixirs.

Improved palatability

Oral solutions can be flavored to improve taste and palatability, making them more acceptable for patients, especially children.

Reduced risk of alcohol exposure

Many spirits and elixirs contain significant amounts of alcohol, which may be undesirable or contraindicated for certain populations. Oral solutions can provide the therapeutic benefits without the alcohol content.

Overall, the advantages of oral solutions make them a more convenient, flexible, and potentially more effective option compared to traditional spirit or elixir formulations in many medical and healthcare applications.

8.2.3 Examples

Pseudoephedrine oral solution

Pseudoephedrine (PSE) is a drug with a long history of medical use; it is helpful in treating symptoms of the common cold and flu, sinusitis, asthma, and bronchitis. Due to its central nervous system (CNS) stimulant properties and structural similarity to amphetamine, it is also used for non-medical purposes. The substance is taken as an appetite reducer, an agent which eliminates drowsiness and fatigue, to improve concentration and as a doping agent.

Chlorpheniramine oral solution

Chlorpheniramine oral solution used for coughing (dry or productive) is the body's way of clearing irritants (like allergens, mucus or smoke) from airways and preventing infection. There are two types of coughs, namely dry cough and chesty cough.

Ferrous sulfate oral solution

This medication is iron supplement used to treat or prevent low blood levels of iron (such as those caused by anemia or pregnancy). Iron is an important mineral that the body needs to produce red blood cells and keep you in good health. Ferrous sulfate works best when you take it on empty stomach. If you can, take it 30 min before eating, or 2 h after eating.

Aluminum subacetate

Prepared by reacting aluminum sulfate sol. with calcium carbonate and acetic acid forming magma.

Magnesium citrate

Prepared by reacting official magnesium carbonate with citric acid, flavoring and sweetening agents, filtering talc and carbonating it by potassium or sodium bicarbonate.

Theophylline (for asthma)

Theophylline is a medication used to treat asthma and chronic obstructive pulmonary disease as a second-line drug. It is a bronchodilator. This activity reviews the indications, action, and contraindications for theophylline as a potential agent in treating asthma and chronic obstructive pulmonary disease.

Cimetidine HCl (for peptic ulcer)

Cimetidine is a gastric acid reducer used in the short-term treatment of duodenal and gastric ulcers. Therefore, the drug effectively manages gastric hypersecretion and is used to manage reflux esophagitis disease and prevent stress-related gastric ulcers.

Ergocalciferol (for vitamin D deficiency)

Vitamin D_2 (ergocalciferol) is a form of vitamin D that is available both over the counter and with a prescription, depending on the dose. It's used to treat vitamin D deficiency (low vitamin D levels), including in various medical conditions, such as hypoparathyroidism (low parathyroid hormone) and rickets.

8.3 Dry mixtures for solution

A variety of medicinal agents, such as antibiotics, exhibit poor solubility in water, necessitating their formulation in dry powdered or granular form for subsequent reconstitution with a specified quantity of purified water prior to immediate dispensing [5]. Dry mixtures encompass all components, including the drug, flavoring agents, colorants, and buffers, except for the solvent.

Once these solutions are prepared, they can typically be stored in refrigerators for 7–14 days, allowing patients easy access during this period. However, if any solution remains beyond the treatment duration, it is advisable to discard it due to potential concerns regarding the stability of the drug.

In order to minimize the risk of microbial contamination and guarantee that all ingredients are of the proper quality, the manufacturing process for liquid preparations intended for oral use should adhere to good manufacturing practice.

8.3.1 Examples

8.3.1.1 Cloxacillin sodium

Cloxacillin is used to treat a wide variety of bacterial infections. This medication is a type of penicillin antibiotic. It works by stopping the growth of bacteria. This antibiotic treats only bacterial infections.

8.3.1.2 Oxacillin sodium

Oxacillin is a penicillin beta-lactam antibiotic used in the treatment of bacterial infections caused by susceptible, usually gram-positive, organisms. The name "penicillin" can either refer to several variants of penicillin available, or to the group of antibiotics derived from the penicillin.

8.3.1.3 Penicillin V/G potassium

Penicillin V potassium is used to treat certain infections caused by bacteria such as pneumonia and other respiratory tract infections, scarlet fever, and ear, skin, gum, mouth, and throat infections.

8.3.1.4 Potassium chloride

This medication is a mineral supplement used to treat or prevent low amounts of potassium in the blood. A normal level of potassium in the blood is important. Potassium helps your cells, kidneys, heart, muscles, and nerves work properly. Most people get enough potassium by eating a well-balanced diet.

8.4 Oral rehydration solutions

The rapid loss of fluids due to diarrhea can result in dehydration, which, in severe cases, may lead to fatality, particularly in patients under the age of 4. During diarrhea, the small intestine excretes an excessive amount of fluid and electrolytes, surpassing the normal absorption capacity of the large intestine [5]. This fluid, along with electrolytes, is absorbed from the extracellular body fluids, causing the patient to enter hypovolemic shock due to reduced blood volume. Oral solutions are typically effective for patients experiencing a body weight loss of 5–10 kg.

To promote optimal fluid absorption, it is imperative to stress that oral rehydration solutions contain a well-balanced combination of glucose and electrolytes. Furthermore, because of their low sodium content and high osmolality, inappropriate fluids such as energy drinks, soft drinks, and juices high in sorbitol should be avoided as they can make diarrhea worse.

In conclusion, treating dehydration in young children caused by diarrhea requires prompt and appropriate rehydration. In severe cases, intravenous fluids may be required, even though oral rehydration solutions are helpful for mild to moderate dehydration. To prevent complications and speed up recovery, careful observation and adherence to prescribed protocols are essential.

8.4.1 Composition of oral hydration solution

Table 8.1 shows the composition of oral rehydration solution.

Tab. 8.1: Composition of oral rehydration solution.

S. no.	Ingredients	Concentration
1	Sodium	45 mEq/L
2	Potassium	20 mEq/L
3	Chloride	35 mEq/L
4	Citrate	25 mEq/L
5	Glucose	25 g/L

This is standard formula for oral rehydrates. Oral rehydration solutions (ORSs) should not be combined with milk or juices because it would disrupt the standard ionic electrolyte concentration by adding electrolytes from these sources. The sodium/glucose co-transport in the gut is optimized with sodium:glucose ratio of 1:1. Addition of electrolytes from milk or juices would interfere with this optimal balance, potentially leading to further fluid loss and dehydration. Oral rehydrates are given in both dry and liquid form.

8.4.2 Uses

It is a systemic alkalinizing agent designed for patients suffering from uric acid and cystine stones in the urinary tract. It proves beneficial as a complement when used alongside uricosuric medications in treating gout, as acidic urine tends to promote the crystallization of urates:

- Glucose helps sodium (and consequently water) to be absorbed in equal molar amounts in the small intestine.
- Sodium and potassium are essential for replenishing the losses of these vital ions during diarrhea (and vomiting).
- Citrate addresses the acidosis that arises from diarrhea and dehydration.

8.4.3 Mechanism

It has been observed that glucose is actively absorbed during diarrhea, despite the fact that it is typically associated with the malabsorption of nutrients. The mechanism of action behind the efficacy of oral rehydration solution is based on the simple principle of co-transport of glucose and sodium across the intestinal membrane. This is the most effective way to promote the absorption of fluids and electrolytes. The co-transport mechanism is activated when glucose binds to the sodium-glucose transporter type 1 (SGLT1), which facilitates the transport of glucose against its concentration gradient. Glucose and sodium are transferred across the intestinal cell into the blood, where sodium binds with the chloride to form salt.

The salt creates an osmotic gradient and attracts water into the blood, thus replenishing fluids and electrolytes. In patients with diarrhea, the co-transport of glucose and sodium is preserved and continues to promote rehydration even when water and electrolytes are lost in stools. An optimal balance of glucose and sodium is critical for an efficient co-transport mechanism, the basis on which oral rehydration solutions are formulated to prevent dehydration.

8.5 Oral colonic lavage solutions

Clinically, oral colonic lavage solutions are used for colonoscopy in the gastrointestinal tract. The procedure typically involves a clear liquid diet for 24–48 h prior to the procedure, followed by the administration of an oral laxative, such as magnesium citrate or bisacodyl and then a morning administration of the laxative. Remember that these types of laxatives are taken at bed time because of their long onset of action. This is done to avoid hospitalization and its associated problems, such as malnutrition and poor oral intake prior to the procedure, which can exacerbate issues. However, the use of these laxatives at home can lead to complications, and patients are advised to follow the instructions provided by their healthcare provider.

As a result, an alternative approach has been implemented, which is less time consuming and involves fewer dietary restrictions. This method entails administering a well-balanced electrolyte solution with polyethylene glycol (PEG) 3350. The solution is prepared before administration. PEG functions as an osmotic agent in the gastrointestinal tract, while the balanced electrolyte concentration leads to minimal net absorption or secretion of ions. Consequently, a significant volume of the solution can be administered without causing substantial changes in water or electrolyte balance.

Since the electrolyte concentration is balanced, there is no further change in their concentration. The recommended dose for adults is 4 L, which is administered systematically in doses of 250 mL every 10 min. Before the procedure, patients are advised to avoid food intake for 2–3 h prior to the first administration. After the procedure, soft drinks can be consumed.

8.5.1 Examples

8.5.1.1 PEG (polyethylene glycol) oral colonic lavage solutions

They are commonly used for the following purposes: (a) Bowel preparation for colonoscopy: PEG-based solutions are widely used to cleanse the colon before a colonoscopy procedure. They help remove stool and other debris from the intestines, allowing for better visualization of the colon during the examination. (b) Constipation management: PEG-based laxatives, such as PEG 3350, are used to treat chronic constipation. They work by drawing water into the intestines, softening stool and facilitating bowel movements. (c) Hepatic encephalopathy: In patients with liver disease and associated hepatic encephalopathy, PEG-based solutions may be used as a part of the treatment regimen. They help reduce the absorption of ammonia and other toxins in the intestines, which can help improve cognitive function. (d) Bowel preparation for other procedures: PEG-based solutions may be used for bowel preparation before other medical procedures, such as surgery or radiological imaging, where a clean colon is essential for better visualization and assessment. (e) Impaction removal: In cases of fecal impaction, PEG-based solutions can be used to soften and evacuate the impacted stool, allowing for resolution of the condition.

It's important to note that the specific formulation, dosage, and instructions for use of PEG-based colonic lavage solutions may vary depending on the intended purpose and the healthcare provider's recommendations. Patients should always follow the instructions provided by their healthcare professionals when using these solutions (Tab. 8.2).

Table 8.2 shows the composition of PEG oral colonic lavage solution [6].

Tab. 8.2: Composition of polyethylene glycol oral colonic lavage solution.

S. no.	Ingredients	Concentration
1	Sodium glycol (PEG) 3350	236 g
2	Sodium sulfate	22.74 g
3	Sodium bicarbonate	6.74 g
4	Sodium chloride	5.86 g
5	Potassium chloride	2.97 g
6	Purified water (q.s.)	4.0 L

8.5.1.2 Magnesium citrate oral colonic lavage solution

Commonly known as citrate of magnesia, this substance is a colorless to slightly yellow clear effervescent liquid with a sweet, acidulous taste and a lemon flavor. It contains magnesium citrate equivalent to 1.55–1.99 g of magnesium oxide in every 100 mL. Magnesium citrate is employed as a pretreatment before colonoscopy when using PEG oral colonic lavage solution. One of its key benefits is the reduction in the amount of PEG needed for colonoscopy, decreasing the volume required from 4 to 2 L [7].

8.5.2 Uses

The PEG electrolyte lavage solution (PEG-ELS), a chronic lavage solution presently used for gut cleansing in preparation for colonic procedures, can be an effective agent in treatment of chronic constipation. It is also employed for unlabeled use in management of acute iron overdose in children. The PEG acts as an osmotic agent in gastrointestinal tract and the balanced electrolyte concentration results in virtually no net absorption or secretion of ions. Thus, a large volume of this solution can be administered without significant change in water or electrolyte balance.

8.6 Syrups

Syrups are concentrates, aqueous preparations of a sugar or sugar-substitute with or without flavoring agents and medicinal substance.

8.6.1 Types of syrups

8.6.1.1 Flavored syrup

Traditionally syrups are flavored syrups containing a flavoring agent. Various flavored syrups, like cherry, are available in the pharmaceutical industry, with options that are free from alcohol and dyes. For instance, Medisca's Oral Syrup is a cherry-flavored liquid designed for standalone use or in conjunction with other products like Medisca's Oral Suspend to enhance the physical stability of suspensions. This formulation is prepared with a slightly acidic pH and includes a balance of preservatives and buffering agents to help minimize common degradation. It is also alcohol and dye-free, catering to patients with dietary restrictions or sensitivities.

In addition to standard flavored syrups, Medisca offers a sugar-free version of their flavored syrup vehicle. This sugar-free, cherry-flavored liquid is formulated to assist compounding pharmacists in the preparation of extemporaneous dosage forms when dietary restrictions are of concern. Like its standard counterpart, it is prepared with a slightly acidic pH and balanced with preservatives and buffering agents to help minimize common degradation. It is also alcohol and dye-free.

These flavored syrup vehicles are designed to assist compounding pharmacists in the preparation of oral extemporaneous dosage forms. They can be used alone as flavoring agents in aqueous solutions or in combination with other products like Medisca's Oral Suspend to enhance the physical stability of suspensions. The incorporation of flavoring agents helps improve patient compliance, especially in pediatric and geriatric populations, by masking the taste of the active pharmaceutical ingredients.

8.6.1.2 Simple syrup

It is composed of an aqueous solution containing 85% w/v sucrose (United States Pharmacopeia – USP). An instance of unflavored syrup within the pharmaceutical field is the brompheniramine maleate, pseudoephedrine hydrochloride, and dextromethorphan hydrobromide syrup. This particular formulation is transparent, sweet, unflavored pink syrup employed for medicinal applications.

Simple syrup, commonly composed of an 85% w/v sucrose solution, serves as a fundamental pharmaceutical excipient. This high concentration of sucrose not only imparts sweetness but also acts as a preservative by inhibiting microbial growth. The syrup's viscous nature ensures uniform distribution of active ingredients in liquid formulations. In the context of combination syrups, such as those containing brompheniramine maleate, pseudoephedrine hydrochloride, and dextromethorphan hydrobromide, simple syrup provides a stable and palatable medium. These formulations are designed to deliver antihistamine, decongestant, and antitussive effects, addressing symptoms associated with colds and allergies. The syrup's composition may include additional excipients like glycerin, citric acid, and preservatives to enhance stability and taste. It is essential to note that such combination syrups should be used cautiously in patients with certain medical conditions, including asthma, glaucoma, gastrointestinal obstruction, and urinary retention, due to the pharmacological actions of the active ingredients

8.6.1.3 Medicated syrup

Medicated syrup is a liquid oral dosage form that contains one or more therapeutic agents dissolved or suspended in a sweetened aqueous base. These syrups are widely used in pediatric and geriatric populations due to their palatable taste and ease of administration. They offer the advantage of masking unpleasant flavors of active ingredients while ensuring accurate dosing. According to pharmaceutical literature, the viscosity and sweetness of the syrup also aid in soothing irritated mucous membranes, making it particularly effective for cough and cold formulations. Medicated syrups are classified based on the type of active ingredient they contain, such as antitussives, antihistamines, expectorants, and antipyretics. They must comply with pharmacopeial standards regarding uniformity, stability, and microbial limits to ensure therapeutic efficacy. Additionally, regulatory guidelines emphasize the importance of using pharmaceutically acceptable excipients and maintaining stringent quality control during manufacturing.

It has a therapeutic agent that may either be directly incorporated into these systems or may be added as the syrup is being prepared. To preserve the integrity of the syrup, it should be kept tightly sealed in a cool and dry environment after use [8]. Medicated syrups are industrially manufactured by combining specific components to create the final product. These components typically include sucrose, purified water, non-medicated syrup (flavoring syrup), colorants, and therapeutic agents. The process involves carefully blending these ingredients to produce the desired medicated syrup formulation for pharmaceutical use.

8.6.2 Preparation methods of syrup

There are four methods used generally for the preparation of syrup [9].

8.6.2.1 Solution by heating

When a quick preparation of syrup is desired without damaging or volatilizing its components, a specific method is employed. To prevent the darkening or amber coloration of the solution due to overheating sucrose, it is crucial to follow a precise procedure.

Initially, sucrose is added to purified water and heated for proper dissolution. Subsequently, heat-stable components are incorporated. If heat-labile or volatile substances like flavoring oils and alcohol need to be included, they are typically added after the sugar has dissolved through heat, followed by rapid cooling to room temperature. These heat-labile components often consist of volatile flavoring agents, including alcoholic products [10].

Inversion of syrup: Sucrose, a disaccharide, can be hydrolyzed into glucose (dextrose) and fructose (laevulose) when exposed to greater amounts of heat. This process is referred to as the inversion of syrup, and the combination of the two monosaccharides is known as invert sugar. When heat is applied during the preparation of sucrose syrup, some inversion of sucrose is almost certain. The extent of inversion is greatly increased by the presence of acids. As the inversion of syrup occurs, the syrup becomes sweeter, as fructose is sweeter than glucose [11].

8.6.2.2 Solution by agitation without heating

To prevent heat-induced inversion of sucrose, syrup may be prepared without heat by agitation. On a small scale, sucrose and other formulative agents can be dissolved in purified water by placing the ingredients in a vessel larger than the volume of syrup to be prepared, allowing for thorough agitation of the mixture. This process is more time-consuming than using heat, but the product has maximum stability.

8.6.2.3 Percolation method

The percolation method can involve percolating sucrose to prepare the syrup or percolating the source of the medicinal component to create an extractive, to which syrup is subsequently added. This method follows the USP guidelines for syrup preparation. An example of a syrup produced using this approach is ipecac syrup. Ipecac syrup is formulated by combining glycerin and syrup with an extractive derived from powdered ipecac obtained through percolation. This syrup serves as an emetic drug utilized in cases of orally ingested poison.

8.6.2.4 Addition of sucrose to medicated liquids

In the preparation of syrups, medicated liquids such as tinctures or fluid extracts are occasionally utilized. These liquids often contain substances soluble in alcohol and are formulated with alcohol or a combination of alcohol and water. When medicinal agents are in alcohol-soluble form but need to be water-soluble, a specific method is employed for this conversion. Conversely, if the alcohol-soluble components are undesired, they can be eliminated by mixing the liquid with water, allowing separation, and subsequent filtration. The resulting medicated liquid is then blended with sucrose to produce the syrup. If the tincture or fluid extract is compatible with water, it can be directly added to a simple or flavored syrup [12].

8.6.3 Components of syrup

Mostly syrups contain the following substances [13]:
–   Sugar usually sucrose
–   Antimicrobial preservatives
–   Flavorants
–   Colorants

Examples of sugars
– Sucrose (85 g/100 mL)
– Glucose
– Fructose
– Invert sugar
– Maltose

Examples of flavorants
– *Cherry:* Frequently used in antibiotics and cough syrups to mask bitterness.
– *Strawberry and raspberry:* Often used in pediatric formulations to improve taste.
– *Anise:* Commonly found in cough syrups and expectorants for its soothing properties.
– *Peppermint:* Often added to cough syrups and lozenges for a refreshing taste.
– *Caramel:* Adds a sweet and slightly toasted flavor, improving the palatability of various syrups.

Examples of antimicrobial preservatives
– Benzoic acid (0.1–0.2%)
– Sodium benzoate (0.1–0.2%)
– Methyl, propyl, and butyl parabens (0.1%)
– Alcohol (15–20%)

Examples of colorants
– *Beetroot red:* Used in fruit syrups and beverages for its vibrant red hue.
– *Turmeric:* Provides a bright yellow color, commonly found in syrups for health drinks and teas.
– *Tartrazine (E102):* A yellow dye frequently used in lemon-flavored syrups and soft drinks.
– *Brilliant blue (E133):* Used to create blue shades in syrups, especially in novelty beverages.
– *Caramel color:* This is a widely used colorant that provides a brown hue, commonly found in cola syrups and various flavored syrups.

8.6.4 Sucrose and non-sucrose-based syrup

Primarily, sucrose serves as the predominant sugar utilized in syrups; however, under specific conditions, it may be substituted wholly or partially by other substances, such as in syrups designed for diabetic patients. Syrups containing sucrose commonly impact oral health and are often associated with dental caries. To enhance oral health, alternatives to sucrose, such as palatinose and sugar alcohols, are employed as substitutes in syrups [14].

Non-glucogenic substances like methyl cellulose or hydroxyethyl cellulose can be utilized in formulations. Polyols such as sorbitol, either alone or in combination with glycerin, are frequently incorporated. Sorbitol solution USP, containing 64% by weight of polyhydric alcohol, is commonly employed either partially or entirely in these preparations.

The emergence of diverse sweeteners has facilitated the creation of syrups that are not sucrose-based. This advancement is particularly significant for diabetic individuals and those mindful of weight management. Products with reduced sugar content can achieve comparable sweetness levels by incorporating intense sweeteners like aspartame and saccharin [15].

8.6.4.1 Antihistamine syrup

Table 8.3 shows the composition of antihistamine syrup.

Tab. 8.3: Composition of antihistamine syrup.

S. no.	Ingredients	Concentration
1	Chlorpheniramine maleate	0.4 g
2	Glycerin	25 mL
3	Syrup	83 mL
4	Sorbitol solution	282 mL
5	Sodium benzoate	1 g
6	Alcohol	60 mL
7	Colorant	q.s.
8	Flavorants	q.s.
9	Purified water (q.s.)	1,000 mL

Antihistamine syrup, such as Claritin D, possesses certain characteristics that may potentially cause erosion. These properties include a low endogenous pH, high titrable acidity, the presence of citric acid, the absence of fluoride and phosphate, and a minimal quantity of calcium in its composition [16].

Antihistamine syrup contains chlorpheniramine maleate as an active ingredient. Chlorpheniramine maleate syrup works by blocking H1-receptor sites on tissues. Chlorpheniramine maleate syrup is used for the treatment, control, prevention, and improvement of the following diseases, conditions and symptoms like common cold, itchy throat/skin, hay fever, hives, allergies, anaphylactic shock, and watery eyes.

8.6.4.2 Ferrous sulfate syrup

Table 8.4 shows the composition of ferrous sulfate syrup.

Tab. 8.4: Composition of ferrous sulfate syrup.

S. no.	Ingredients	Concentration
1	Ferrous sulfate	40 g
2	Citric acid, hydrous	2.1 g
3	Peppermint spirit	2 mL
4	Sucrose	825 g
5	Purified water (q.s.)	1,000 mL

A joint consultation by UNICEF and USAID has advised that the most suitable iron supplement for infants and young children should be an aqueous solution containing a
soluble ferrous salt, like ferrous sulfate (FS), or a ferric complex, such as iron polymaltose (IPC) [17].

Ferrous sulfate syrup contains ferrous sulfate as an active ingredient. Ferrous sulfate syrup works by completing the need of iron in the body. It is used:
- as iron supplements are indicated in patients with diseases caused by iron deficiency;
- in the treatment of iron deficiency anemia, prophylaxis for iron deficiency in
pregnancy; and
- in precaution if sedation or general anesthesia is required; risk of the hypotensive
episode.

8.6.5 Advantages of syrups

The advantages of syrups are as follows:

Palatability
Syrups are often sweetened, making them more palatable than pills or capsules, especially for children and those who have difficulty swallowing. This can enhance patient
compliance with medication regimens.

Versatile delivery
They can serve as vehicles for a variety of medicinal compounds, including antibiotics,
antihistamines, and cough suppressants, allowing for effective delivery of these drugs.

Soothing effect

Syrups can coat the mucous membranes of the throat, providing relief from irritation and inflammation. This is particularly beneficial in cough syrups, where the sugar content helps to soothe a sore throat by attracting moisture and cooling inflammation.

Nutritional benefits

Some syrups, especially those made from natural sources like maple syrup, contain antioxidants and minerals that may contribute to overall health. For instance, maple syrup has been shown to have anti-inflammatory properties and may help protect against oxidative stress.

Precision in dosing

Syrups allow for more precise dosing, especially in pediatric medicine, where accurate dosing is critical. The liquid form can be easily measured using syringes or dosing cups.

8.6.6 Difference of simple syrup in BP and USP

8.6.6.1 British Pharmacopoeia

In the British Pharmacopoeia, syrups are formulated in weight/volume (W/V) measurements. Specifically, 66.7 g of sucrose is incorporated into an adequate amount of water to achieve a total weight of 100 g for the syrup preparation.

8.6.6.2 USP

In the USP, syrups are typically formulated in percentage weight/volume (% W/V) measurements. Specifically, 85 g of sucrose is combined with an appropriate amount of water to achieve a total volume of 100 mL for the syrup preparation.

8.6.7 Uses of syrup

This form of medication is swiftly absorbed by the body, offering a more soothing experience during swallowing [18]. Syrups are widely recognized as a prominent delivery vehicle for antitussive medication due to their ease of ingestion compared to tablets and capsules.

Functions as an antioxidant

Slows down oxidation by converting sugar into dextrose and levulose, reducing the sugar content, thereby preventing the breakdown of various substances without requiring additional preservatives.

Acts as a preservative
Creates significant osmotic pressure, inhibiting the growth of microorganisms like bacteria, fungi, and molds.

Serves as a pleasant sweetener
Enhances taste by masking bitterness or unpleasant flavors, making it a suitable carrier for bitter or nauseating substances.

Helps preserve vegetable drugs
Aids in preventing the deterioration of numerous plant-based medicines.

8.7 Elixirs

The *British Pharmacopoeia* (2013) states that "elixirs are clear aromatic solutions for oral use that contain one or more active ingredients dissolved in a solvent, usually containing a high proportion of sucrose, or in a suitable polyatomic alcohol or alcohols and can also contain ethanol (96%) or a dilution" [19].

The *USA Pharmacopeia* (39th Ed.) gives only the definition of the DF in the monograph *Pharmaceutical Dosage Forms* (1151), in which it states that "elixirs, as a rule, are clear, aromatic, sweetened hydroalcoholic liquids intended for oral use."

Just like syrups, elixirs are also of two types:
1. Non-medicated elixirs: used as a vehicle.
2. Medicated elixirs: used therapeutically.

8.7.1 Method of preparation

Elixirs are typically prepared by initially dissolving alcohol and water-soluble ingredients separately in alcohol and purified water, respectively. Subsequently, the alcohol solution is added to the water, not vice versa, to maintain the highest possible alcoholic strength consistently and minimize the separation of alcohol-soluble ingredients. Following this step, the final volume is adjusted. If the elixir appears somewhat turbid at this stage, a small amount of talc can be added to address this issue. After adding talc, allow some time for settling before separating it. This turbidity often arises from the separation of some flavoring oil from the elixir due to the reduced alcohol concentration [20].

8.7.2 Medicated elixirs

Medicated elixirs are primarily employed to exhibit therapeutic effects, with most elixirs containing a single therapeutic agent. The key advantage of using a single therapeutic agent is the flexibility to adjust the dosage by increasing or decreasing the amount of elixir administered. However, when multiple therapeutic agents are combined in an elixir, adjusting the dosage becomes challenging as it can impact the effects of all the therapeutic agents simultaneously [21].

Elixirs are typically alcohol-based solutions that help water-soluble and alcohol-soluble components dissolve, increasing the stability and bioavailability of the active ingredients. Alcohol also acts as a preservative, extending the product's shelf life. Additionally, sweeteners and flavorings are commonly added to elixirs to mask the unpleasant taste of the medication, making them more palatable, especially for older and pediatric populations. Because they are liquids, elixirs have a rapid onset of action and facilitate quicker absorption in the gastrointestinal tract. Nevertheless, despite their benefits, elixirs should be used carefully around people who have liver problems, are intolerant to alcohol, or are taking drugs that could interfere with alcohol. In order to strike a balance between patient adherence, safety, and efficacy, formulation scientists carefully select the ingredients and their concentrations.

8.7.2.1 Examples of medicated elixirs

Examples of medicated elixirs are as follows [22]:
- Phenobarbital elixir (sedative, 0.4% elixir)
- Chlorpheniramine elixir
- Paracetamol elixir (analgesic, antipyretic)
- Decadron elixir (dexamethasone elixir)

8.7.3 Non-medicated elixirs

Non-medicated elixirs may be used for the following purposes:
1. Addition of a therapeutic agent for making medicinal elixirs
2. Dilution of an existing medicated elixir

When dealing with a liquid therapeutic agent, it is essential to assess its solubility and stability in both water and alcohol. Diluting a pre-existing medicated elixir requires the use of a non-medicated elixir. Moreover, the alcohol base in the diluent should match that of the medicated elixir to prevent adverse effects on the solubility of already soluble substances. Additionally, the color and flavor of the diluent should closely resemble those of the medicated elixir to avoid potential physical and chemical incompatibilities.

As a standardized medium, non-medicated elixirs ensure consistency and uniformity in drug formulation. To maintain their organoleptic properties, they usually contain flavorings, sweeteners, and preservatives. Potential interactions with future active ingredients must also be considered when selecting ingredients for non-medicated elixirs. Preventing problems like phase separation, precipitation, and microbial contamination during storage requires proper formulation. Furthermore, in pharmacy practice, non-medicated elixirs are especially helpful when it comes to impromptu compounding. These elixirs make it easier to prepare customized dosages that meet each patient's unique needs.

8.7.3.1 Examples of non-medicated elixirs

Examples of non-medicated elixirs are as follows [23]:
– Aromatic elixirs
– Compound benzaldehyde elixir
– Iso-alcoholic elixir

8.7.4 Differences between elixir and syrup

Differences between elixirs and syrups are given in Tab. 8.5.

Tab. 8.5: Differences between elixirs and syrups.

S. no.	Elixir	Syrup
1	Less viscous	More viscous
2	Is less sweet	Is sweeter
3	Contains alcohol from 4% to 40%	No limit alcohol concentration if present
4	Are cleared, sweetened, pleasantly flavored hydroalcoholic solutions	Are concentrates, aqueous preparations of a sugar or sugar-substitute

8.7.5 Composition of elixirs

The alcohol content in elixirs typically ranges from 4% to 40%. Elixirs formulated for alcohol-soluble ingredients have a higher alcohol concentration, and vice versa. Glycerin is a common component in most elixirs, sometimes accompanied by propylene glycol. While many elixirs are sweetened with sucrose syrup, saccharin is used as a sweetening agent when the alcohol concentration is high because sucrose syrup is less soluble in high alcohol concentrations. Flavoring agents and colorants are added to enhance taste and appearance. Elixirs with alcohol concentrations between 10% and 21% are self-preserving and do not require additional antimicrobial agents [24].

8.7.6 Uses of elixirs

Elixirs are used for various purposes as follows:

Oral medications

Elixirs are commonly used for medications that are soluble in water but have an unpleasant taste. They are flavored to improve patient acceptance and are available in various flavors, making them more palatable.

Preservation

The use of alcohol in elixirs helps preserve the medication and prevent bacterial growth, ensuring the stability of the pharmaceutical preparation.

Therapeutic use

Elixirs can be utilized for a variety of medications, including cough syrups, antacids, and anti-nausea medications, providing therapeutic benefits to patients.

Administration

Elixirs are easy to administer due to their liquid consistency, making them easy to swallow and manipulate for patients.

Antihistaminic treatments

Medicated elixirs such as chlorpheniramine maleate and diphenhydramine HCl are used to relieve allergy symptoms, including sneezing, itching, and runny nose.

Sedative and hypnotic effects

Elixirs containing barbiturates or other sedatives are used to induce drowsiness or sleep, making them useful for treating insomnia or anxiety.

Pediatric applications

Certain medicated elixirs, like chloral hydrate, are specifically formulated for children, providing a palatable option for administering medications.

Expectorants

Elixirs such as terpin hydrate are used to facilitate productive coughs, helping to clear mucus from the respiratory tract.

Pain relief

Some medicated elixirs, like acetaminophen elixir, are formulated to provide analgesic effects, offering relief from pain and discomfort.

Hormonal treatments

Dexamethasone elixir, containing a synthetic corticosteroid, is used for its anti-inflammatory and immunosuppressive properties.

8.7.7 Benefits

Vehicle for medications
They act as carriers for active pharmaceutical ingredients, facilitating the solubility and delivery of medications in elixir formulations.

Flavoring agents
Non-medicated elixirs are often flavored to improve palatability, making them more acceptable to patients, particularly children.

Stability and preservation
The alcohol content in these elixirs helps preserve the solution and maintain the stability of both water-soluble and alcohol-soluble components, extending shelf life.

Dilution
They can be used to dilute medicated elixirs, allowing for precise adjustments in dosage without compromising therapeutic effects.

Ease of preparation
Non-medicated elixirs are relatively simple to prepare, making them efficient for pharmaceutical compounding.

Extemporaneous compounding
Pharmacists utilize them in the preparation of customized medications tailored to individual patient needs.

Hydration and nutritional supplementation
Some formulations serve as hydration solutions or nutritional supplements, providing essential nutrients in a palatable form.

8.7.8 Examples

8.7.8.1 Phenobarbitol elixir

The ingredients of phenobarbital elixir are given in Tab. 8.6.

It is used as a sedative and hypnotics, for the short-term treatment of insomnia, since they appear to lose their effectiveness for sleep induction and sleep maintenance after 2 weeks. It is used as long-term anticonvulsants for the treatment of generalized tonic-clonic and cortical local seizures, and in the emergency control of certain acute convulsive episodes, e.g., those associated with status epilepticus, cholera, eclampsia, meningitis, tetanus, and toxic reactions to strychnine or local anesthetics.

Tab. 8.6: Composition of phenobarbital elixir.

S. no.	Ingredients	Concentration
1	Phenobarbital	4 g
2	Orange oil	0.25 mL
3	Propylene glycol	100 mL
4	Alcohol	200 mL
5	Sorbitol solution	600 mL
6	Colorant	q.s.
7	Purified water (q.s.)	1,000 mL

8.7.8.2 Paracetamol elixir

The ingredients of paracetamol elixir are given in Tab. 8.7.

Tab. 8.7: Composition of paracetamol elixir.

S. no.	Ingredient	Concentration
1	Paracetamol	2.5 g
2	Ethanol 96%	10mL
3	Propylene glycol	10 mL
4	Chloroform spirit	2 mL
5	Amaranth solution	q.s.
6	Glycerin	100 mL
7	Simple syrup	30.0 mL

It is used for the treatment of mild to moderate pain, anti-pyretic and post immunization pyrexia. Paracetamol elixir is used in babies and children for the treatment of mild or moderate pain and fever, and also in babies who develop fever after vaccination. It is used for temporary relief of fever and pain associated with teething, immunization, earache, headache, symptoms of colds and flu. It is used as an analgesic and antipyretic in patients sensitive to or unable to take aspirin.

8.8 Spirits

Spirits also referred to as essences are alcoholic or hydroalcoholic solutions containing volatile substances with alcohol contents exceeding 60%. The preparation typically involves the use of 90% alcohol. The active ingredient in spirits can exist in solid, liquid, or gaseous form. These solutions are administered through inhalation, orally, or externally. Primarily utilized as flavoring agents, spirits excel in retaining water-insoluble volatile oils in solution more effectively than tinctures. The production of spirits involves several key stages: fermentation, distillation, and maturation. Fruit spirits are a significant type of spirit, with a minimum ethanol content of 37.5% by volume. Plum spirit, derived from fermented plums, is a widely popular fruit spirit, particularly produced on a large scale in central and Eastern Europe. Cherry spirits are another favored variety, commonly manufactured in southern regions of Germany, France, Switzerland, and Serbia. Additionally, fruit spirits crafted from fermented apple and pear juices are relatively prevalent in the market [25].

8.8.1 Uses of spirits

The uses of spirits in pharmacy are varied and essential for pharmaceutical preparations. Spirits, which are alcoholic or hydroalcoholic solutions of volatile substances, serve several purposes in pharmacy.

Flavoring agent
Spirits are commonly used as flavoring agents in pharmaceutical preparations to impart the flavor of their solute to other medications.

Medicinal purposes
Spirits are utilized for their therapeutic value, both orally, externally, or through inhalation, depending on the specific preparation. They can act as carminatives, antacids, and mild reflex circulatory stimulants, depending on their composition.

Official spirits
Some spirits are officially recognized in pharmacopeias like the USP/NF, such as aromatic ammonia spirit, camphor spirit, compound orange spirit, and peppermint spirit. These official spirits have defined pharmaceutical uses and are part of standard pharmaceutical practice.

In summary, spirits play a crucial role in pharmacy as flavoring agents, for medicinal purposes, and as officially recognized pharmaceutical preparations with specific therapeutic effects.

8.8.2 Methods of preparation of spirits

Methods of preparation of spirits are discussed further.

8.8.2.1 Simple solution

A definite amount of volatile oil is dissolved in alcohol simply. The ingredients of compound spirit of orange are given in Tab. 8.8.

Tab. 8.8: Compound spirit of orange.

S. no.	Ingredients	Concentration
1	Oil of orange terpeneless	6 mL
2	Oil of lemon terpeneless	3 mL
3	Anethol	4 mL
4	Oil of coriander	12 mL
5	Alcohol	q.s.
6	Distilled water	q.s.
7	Purified talc	15.0 g

To prepare the spirit, dissolve the oils and anethol in alcohol, and then add sufficient distilled water to achieve the desired quantity. Add purified talc and shake vigorously for several minutes. Filter the mixture through paper in a well-covered funnel, washing the filter with diluted alcohol to obtain the required yield [26].

The uses of compound orange spirit are as follows [27]:
- Used as a flavorant
- Used in aromatherapy
- Used in food and beverages industry

The ingredients of compound spirit of cardamom are given in Tab. 8.9.

Tab. 8.9: Compound spirit of cardamom.

S. no.	Ingredients	Concentration
1	Oil of cardamom	20 mL
2	Oil of orange	20 mL
3	Oil of cinnamon	2 mL
4	Oil of clove	1 mL
5	Oil of caraway	0.2 mL
6	Anethol	1 mL
7	Alcohol	200 mL

Mix the oil and anethol with sufficient alcohol to make the product measure 200 mL.

The uses of compound spirit of cardamom are [28]:
- As a flavoring agent
- As a carminative
- As an anti-spasmodic agent

8.8.2.2 Maceration

In this method, the crude drug is macerated in purified water to extract water-soluble materials. Subsequently, the moist water-macerated leaves are combined with a specific quantity of alcohol, and the final volume is adjusted by adding volatile oil. The following is an example of a spirit prepared using this approach.

Peppermint spirit: The ingredients of peppermint spirit are [29] peppermint oil, alcohol, methyl acetate and methanol. Peppermint spirits are used as carminative agent, for aromatherapy, in treating irritable bowel syndrome (IBS), as potential treatment for non-ulcer dyspepsia. Additionally, when applied topically, peppermint oil has been utilized as a remedy for tension headaches.

Peppermint, a plant that thrives in shaded and humid environments, is the source of peppermint oil. The oil is extracted through a process that involves the following steps [30]. (a) Maceration: Macerate the peppermint leaves with water for 1 h to remove water-soluble substances. Add the moistened leaves in alcohol and macerate for 6 h. (b) Straining and pressing: Strain the mixture to separate the liquid from the solid residue (marc). Press the marc to extract any remaining liquid. (c) Combining and filtering: Combine the strained and expressed liquids and let them sit for some time then filter the mixture to remove impurities. (d) Final touch: Add peppermint oil to adjust the volume of the macerate. This process results in a peppermint macerate that can be used in various skincare, haircare, and hygiene products due to its anti-inflammatory, refreshing, purifying, and calming properties. (e) Storage: Store in well-closed, well-filled container to avoid the loss of volatile oil and alcohol [31].

8.8.2.3 Chemical reaction

Unofficial spirit of ethyl nitrate can be prepared by this process.

Reaction of NaNO$_2$ with alcohol: React sodium nitrite (NaNO$_2$) with alcohol in a cold solution.

Reaction of alcohol and H$_2$SO$_4$: In a separate step, react the alcohol with sulfuric acid (H$_2$SO$_4$) in a cold solution.

Preparing ethyl nitrate spirit: The resulting mixture can be used to prepare ethyl nitrate spirit.

This process is not officially recognized and may not produce a safe or effective product. It is essential to exercise caution when attempting to prepare such substances [32].

8.8.2.4 Distillation

Good-quality spirits are prepared by this process. Following is the example of spirit prepared by this method;

The ingredients of aromatic ammonia spirit are given in Tab. 8.10.

Tab. 8.10: Composition of aromatic ammonia spirit.

S. no.	Ingredients	Concentration
1	Ammonium bicarbonate	3–4%
2	Strong ammonia solution	9%
3	Aromatic oils (lemon, lavender, and nutmeg)	q.s.
4	Alcohol	60%

It is a nearly colorless liquid when freshly prepared, but gradually acquires a yellowish color upon standing. The aromatic ammonia spirit can be administered by inhalation to shorten the post anesthetic recovery time. The aromatic ammonia spirit being a reflex respiratory and circulatory stimulant may also be used along with anesthetics, Inhalation of the vapors of aromatic ammonia tends to decrease shuck and vomiting in spinal and caudal anesthesia [33].

The procedure for the preparation of aromatic spirit of ammonia consists of the following steps. (a) Distil volatile oil, alcohol, and water and get 875 mL distillate and keep it separate in stoppered bottle as reserved portion. (b) Carry on distillation further and collect 55 mL of distillate and add ammonium bicarbonate and strong ammonia solution. To this portion, heat on water bath at 60 °C. In a sealed bottle, shake occasionally until solution is complete. Cool the solution and filter through cotton wool. (c) Mix this filtrate with reserved proportion of distillate add sufficient amount of purified water to make the volume and mix thoroughly [34].

References

[1]   Lein A, Ng SW. Oral liquids. In: Practical pharmaceutics: An international guideline for the preparation, care and use of medicinal products: Springer; London England, 2023. pp. 277–98.

[2]   Salunke S, O'Brien F, Tan DCT, Harris D, Math M-C, Ariën T, et al. Oral drug delivery strategies for development of poorly water soluble drugs in paediatric patient population. Advanced Drug Delivery Reviews. 2022; 190(114507): 1–23.

[3]   Vinoth R, Rani S, Manikandan V Formulation development, evaluation and technology transfer for lacosamide oral syrup. Recent trends of innovations in chemical and biological sciences volume-I.126.

[4]   Klie GC. Preparation of syrups by percolation. American Journal of Pharmacy. 1835–1907; 1881: 1.

[5]   Davenport R, Slinn J. Glaxo: A history to 1962: Cambridge University Press; Shaftesbury Road, Cambridge, CB2 8EA, United Kingdom, 1992.

[6]   Morotomi M, Guillem J, Pocsidio J, LoGerfo P, Treat M, Forde K, et al. Effect of polyethylene glycol-electrolyte lavage solution on intestinal microflora. Applied and Environmental Microbiology. 1989; 55(4): 1026–28.

[7]   Sharma VK, Chockalingham SK, Ugheoke EA, Kapur A, Ling P, Vasudeva R, et al. Prospective, randomized, controlled comparison of the use of polyethylene glycol electrolyte lavage solution in four-liter versus two-liter volumes and pretreatment with either magnesium citrate or bisacodyl for colonoscopy preparation. Gastrointestinal Endoscopy. 1998; 47(2): 167–71.

[8]   Ansari ARM, Mulla SJ, Pramod GJ. Review on artificial sweeteners used in formulation of sugar free syrups. International Journal of Advances in Pharmaceutics. 2015; 4(2): 5–9.

[9]   Desu HR, Narang AS, Kumar V, Thoma LA, Mahato RI. Liquid dosage forms: Pharmaceutics: Elsevier; 2024. pp. 271–318.

[10]   Djohan YA, Meenune M Effect of heating conditions on physical and chemical characteristics of sugar syrup.

[11]   Oroian M, Paduret S, Ciursa P, Pauliuc D. Influence of different adulteration agents (Glucose, fructose, inverted sugar, hydrolysed syrup and malt wort syrups) on honey textural properties. International Multidisciplinary Scientific GeoConference: SGEM. 2019; 19(6.3): 111–18.

[12]   Allen L, Ansel HC. Ansel's pharmaceutical dosage forms and drug delivery systems: Lippincott Williams & Wilkins; Philadelphia, Pennsylvania, 2013.

[13]   Mulla SJ, Swain K, Deshmukh S, Ansari ARM, Ahale D, Prasanjit I Formulation of multicomponent cold and cough syrup. 2015.

[14]   Matsukubo T, Takazoe I. Sucrose substitutes and their role in caries prevention. International Dental Journal. 2006; 56(3): 119–30.

[15]   Chetana R, Krishnamurthy S, Yella Reddy SR. Rheological behavior of syrups containing sugar substitutes. European Food Research and Technology. 2004; 218: 345–48.

[16]   Costa C, Almeida I, Costa Filho L. Erosive effect of an antihistamine-containing syrup on primary enamel and its reduction by fluoride dentifrice. International Journal of Paediatric Dentistry. 2006; 16(3): 174–80.

[17]   Mahmood T, Khan TM, Mahmood NKT, Khan TM, Khizar N. Comparison of ferrous sulphate with iron polymaltose in treating iron deficiency anaemia in children. Journal of Rawalpindi Medical College. 2017; 21(4): 376–379.

[18]   Jadhao AG, Sanap MJ, Patil PA. Formulation and evaluation of herbal syrup. Asian Journal of Pharmaceutical Research and Development. 2021; 9(3): 16–22.

[19]   Sakanyan E, Lyakina M, Alekseeva A, Shishova L, Shemeryankina T. Elixir dosage form in pharmaceutical practice. Pharmaceutical Chemistry Journal. 2017; 50: 764–67.

[20]   Pregadio F. Elixirs and alchemy. In: Daoism handbook: Brill; Leiden, Netherlands, and Boston, USA, 2000. pp. 165–95.

[21] Dahab AA. Drug formulations. In: Understanding pharmacology in nursing practice: Springer; London England, 2020. pp. 57–88.

[22] Pathak Y, dos Santos MA, Zea L. Handbook of space pharmaceuticals: Springer; London England, 2022.

[23] Mtonga T, Douglas GP. Standardizing representation of medication in LMICs: Case of Malawi and RxNORM. Journal of Health Informatics in Africa. 2019; 6(2): 51–56.

[24] Prountzos D, Manevich R, Pingali K, editors. Elixir: A system for synthesizing concurrent graph programs. In: Proceedings of the ACM international conference on Object oriented programming systems languages and applications: New York, NY, United States, 2012.

[25] Egea T, Signorini MA, Bruschi P, Rivera D, Obón C, Alcaraz F, et al. Spirits and liqueurs in European traditional medicine: Their history and ethnobotany in Tuscany and Bologna (Italy). Journal of Ethnopharmacology. 2015; 175: 241–55.

[26] Jones ER. Soluble compound spirit of orange and a simplified process for aromatic elixir. Journal of the American Pharmaceutical Association. 1922; 11(4): 277–78.

[27] Da Porto C, Pizzale L, Bravin M, Conte LS. Analyses of orange spirit flavour by direct-injection gas chromatography–mass spectrometry and headspace solid-phase microextraction/GC–MC. Flavour and Fragrance Journal. 2003; 18(1): 66–72.

[28] Council GM. The british: Pharmacopœia: Spottiswoode; 1899.

[29] Thompson HL. The assay of the spirit of peppermint: The University of Nebraska-Lincoln; Lincoln, Nebraska, 1914.

[30] Mughal SS. Peppermint oil, its useful, and adverse effects on human health: A review. Authorea Preprints. 2022; 8(6): 1–4.

[31] Safaeian Laein S, Khanzadi S, Hashemi M, Gheybi F, Azizzadeh M. Improving quality of trout fillet using gelatin coating-contain peppermint essential oil loaded solid lipid nanoparticles (PEO-SLN). Journal of Food Measurement and Characterization. 2024; 18(1): 345–356.

[32] Palmieri S, Pellegrini M, Ricci A, Compagnone D, Lo Sterzo C. Chemical composition and antioxidant activity of thyme, hemp and coriander extracts: A comparison study of maceration, Soxhlet, UAE and RSLDE techniques. Foods. 2020; 9(9): 1221.

[33] Stover OH. A preliminary report on the postoperative use of aromatic spirits of ammonia*. Anesthesia & Analgesia. 1928; 7(2): 77–79.

[34] Bessling B, Löning J-M, Ohligschläger A, Schembecker G, Sundmacher K. Investigations on the synthesis of methyl acetate in a heterogeneous reactive distillation process. Chemical Engineering & Technology. 1998; 21(5): 393–400.

Muhammad Yasir Ali, Shams ul Hassan, Ijaz Ali, Saeed Ahmad,
Udo Bakowsky, Nisar ur Rahman, Malik Saadullah, Ali Moghadam

9 Emulsions

9.1 Introduction

An emulsion is a dispersion in which the dispersed phase is composed of small globules of a liquid distributed throughout a vehicle in which it is immiscible (Fig. 9.1) [1]. The liquid that is dispersed into globules is called the dispersed phase or internal phase, and the system in which globules are dispersed is called the dispersion medium or external phase [2]. The internal phase and external phase are also called discontinuous and continuous phases, respectively [3]. The globules remain dispersed only for a short time, and separation takes place quickly upon standing [4]. So, a third system is required for stability, called an emulsifying agent or emulsifier [5].

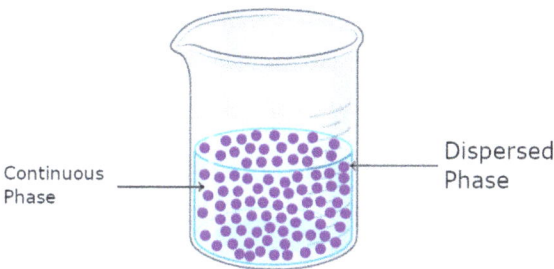

Fig. 9.1: Schematic illustration of a general emulsion.

It is a thermodynamically unstable system of two liquids in which one, called dispersed phase, is continuously distributed into another called dispersion medium, and the system is stabilized by the presence of a third substance called emulsifier or emulsifying agent [3]. Emulsions can be formulated for all routes of administration, but most commercial products are developed for oral, topical, and parenteral routes. An emulsion is manufactured to mask the bitter taste of the drug and also to increase the rate of absorption. Many drugs have a bitter taste, which can be unpleasant for patients. By incorporating the drug into an emulsion, taste can be masked by other ingredients in the formulation, such as flavoring agents or sweeteners. The oil phase of the emulsion can coat taste buds and reduce the perception of bitterness, making the drug more palatable to the patient. This can improve patient compliance with medication regimens as patients are more likely to take medications that taste better.

Emulsions can also enhance the absorption of drugs in the body. The oil droplets in the emulsion can help solubilize lipophilic (fat-soluble) drugs, improving their bioavailability. Additionally, emulsions can increase the surface area of the

https://doi.org/10.1515/9783111438108-009

drug particles, facilitating their interaction with the body's tissues, and increasing absorption rates. This can lead to faster onset of action and more consistent therapeutic effects compared to drugs administered in other dosage forms, such as tablets or capsules. For example, let's say we have a bitter-tasting drug that is poorly absorbed when administered alone. By formulating it into an emulsion, we can improve its palatability while also enhancing its absorption rate. The emulsion acts as a vehicle for the drug, optimizing its delivery and bioavailability.

In summary, emulsions are versatile dosage forms that can be tailored to meet specific pharmaceutical needs, including taste masking and enhancing drug absorption [6].

Macromolecules, e.g., insulin and heparin, are dissolved in oil, and then an emulsion is formed [7]. Also, lotions, creams, liniments, and ointments are all emulsions [8]. Some emulsions are used for clinical investigation, e.g., radio-opaque x-ray examination [6].

9.2 Types of emulsions

Emulsions are of two types [3]:
1. Water-in-oil (W/O)
2. Oil-in-water (O/W)

9.2.1 Water-in-oil emulsion

A water-in-oil (W/O) emulsion is characterized by having a hydrophobic material, typically oil, as the continuous phase, while water is dispersed throughout the internal phase (Fig. 9.2). This type of emulsion predominates in crude oil emulsions found in oil fields, accounting for over 95% of cases. In W/O emulsions, three main components are present: a solvent, a surfactant, and water.

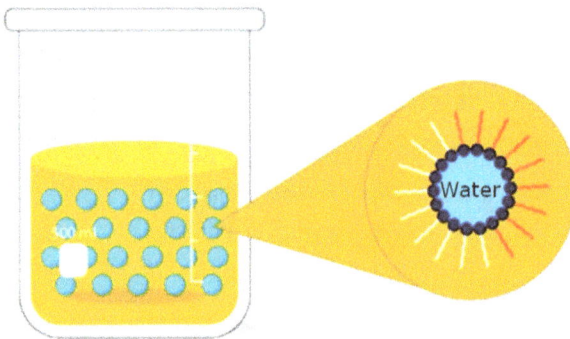

Fig. 9.2: Magnified droplet of water-in-oil (W/O) emulsion.

The composition of W/O emulsions plays a crucial role in their formation. Numerous studies have highlighted stability as the paramount characteristic of W/O emulsions, often achieved through the use of natural surfactants. Stable W/O emulsions typically exhibit a brown hue and contain a water content ranging from 60% to 80%. Mesostable emulsions, appearing brown or black, possess properties that fall between stable and unstable states, resembling O/W emulsions. Unstable emulsions swiftly separate into two distinct phases of water and oil shortly after formation. The entrained water in unstable emulsions initially appears black and contains around 30–40% water content for a brief period, reducing to approximately 10% within a week. Notably, only stable and mesostable emulsions are considered distinct from the other two states and can be categorized as true emulsions.

9.2.2 Oil-in-water emulsion

An O/W emulsion is characterized by the presence of oil as the dispersed phase and water as the continuous phase [9–11] or dispersion medium (Fig. 9.3). In the petroleum industry, both W/O and O/W emulsions can lead to significant financial losses if not appropriately treated. However, W/O emulsions are more prevalent than O/W emulsions, making O/W emulsions often referred to as reverse emulsions.

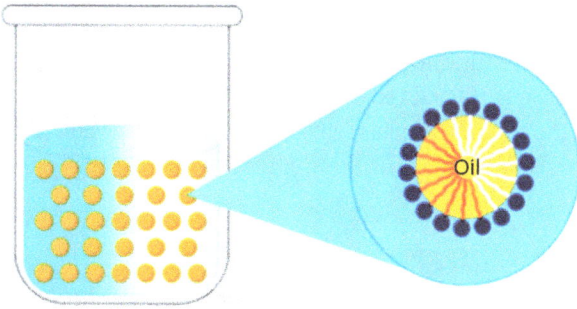

Fig. 9.3: Magnified droplet of oil-in-water (O/W) emulsion.

9.2.3 Multiple emulsions

In certain cases, multiple emulsions such as water-in-oil-in-water (W/O/W) and oil-in-water-in-oil (O/W/O) can be observed. Typically, multiple emulsions are stabilized using a combination of hydrophilic and hydrophobic surfactants. These emulsions are more intricate, comprising very small droplets suspended within larger droplets, all dispersed within a continuous phase. For example, W/O/W emulsions consist of water droplets encapsulated within larger oil droplets, which are then suspended in a continuous water phase. Moreover, these emulsions necessitate the presence of at least two types of emulsifiers in the system: one with a low hydrophilic-lipophilic balance (HLB) value and another with a high HLB value [12].

9.3 Tests for identification of emulsions

The following test can be performed for the identification of emulsion type.

9.3.1 Miscibility test

In this test, add a large amount of water to the emulsion. If emulsion is broken down by the water addition, then it is W/O, and if not, then it is O/W emulsion. Remember that by diluting with an outer or external phase, there is no separation of phases [1].

9.3.2 Staining test

Take a drop of emulsion on the glass slide and a drop of scarlet red. Then observe under the microscope. If dispersed globules are stained red, then it will be O/W emulsion. If the continuous phase is red, then it is W/O-type emulsion because scarlet red is soluble in oil, not in water [1]. Amaranth can also be used for staining tests, a water-soluble dye. If the continuous phase is stained red, then it is O/W emulsion, but if dispersed globules appear red and the continuous phase colorless, it is W/O emulsion [13].

9.3.3 Conductivity test

As water can conduct electricity, but oil cannot so this property can be used for testing whether emulsion is water in oil or oil in water. If water is present in the continuous phase, then electricity is passed, but if the water is in dispersed globular form, then electricity is not passed [1].

9.3.4 Cobalt chloride filter paper test

A filter paper impregnated with cobalt chloride and then dried (originally blue). It is then immersed in an emulsion, and its color changes from blue to pink. This signifies that the emulsion is of the O/W type. It is important to note that this test may not yield accurate results if the emulsion is unstable because of the presence of electrolytes [13].

9.3.5 Fluorescence test

When oils come into contact with UV rays, they emit fluorescence. O/W emulsions show irregular patterns, while W/O emulsions fluoresce uniformly across the entire field. However, it's important to recognize that this technique may not be applicable. Out of these tests, the first three are applicable and hence commonly used [13].

9.4 Emulsion theories

Emulsion theories have been developed to elucidate how emulsifying agents facilitate emulsification and uphold the stability of the resulting emulsion [3].

9.4.1 Surface tension theory

Interfacial tension refers to the force that prevents each liquid from breaking up into smaller particles. Surfactants aid in reducing this resistance. According to the surface tension theory, the presence of surfactants reduces the interfacial tension between immiscible liquids by disrupting the cohesive forces at the interface. The hydrophobic portion of the surfactant molecule tends to interact with the hydrophobic phase (such as oil), while the hydrophilic portion interacts with the hydrophilic phase (such as water). By doing so, surfactants essentially "coat" the interface between the two liquids, reducing the resistance to breaking up into smaller particles. This allows for better mixing and dispersion of one liquid phase into another, leading to the formation of stable emulsions or suspensions. In summary, the surface tension theory explains how surfactants reduce the resistance between immiscible liquids by lowering the interfacial tension, thereby promoting better mixing and dispersion, which is crucial in various industrial and pharmaceutical processes.

9.4.2 Surface orientation theory

An emulsifying agent with a stronger hydrophilic character than hydrophobic character will encourage the formation of an O/W emulsion. The surface orientation theory of emulsions suggests that surfactant molecules, which have both hydrophilic and hydrophobic parts, align themselves at the interface between immiscible liquids in a way that minimizes the free energy of the system. Hydrophobic tails of surfactants tend to orient themselves towards the oil phase, while hydrophilic heads face the aqueous phase. This arrangement reduces the interfacial tension between the two liquids, allowing for stable emulsions to form. Additionally, the orientation of surfactant molecules prevents the coalescence of droplets by creating a protective barrier around them, enhancing the stability of the emulsion over time.

9.4.3 Plastic or interfacial film theory

The emulsifying agent forms a thin film surrounding the droplets of the internal phase, preventing them from coming into contact and coalescing with each other. The plastic or interfacial film theory of emulsions posits that emulsions are stabilized by a thin layer of adsorbed surfactant molecules forming a flexible film at the interface between immiscible liquids. According to this theory, the surfactant molecules create a plastic-like barrier that envelops the dispersed droplets, preventing them from coalescing or merging with each other. This film acts as a protective layer, minimizing the contact between the dispersed phase and the continuous phase, thus reducing the interfacial tension and enhancing the stability of the emulsion. The flexibility of the film allows it to adjust to changes in the system, such as agitation or temperature variations, maintaining the integrity of the emulsion over time. Overall, the plastic or interfacial film theory provides insights into the mechanisms by which surfactants stabilize emulsions, crucial for various industrial and pharmaceutical applications.

9.5 Ingredients of emulsion

9.5.1 Oil phase

Oils utilized in emulsions either serve as medicaments or act as carriers of water-insoluble drugs. Important parameters that help select the oil phase may include physical characteristics of emulsion, miscibility of aqueous and oil phases, drug solubility in oil phase, and required consistency of emulsion. Oil from vegetable origin gets rancid, so it needs incorporation of antioxidants. Liquid paraffin, soft paraffin, hard paraffin, silicone oils, turpentine oil, and benzoyl benzoate are commonly used oils for externally applied emulsions. Commercial parenteral emulsion utilizes purified soya bean or safflower oils. Purified oleic, linoleic, palmitic, and stearic acids are also used. Emulsions for parenteral administration have extremely limited options for oil because they are naturally unsafe. Oral emulsions mostly contain castor oil, liquid paraffin oil, or fish liver oils.

9.5.2 Emulsifying agents

There are three classes of emulsifiers [14]: (a) surfactants, (b) hydrophilic colloids, and (c) gums. Surfactants, also known as surface-active agents, have both hydrophilic and lipophilic portions [15]. If hydrophilic value is increased, then they are soluble in water, and if lipophilic value is increased, then they are soluble in lipid. When the hydrophilic value of a surfactant is increased, it means that the proportion of the hydrophilic portion of the molecule relative to the lipophilic portion is higher. In other words, the surfactant molecule becomes more polar or water-loving. As a result, surfactants with increased hydrophilic values are more soluble in water because they can form favorable interactions with water molecules, such as hydrogen bonding. Conversely, when the lipophilic value of a surfactant is increased, it means that the proportion of the lipophilic portion of the molecule relative to the hydrophilic portion is higher. In this case, the surfactant molecule becomes more non-polar or lipid-loving. Surfactants with increased lipophilic values are more soluble in lipid phases because they can form favorable interactions with oil molecules, such as van der Waals forces. Overall, the balance between the hydrophilic and lipophilic properties of surfactants determines their solubility in water or lipid phases. By adjusting these properties, surfactants can be tailored to suit specific applications, such as forming stable emulsions, dissolving oils in water (or vice versa), or enhancing the wetting and spreading of liquids on surfaces [16].

9.5.3 Classification of emulsifying agents

Emulsifying agents are classified as shown in Tab. 9.1.

Tab. 9.1: Classification of emulsifying agents.

S. no.	Class	Examples
1.	Natural emulsifiers from plant sources	– Acacia – Tragacanth – Chondrus – Pectin – Starch
2.	Natural emulsifier from animal source	– Gelatin – Egg yolk – Wool fat
3.	Synthetic emulsifier	– Anionic – Cationic – Non-ionic – Amphoteric

Tab. 9.1 (continued)

S. no.	Class	Examples
4.	Semisynthetic emulsifier	– Methylcellulose – Sodium carboxymethyl cellulose
5.	Finely divided solids/inorganic emulsifying agents	– Bentonite – Magnesium oxide – Magnesium trisilicate – Magnesium aluminum silicate – Milk of magnesia

9.5.3.1 Natural emulsifier from plant source

Emulsifying agents derived from plant sources encompass carbohydrates, including gums and mucilaginous substances. Possessing anionic properties, these agents facilitate the formation of O/W emulsions. Some function as true emulsifiers, termed primary emulsifying agents, while others serve as emulsion stabilizers, referred to as secondary emulsifying agents. They exhibit the ability to emulsify a wide range of substances; however, to ensure the preservation of resulting emulsions, it becomes necessary to introduce a suitable preservative. Common preservatives include alcohol, sodium benzoate, benzoic acid, or a combination of methyl paraben and propyl paraben, as carbohydrates provide an optimal medium for bacterial growth [13].

Acacia: Acacia is the best emulsifying agent for the extemporaneous preparation of oral emulsion. The emulsions prepared by this are attractive and quite palatable. They are stable over a wide range of pH, i.e., 2–10. These emulsions are of low viscosity; therefore, creaming occurs rapidly, which can be prevented by increasing the viscosity by incorporating tragacanth, agar, or pectin along with acacia. They are also susceptible to bacterial growth; therefore, they must be suitably preserved [13].

Tragacanth: It is an auxiliary type of emulsifier, because it does not reduce interfacial tension, and thus, the oil globules are usually large. It produces coarse and thick emulsion, and sometimes viscosity increases to such an extent that pouring becomes difficult. A very stable emulsion is produced if both acacia and tragacanth are used as emulsifying agents. The amount of tragacanth needed for this purpose is one-tenth of the quantity of acacia utilized [13].

Agar: It is not a good emulsifier as it produces very coarse and viscous emulsions. It is commonly used as a thickening agent along with acacia for the emulsification of mineral oils [13]. In the context of mineral oil emulsions, agar might be included in the formulation alongside other emulsifiers to help create a stable emulsion with the desired viscosity and texture. The combination of agar with other emulsifiers can help to overcome the limitations of agar's emulsifying properties and achieve the desired emulsion characteristics. Overall, while agar is not typically used as a standalone emulsifier, it can still contribute to the emulsification process when incorporated into formulations alongside other emulsifying agents and stabilizers.

Chondrus (Irish moss): It is primarily valued for its carrageenan content, which is extracted and used in various applications, including emulsification.

Extraction, characterization, and use of carrageenan: Carrageenan, derived from *Chondrus crispus* and other red seaweeds, is a polysaccharide with excellent thickening, stabilizing, and emulsifying properties. It is dried seaweed. It is not very suitable for small-scale emulsification because the preparation of mucilage is time-consuming, and its emulsion must be homogenized. A very stable emulsion can be made by mechanical methods, and Chondrus has been used industrially as an inexpensive emulgent for fixed oils, particularly for cod liver. It is also a useful emulsion stabilizer [17].

Pectin: Pectin is obtained from the inner rind of citrus fruit or apple pulp remaining after cider making. It is extracted with dilute acids and then purified. It has been used in the proportion of 0.1 g/g of acacia to replace acacia, completely or particularly, in internal emulsion. It is also employed for stabilizing pharmaceutical and cosmetic lotions and creams. To prevent clumping on addition of water, it should be previously wetted with alcohol, glycerol, or syrup. Pectin emulsions are incompatible with alkalis, strong alcohol, heavy metals, and tannic and salicylic acids [13].

Starch: The utilization of starch as an emulsifying agent is infrequent, with its application limited to preparations intended for use as enemas [13]. There's how starch can contribute to emulsification: (a) As a thickening agent: Starches can absorb water and swell when heated in the presence of water, forming a viscous gel. Physicochemical and functional aspects: This gel-like consistency helps to increase the viscosity of the liquid phase in an emulsion, making it thicker and more stable. Octenyl succinate quinoa starch granule-stabilized pickering emulsion gels: preparation, microstructure, and gelling mechanism. By thickening the continuous phase (usually water) of the emulsion, starch helps to prevent the separation of the dispersed phase (usually oil) and maintain the uniformity of the emulsion. W/O emulsions are stabilized by surfactants, biopolymers, and/or particles. (b) As a stabilizer: In addition to thickening, starches can also contribute to the stability of emulsions by forming a network or matrix that traps and immobilizes the dispersed phase (oil droplets). Double emulsions relevant to food systems: preparation, stability, and applications. This network helps to prevent the coalescence and aggregation of oil droplets, thereby inhibiting phase separation and maintaining the integrity of the emulsion over time. Factors affecting the stability of emulsions are stabilized by biopolymers.

9.5.3.2 Synthetic emulsifiers

There are four types of these emulsifying agents (surfactants), i.e. [18],
- Anionic
- Cationic
- Non-ionic
- Amphoteric

Anionic surfactants

There are five relevant subgroups of this class. (a) Alkali metal and ammonium soap: These are the sodium, potassium, or ammonium salts of long-chain fatty acids such as oleic, stearic, and ricinoleic acids. When they are dissolved in media, they are dissociated and give anions [19]. Their physical action and unpleasant taste make them unsuitable for internal emulsion, and because of their alkaline nature, they should not be used in preparations for broken skin [20]. They are resistant to attack by microbes [21]. (b) Soaps of divalent and trivalent metals: Although calcium, magnesium, and zinc salts of fatty acids are W/O emulsifying agent, only the calcium soaps are commonly used as such [25]. Like alkali soap emulsions, they cannot be used in internal emulsion, but are less alkaline and less sensitive to acids, e.g., a-oleate [26]. Metallic soaps, due to their amphiphilic nature (having both hydrophilic and hydrophobic parts), can act as effective emulsifiers. Production of green surfactants: Market prospects. The hydrophobic part of the metallic soap molecule is attracted to oil, while the hydrophilic part is attracted to water. This allows the metallic soap molecules to surround and stabilize the oil droplets in the water phase, forming a stable emulsion. (c) Amine soap: Due to structural resemblance with ammonia, they form soap by reacting with fatty acids. They form an O/W emulsion and can be applied on broken skin but are not suitable for internal use [27]. One example is benzyl benzoate lotion in which they are used which is for itching and scabbing [28]. In emulsification, amine soaps function by adsorbing onto the oil-water interface, with their hydrophobic tails oriented towards the oil phase and their hydrophilic heads facing the water phase. This arrangement creates a barrier around the dispersed oil droplets, preventing them from coalescing and forming larger oil droplets or separating from the continuous water phase. (d) Sulfate and sulfated soap: They are esters of fatty alcohol and sulfuring or sulfuric acid Na-lauryl sulfate (ester of H_2SO_4 and lauryl alcohol) is not used alone and is used with cetostearyl alcohol Na-dodecyl sulfosuccinate (ester of sulfonic acid) and all other sulfurated soap are not used as emulsifier but used as wetting agents. Sulfates are susceptible to hydrolysis [3]. Emulsions with these emulsifiers are compatible with calcium and magnesium ions [3]. Therefore, with water also, so are water-soluble but are incompatible with cationic emulsifiers, such as cetrimide and with cationic medicaments, such as crystal violet and brilliant green [22]. When heated becomes acidic due to the breakage of the ester bond, produce O/W emulsion [23]. (e) Alkyl phosphates: They are similar to alkyl sulfates, but the alcohols are phosphated instead of sulfated. Like alkyl sulfates, they are used in combination with fatty alcohols [31]. They also produce O/W emulsion [32]. Alkyl phosphates are particularly useful in formulating W/O emulsions, where water droplets are dispersed within the continuous oil phase. Emulsions and emulsifiers: They help maintain the stability of W/O emulsions by providing a barrier between water droplets and preventing their aggregation. A critical review of emulsion stability and characterization techniques in oil processing is performed.

Cationic surfactant

When used, these surfactants give cations. Although they are mainly used for their disinfectant and preservative properties, they are also O/W emulsifying agent [24]. The quaternary ammonium compounds comprise the most important group of this class [25]. The most useful quaternary ammonium compound is cetrimide, which consists of tetradecyl trimethyl ammonium bromide [25]:

Cetrimide $CH_3(CH_2)_{15}NH^+(CH_3)_3$

Cetyltrimethyl bromide: $CH_3(CH_2)_{15}N^+(CH_3)_3 Br^-$

Both anionic and cationic surfactants are cheaper in cost.

Non-ionic surfactant

They are comparatively stable than both anionic and cationic surfactants. They are more compatible with other substances [26]. They may be used for oral preparation because they are more stable. They are comparatively expensive [27].

They have two parts, i.e.,

– hydrophilic part
– hydrophobic part

Hydrophilic part contains hydroxyl group, polyethylene groups, or oxyacetylene group. The hydrophobic part consists of fatty acids or fatty alcohols [28]. If there is balance between the two parts, then they will be at the interface. But if the molecule is predominantly hydrophilic, the emulsion formed is O/W because they are in the water phase, and if the molecule is predominantly hydrophobic, the emulsion formed is W/O because they go in oily phase [29].

Most of the surfactants are derived from:

1. Fatty acid (usually with 12–18 carbon atoms) [30]
2. An alcohol and/or ethylene oxide [27]

We have a basic nucleus (Fig. 9.4), which is obtained from sorbitol by removal of one water molecule, used in most of this class of surfactants [31].

a) Spans: Sorbitan esters (commonly known as Spans) are amber or yellow oily liquids or amber waxy solids in which the hydrophobic group is slightly predominant, so they form W/O emulsion [32]. General formula is shown in Fig. 9.5.

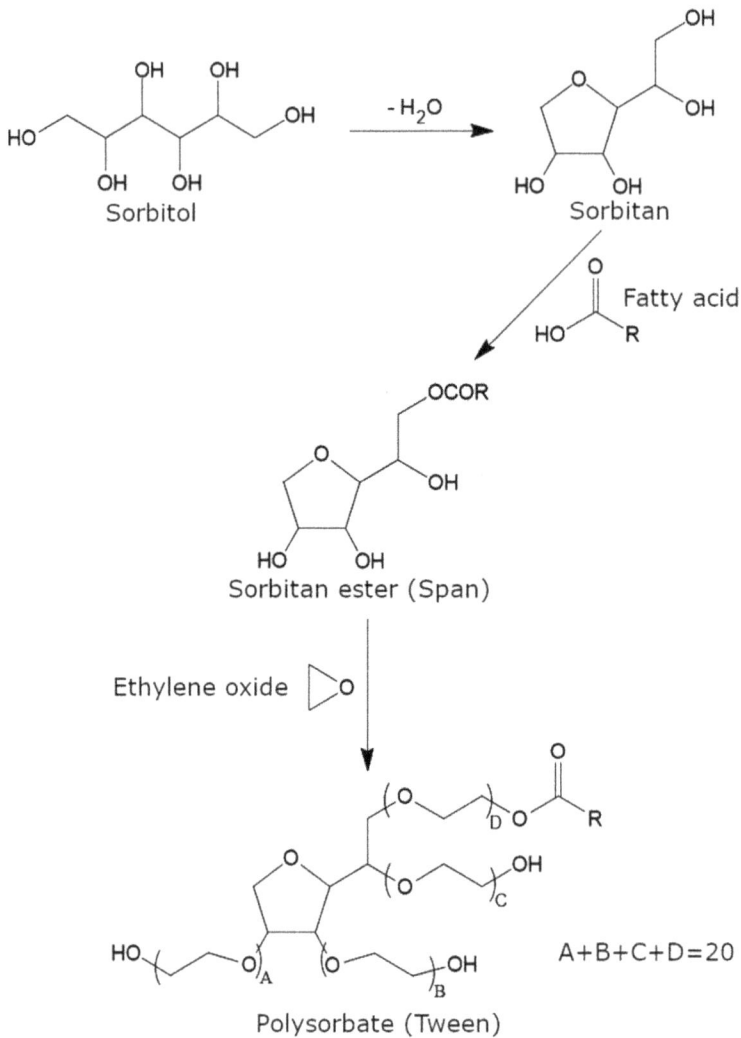

Fig. 9.4: Conversion of sorbitol to sorbitan and polysorbate.

They, as mentioned earlier, alone form W/O emulsion, creams, and ointments, but with polysorbates, they may form W/O or O/W emulsion [33]. Span creams are fine-textured [34]. These are relatively insensitive to high concentrations of electrolytes [35].

b) Polysorbates: When sorbitan esters are treated with ethylene oxide and polyethylene glycol, then polysorbates are obtained [36]. General formula is shown in Fig. 9.4.

By changing oxyethylene groups and by acid changing a different polysorbate, we obtained [37]:

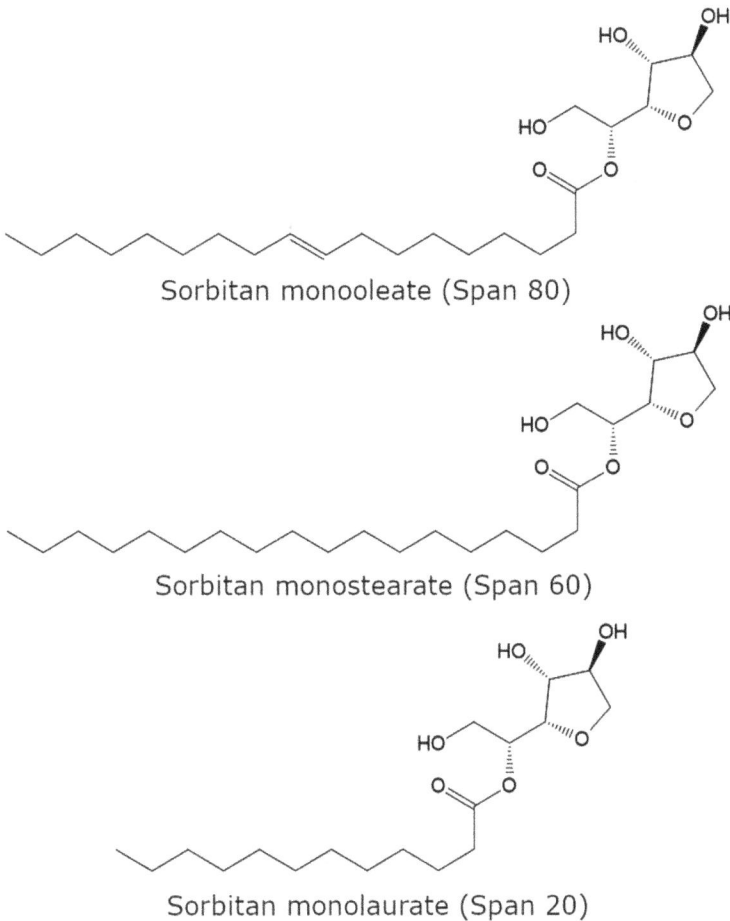

Sorbitan monooleate (Span 80)

Sorbitan monostearate (Span 60)

Sorbitan monolaurate (Span 20)

Fig. 9.5: Examples of sorbitan esters (Spans).

- Polyoxyethylene (20): This is obtained from sorbitan monolaurate and also called polysorbate 20 or Tween 20.
- Polyoxyethylene (60): This is obtained from sorbitan monostearate and also called polysorbate 60 or Tween 60.
- Polyoxyethylene (80): This is obtained from sorbitan monooleate and also called polysorbate 80 or Tween 80.

They produce fine-textured O/W emulsion and are stable to high concentrations of electrolytes. They are suitable for internal emulsions. Ointments are water-soluble [38].
c) Fatty alcohol polyethylene glycol ethers: They are also called macrogol ethers [39]. These are obtained by condensation of polyethylene glycol and fatty alcohols, usually cetyl or cetostearyl [40]. Ethers with less than 10 oxyethylene groups are oil sol-

uble and form W/O emulsions, and those of more than 10 oxyethylene groups are water soluble and form O/W emulsions [41]. Ceto-macrogol 1000 (polyethylene glycol 1000 monocetyl ether) [42]. The ceto-macrogol emulsifier contains 80% of acetos-tearyl alcohol and 20% of ceto-macrogol [43]. Macrogol ethers reduce the antimicrobial activity of quaternary ammonium compounds if the molar ratio of macrogol ether to antimicrobial agent is high [44].

d) Fatty acid polyethylene glycol esters: They are also called macrogol ester [39]. The macrogol esters are polymers of ethylene oxide and therefore contain many hydrophilic (oxyethylene) groups. So, they are O/W [45], for example, polyoxymethylene 40 monostearate [46]. Because of ester linkage, they are susceptible to hydrolysis. This may cause cracking of emulsion. Emulsions are stable to electrolytes [47].

e) Aliphatic alcohols: Aliphatic alcohols can also influence the viscosity, texture, and sensory properties of emulsion-based products. They may contribute to the formation of smoother textures, improved spreadability, and a more pleasant skin feel in cosmetic formulations. These are higher members of the series of saturated aliphatic monohydric alcohol [58]. They are sterilizers of O/W emulsion (auxiliary) or form W/O emulsion [59], e.g.:

1. Cetyl alcohol : $C_{16}H_{34}O$
2. Stearyl alcohol : $C_{18}H_{38}O$
3. Cetostearyl alcohol: $C_{56}H_{114}O_{21}$

Amphoteric surfactant

They can give both cationic and anionic surfactants. The emulsifying capabilities diminish as the pH approaches the isoelectric point of the surface-active agent. They are not commonly used, e.g., lecithins. Lecithin finds application in emulsions, specifically for intravenous and intramuscular administration, as well as creams, where it functions as an O/W emulsifying agent [18].

9.5.3.3 Semisynthetic polysaccharides

They are used as emulsifying agents and/or stabilizers in O/W emulsions for both internal and external use. Semisynthetic polysaccharides offer several advantages in emulsification, including their biocompatibility, versatility, and availability. Applications of natural, semisynthetic, and synthetic polymers in cosmetic formulations: They are commonly used in various industries, including food, pharmaceuticals, cosmetics, and personal care products to formulate stable emulsions with desired properties. Renewable polysaccharides micro/nanostructures are used for food and cosmetic applications.

Methylcellulose

Methylcellulose, a synthetic derivative of cellulose, sees extensive use in the pharmaceutical industry, serving as a suspending, thickening, and emulsifying agent. It is offered in various forms, including methylcellulose 20, methylcellulose 2500, and methylcellulose 4500 [9]. In emulsification, methylcellulose can contribute to the stability and texture of emulsions by forming a viscous gel-like network that helps to suspend and stabilize the dispersed phase (oil droplets) within the continuous phase (water). The leading role of cellulose, while methylcellulose itself may not possess strong emulsifying properties, can enhance the stability of emulsions formulated with other emulsifiers or surfactants.

Sodium carboxymethyl cellulose (Na-CMC)

In syrups for diabetic patients, we can use methylcellulose. Methylcellulose is not employed as a genuine emulsifier but functions as an emulsion stabilizer at a concentration of 0.5–1%. It exhibits solubility in both cold and hot water [9]. Na-CMC can also contribute to the viscosity and rheological properties of emulsions, enhancing their stability and texture. Additionally, it can improve the suspension properties of dispersed particles or ingredients in emulsions, leading to uniform distribution and improved product performance. Extensional viscosity of O/W emulsion stabilized by polysaccharides is measured on the opposed-nozzle device.

Finely divided solids/inorganic emulsifying agents

Bentonite, magnesium oxide, magnesium trisilicate, magnesium aluminum silicate, and milk of magnesia are used in the preparation of pharmaceutical emulsions. They usually produce O/W emulsion, but bentonite may be used to prepare either O/W or W/O emulsions depending on the order of mixing. A 5% suspension of bentonite is employed as an emulsifying agent [9]. These finely divided solids offer advantages such as improved stability, enhanced texture, and reduced sensitivity to environmental factors like temperature and pH changes. Fundamentals and applications of particle-stabilized emulsions in cosmetic formulations: However, it's essential to carefully select and control the particle size, surface chemistry, and concentration of these solid particles to optimize their emulsifying properties and ensure product performance.

9.5.3.4 Advantages of natural emulsifiers over synthetic emulsifiers

The statement underscores the potential dangers linked with synthetic emulsifiers and detergents frequently utilized in various products, including cosmetics and household items. These artificial compounds, such as betaine, carbomer, and Diethanolamine-based (DEA-based) detergents, may pose health risks, such as skin and eye irritation, as well as the formation of carcinogenic nitrosamines. In contrast, natural emulsifiers are portrayed as safer options due to their compatibility with biological systems, ready availability, and ability to operate at the molecular level within living organisms. Natural substances, often resembling macromolecular structures found in nature, can be metabolized and recognized by the body, thus decreasing the likelihood of adverse reactions and inflammatory responses. The shift towards natural emulsifiers underscores the increasing importance of prioritizing safety in product formulations. However, it is essential to ensure that natural alternatives undergo comprehensive testing to verify their safety and effectiveness. Regulatory agencies like the National Toxicology Program (NTP) play a crucial role in evaluating the potential risks associated with synthetic compounds and establishing standards to safeguard consumer health. Overall, the statement emphasizes the necessity for heightened awareness and regulation concerning the use of synthetic emulsifiers and detergents, as well as the growing demand for safer, natural substitutes.

9.6 Methods of preparation

9.6.1 Trituration method

A mortar and pestle can be employed on a small scale, but its effectiveness is restricted. Small electric mixers can be used to get around these limitations, but caution must be used to prevent excessive air entrapment.

We have the following three types of methods for the preparation of emulsion [48]:

9.6.1.1 Dry gum method

Measure the oil very accurately. Pour it into flat-bottomed, perfectly dry mortar completely. Measure the aqueous vehicle. Weight the acacia gum. Place the gum on the oil and mix gently for just long enough to disperse the lumps. Add water at once. Stir continuously but lightly in one direction until the mixture thickens under the pestle. Now triturate heavily. When cracking sounds are heard, the emulsion is called primary emulsion, but triturating is done for 2–3 min until a white emulsion is obtained. Finally, remaining aqueous media is added to make up the volume.

9.6.1.2 Wet gum method

This is also known as English methods. It is older, slower, and not much reliable technique. The following steps are performed while going through this method. Triturate the acacia with water to form a mucilage. Add oil in small amounts with constant rapid and light trituration. Triturate vigorously when all the oil has been added for a few minutes. Final volume is made up of water. The quality of an emulsion made using the wet gum method or the dry gum approach may usually be improved by homogenizing it with a hand homogenizer.

9.6.1.3 Bottle method or Forbes method

For the preparation of emulsion of volatile and other non-viscous oils, this method is frequently applied. This method is merely a variation of the dry gum or wet gum method except that a large bottle or flask is used as the mixing device.

9.6.2 Low-energy emulsification

The low-energy emulsification method utilizes the internal chemical energy constituting the system. Emulsion formation takes place by changing the temperature or composition of the emulsion.

9.6.2.1 Phase inversion method

Phase inversion is an energy-saving, room-temperature emulsification technique that produces an emulsion that is quite stable and is believed to contain a finely dispersed internal phase. Phase inversion refers to a phenomenon that occurs when agitated oil in water emulsion reverts to water in oil and vice versa. A phenomenon always accompanying phase inversion is the spontaneous change in the surfactants' arrangements at the oil-water interface. Phase inversion can be brought about by changing the temperature of the system, by changing the volume fraction of the phases, by adding salts, or by imposing particular flows, e.g., extensional flow. This process is defined by some authors as a catastrophic event because it appears to be a sudden and dramatic change in morphology caused by a gradual change in experimental conditions. Achieving phase inversion by changing temperature is the so-called phase inversion temperature (PIT) method, where miscibility of the two phases plays a key role. Another way is to gradually increase the amount of the minor phase up to the emulsion inversion point (EIP) (also known as phase inversion concentration, PIC). Emulsification via phase inversion is widely used in the fabrication of cosmetic products, pharmaceutical products (e.g., vesicles for drug delivery), foodstuffs, and detergents.

Other low-energy emulsification methods are the membrane emulsification method, spontaneous emulsification method, and solvent displacement method. The main drawback of low low-energy emulsification method is its restriction to only specific oils and emulsifiers.

9.6.3 High-energy emulsification methods

High-energy emulsification methods are commonly in vogue for emulsion production. Various types of equipment are available to affect droplet breakup and emulsification either in the laboratory or in production. These devices are mechanical stirrers, homogenizers, ultrasonifiers, microfluidizers, and colloid mills. The most important factor involved in the preparation of an emulsion is the degree of shear and turbulence required to produce a given dispersion of liquid droplets. The amount of agitation required depends on the total volume of liquid to be mixed, the viscosity of the system, and the interfacial tension at the oil-water interface. The latter two factors are determined by the emulsion type, the phase ratio, and the type and concentration of emulsifiers. For this reason, no single method of dispersion can be used for all emulsions, and conversion from one method to another is difficult.

9.7 HLB system

Generally, each emulsifying agent has a hydrophilic portion and a lipophilic portion with one or the other being more or less predominant and influencing the type of emulsion. If an emulsifying agent is predominantly hydrophilic, it tends to form an O/W emulsion. If it is predominantly lipophilic, it will form a W/O emulsion. The HLB of emulsifying agents has been expressed in terms of a number called HLB (HLB) scale or HLB system. These values are assigned between 1 and 20.

If HLB value is less, e.g.,

Span 80 = 4.3
Span 60 = 4.7
Span 20 = 5

It means that the lipophilic part is dominant. As more lipophilic and less hydrophilic parts or characters are more soluble in oils, so they form W/O emulsion [49].

But if HLB value is more, e.g.,

Tween 80 = 15.0
Tween 60 = 14.9
Tween 20 = 16.7

It means that the hydrophilic part is more than the lipophilic part, and the emulsifier is more soluble in water; thus, O/W emulsion is formed [49]. Usually for the preparation of W/O emulsion, the emulsifying agent should have an HLB of 3–8 [50]. An O/W emulsion, on the other hand, is formed with an HLB is from 8 to 18 [51]. But if the value is b/w 0–3, then these emulsifying agents are antifoaming agents. As soaps, when used as emulsifiers, form foam, so we use a very small number of antifoaming agents that will dissolve foam. If HLB value is between 10 and 18, then the emulsifiers are used as stabilizers (Tab. 9.2). In practice, we usually use a combination of Span and Tween so that we may have the required HLB value for the stability of the emulsion (Tab. 9.3).

Tab. 9.2: Activity and HLB value of various surfactants.

S. no.	Activity	HLB value
1.	Antifoaming agents	0–3
2.	Emulsifiers (W/O)	3–6
3.	Wetting agents	7–9
4.	Emulsifiers (O/W)	8–18
5.	Solubilizers	15–20
6.	Detergents	13–16

This can be explained by the following example. Suppose an emulsion containing:
Liquid paraffin 35%
Wool fat 01%
Wool alcohol 01%
Emulsifier 07%
Water (q.s.) 100%

HLB values of liquid paraffin wool fat and wool alcohol are 12.1, 11, and 15, respectively. In the formulation of emulsion, total oily phase is 37% (35% + 1% + 1%).
But liquid paraffin out of this is:

$$\text{Liquid paraffin} = \frac{35}{37} \times 100 = 94.5\%$$

So, wool fat and wool alcohol both are 2.7% each. Now HLB values can be calculated as follows:
First actual values are:
Liquid paraffin = 12.1
Wool fat = 11
Wool alcohol = 15
But now we have:

$$\text{Liquid paraffin} = \frac{94.6}{100} \times 12.1 = 11.4$$

$$\text{Wool fat} = \frac{2.7}{100} \times 11 = 0.3$$

$$\text{Wool alcohol} \quad = \frac{2.7}{100} \times 15 = 0.4$$

So, required HLB value for the emulsifier will be $11.4 + 0.3 + 0.4 = 12.1$.

For this, we have to select a pair of emulsifiers.

Span 80 has an HLB value = 4.3

Tween 80 has an HLB value = 15

Let suppose we use span X g

Then, Tween should be $(1–X)$ g. Now:

$$12.1 \quad = \quad 4.3x(\text{Span}) + 15(1 - x)\text{Tween}$$

$$12.1 \quad = \quad 4.3x(\text{Span}) + 15(1 - x)\text{Tween}$$

$$12.1 \quad = 15 = 4.3x - 15x$$

$$+2.9 \quad = \quad +10.7x$$

$$x \quad = \quad \frac{2.9}{10.7} = 0.2$$

So, for Tween = $\quad 1 - x = 1 - 0.2 = 0.8$

So, Span 80 should be 20% and Tween 80 should be 80% for a 12.1 HLB value.

But the total amount of emulsifier is 7%; therefore, amounts of Tween and Span are:

For Span:

$$\frac{7}{100} \times 20 = 1.4\% \text{ of total formulation}$$

For Tween:

$$7 = 1.4 = 5.6\% \text{ of total formulation}$$

9.8 Stability of emulsions

One crucial factor affecting the stability of emulsion-based products is their emulsion stability, which is challenging to assess easily. When an emulsion experiences the phenomenon of its internal phases forming clusters of droplets, it is deemed to be physically unstable. When sizable clusters or conglomerates of droplets ascend or descend within the emulsion, they gather to create a dense layer of the internal phase either at the top or bottom. If any portion or the entirety of the liquid from the internal phase separates and appears unemulsified at either the top or bottom of the, emulsion stability is decreased. Emulsion instability can be categorized into four phenomena: flocculation, creaming, coalescence, and breaking (Fig. 9.6).

Tab. 9.3: HLB values of various emulsifiers.

S. no.	Agents	HLB value
1.	Ethylene glycol	1.5
2.	Span tristearate (Span 65)	2.1
3.	Propylene glycol monostearate	3.4
4.	Sorbitan monooleate (Span 80)	4.3
5.	Sorbitan monostearate (Span 60)	4.7
6.	Acacia	8.0
7.	Tragacanth	13.2
8.	Polyoxyethylene sorbitan monostearate (Tween 60)	14.9
9.	Polyoxyethylene sorbitan monooleate (Tween 80)	15.0
10.	Polyoxyethylene sorbitan monolaurate (Tween 20)	16.7
11.	Sodium lauryl sulfate	40.0

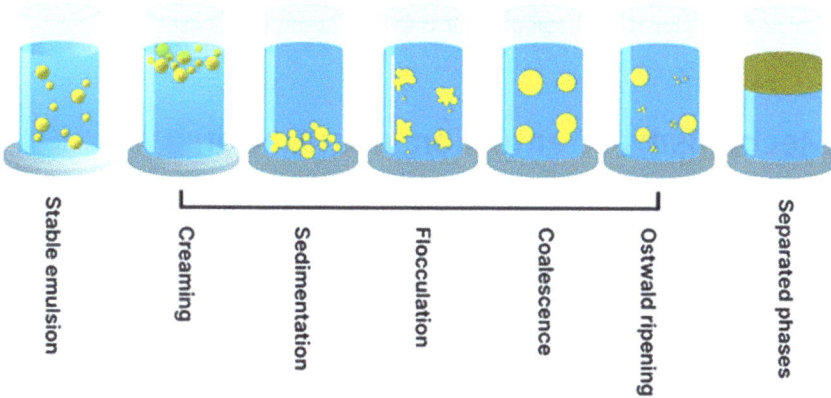

Fig. 9.6: Various forms of emulsion instability.

9.8.1 Flocculation

Flocculation refers to the gathering of small particles into larger aggregates that can be dispersed again through agitation. It's a reversible process where droplets remain intact. Flocculation is seen as the initial stage of coalescence because an abundance of surfactant in the continuous phase can cause emulsion droplets to merge. The depletion effect, primarily caused by excess surfactant, is the main driver behind droplet flocculation. This effect can be described as a system containing surplus surfactant forming micelles.

9.8.2 Creaming

Creaming occurs when the dispersed phase separates from the system, rising to form a layer atop the continuous phase. During creaming, the dispersed phase maintains its globular form, allowing it to be re-dispersed through vigorous shaking. Increasing the viscosity of the continuous phase can help mitigate creaming. O/W emulsions typically experience upward creaming when the dispersed phase globules are less dense than the continuous phase. Conversely, W/O emulsions encounter downward creaming when the dispersed phase globules are denser than the continuous phase.

9.8.3 Coalescence (synonyms: breaking or cracking)

A more nuanced form of emulsion instability, coalescence arises when the mechanical or electrical barrier fails to hinder the amalgamation of increasingly larger droplets. To prevent coalescence, stabilization can be attained by incorporating high boiling point or high-molecular-weight components into the continuous phase.

9.9 Mechanisms of emulsion stabilization

Emulsions are present across diverse industries, spanning from food and pharmaceuticals to petroleum production and refining. Their stability is attributed to the presence of an interfacial barrier that prevents the coalescence of dispersed water droplets. Typically, this stability is governed by the interface film and the mechanism of surfactant adsorption. Four primary mechanisms stabilize emulsions: electrostatic repulsion, steric repulsion, the Marangoni-Gibbs effect, and thin film stabilization. Subsequent sections delve into the details of each of these mechanisms.

9.9.1 Electrostatic repulsion

Electrostatic force arises from the interaction between the electrical double layers surrounding charged droplets, which serves to hinder droplet contact (Fig. 9.7). This mechanism primarily occurs through the adsorption of an ionic surfactant. However, electrostatic repulsion does not significantly contribute to the stabilization of W/O emulsions due to the low dielectric constant of the continuous phase. Instead, the stability of the emulsion stems from the robustness of the interfacial film enveloping the dispersed droplets, rather than from electrostatic forces.

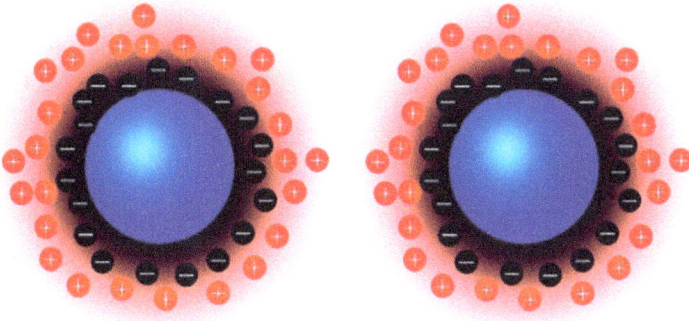

Fig. 9.7: Electrostatic stabilization of emulsion.

9.9.2 Steric repulsion

Steric stabilization is commonly observed in systems stabilized by non-ionic surfactants and polymers. In this mechanism, the dispersed water droplets are coated with surfactant molecules, and the surfactant tails adsorbed onto the particle surface hinder close droplet contact. This type of stabilization is primarily responsible for stabilizing water-in-oil (W/O) emulsions, as these stabilizers typically consist of a non-polar segment with a high affinity for oil and a polar segment with a strong affinity for the aqueous phase or water (Fig. 9.8).

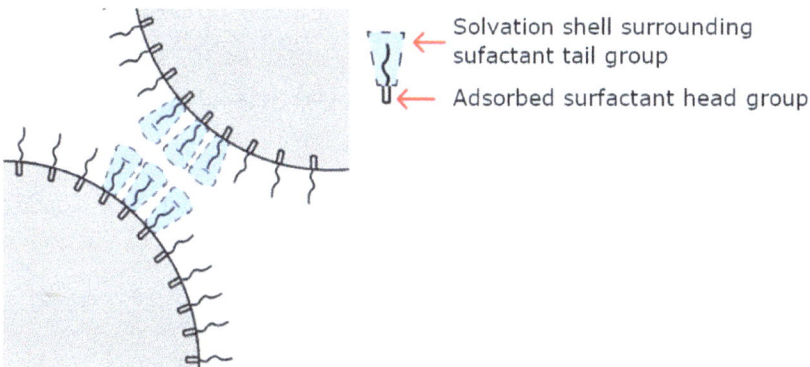

Solvation shell surrounding sufactant tail group

Adsorbed surfactant head group

Fig. 9.8: Schematic representation of steric stabilization.

9.9.3 Marangoni-Gibbs effect

The Marangoni-Gibbs effect stabilizes emulsions by impeding the drainage of the continuous phase between two adjacent droplets. This phenomenon arises from the alteration of droplet surface area as they draw closer. When droplets converge, they create a parallel surface, prompting the film layer to attempt drainage. Furthermore, the stability of the emulsion in this process is attributed to the mechanism of surfactant adsorption and the interfacial film.

9.9.4 Thin film stabilization

During this stage, a sturdy and viscoelastic film envelops the water droplets, halting their coalescence. This process can be intricate, contingent upon asphaltene chemistry, solvency, and the dynamics of diffusion and adsorption.

9.10 Emulsion stability assessment

The stability of emulsions needs to be considered concerning both the physical stability of the emulsion system under examination and the chemical stability of its components.

9.10.1 Macroscopic examination

The rate of creaming or coalescence over a specific period provides an evaluation of the physical stability of the emulsion. This is determined by dividing the volume of the separated creamed part of the emulsion by the total volume of the product.

9.10.2 Determination of particle size and particle count/globule size analysis

Examining alterations in the average particle size is a key parameter for evaluating emulsion stability. This analysis is conducted using optical microscopy, the Andresen apparatus, and the Coulter counter apparatus.

9.10.3 Determination of viscosity/viscosity changes

Alterations in globule size or quantity, or the migration of emulsifying agents during aging, can be identified through variations in apparent viscosity. Emulsions exhibit non-Newtonian flow behavior; hence, in O/W emulsions, this leads to an immediate rise in viscosity.

9.10.4 Determination of electrophoretic properties

Zeta potential serves as a vital parameter in evaluating emulsion stability because particle electric charges influence flocculation rates. Electrostatically, emulsion stabilization occurs through mutual repulsion between the electrical double layers of both phases. This form of stability is highly sensitive to the ionic strength of solutions; as the electrolyte concentration rises, the electrical double layer compresses and the distance of electrostatic repulsion decreases, leading to flocculation.

9.11 Microemulsions

Microemulsions are thermodynamically stable, transparent, isotropic mixtures of oil, and water, which are stabilized by surfactant [52].

Size range of globules is 10–100 nm or 100–1,000 Å. If we are using an emulsifying agent of HLB 15–18, they form an O/W emulsion. However, if HLB value is low, then we have W/O emulsion. The essential difference between emulsion and microemulsion is particle size and their stability; the emulsions are kinetically stable, whereas the microemulsions are thermodynamically stable [53].

9.11.1 Advantages of microemulsions

The following are a few advantages of microemulsions (Tab. 9.4):
- They are more rapidly and efficiently absorbed.
- Physically look transparent.
- Can be given orally and also topically.
- If given transdermally, absorption is enhanced.
- Can be given parentally.
- Can be used as artificial RBC$_S$.
- Target the cancerous cells because in the ordinary emulsion, globule size varies, but here, we have nearly constant size.

Tab. 9.4: Comparison between emulsions and microemulsions.

Properties	Macroemulsions	Microemulsions
Appearance	Cloudy	Transparent
Structure	Static	Dynamic
Particle size	1–100 µm	10–100 nm
Stability	Kinetically stable	Thermodynamically stable
Phases	Biphasic	Monophasic
Viscosity	Highly viscous	Low viscosity with Newtonian behavior
Co-surfactant	No	Yes
Contact position	Direct oil/water contact at the surface	No direct oil-in-water contact surface

9.11.2 Methods of preparation of microemulsions

To prepare a microemulsion system, the drug is dissolved in the lipophilic portion [11, 54]. The oil and water phases are combined by slowly adding surfactant and co-surfactant while stirring continuously. The amount of surfactant and co-surfactant to be applied, as well as the percentage of oil phase that can be assimilated (Tab. 9.5), are calculated using a pseudo-ternary phase diagram [12]. Using an ultrasonicator can help obtain the necessary size range for scattered globules. It is then left to equilibrate. To make gel, add a gelling agent to the microemulsion, as described above. Carbomers (cross-linked polyacrylic acid polymers) are the most often utilized gelling agents.

Tab. 9.5: Example of the content of various ingredients in microemulsions.

S. no.	Compound	Function	Content in microemulsions (%) O/W	Content in microemulsions (%) W/O
1.	Sodium lauryl sulfate	Surfactant	13	10
2.	1-Pentanol	Co-surfactant	8	25
3.	Xylene	Oil	8	50
4.	Water	Lipid	71	15

9.12 Nanoemulsions

A thermodynamically unstable colloidal dispersion of two immiscible liquids, one of which is distributed as tiny, spherical droplets with a radius, less than 100 nm in size, referred to as nanoemulsions.

Particles in typical emulsion usually have mean radii of between 100 nm and 100 mm. Conventional emulsions and nanoemulsions constitute metastable systems, which means that they will eventually degrade as a result of several destabilization mechanisms, such as coalescence, flocculation, gravity separation, and Ostwald ripening (Tab. 9.6). Emulsions, on the other hand, can be made to stay stable for a specific duration of time. Considering the significant influence of molecular interactions on structural and rheological behavior, the addition of stabilizers and co-adjuvant molecules may have a significant impact on the physical characteristics and stability of nanoemulsions [55–57].

Tab. 9.6: Comparison between macroemulsions, microemulsions, and nanoemulsions.

Types of emulsion	Macroemulsion	Microemulsion	Nanoemulsion
Shape	Spherical	Spherical, lamellar	Spherical
Size	1–100 µm	10–100 nm	20–500 nm
Stability	Thermodynamically unstable and weakly kinetically stable	Thermodynamically stable	Thermodynamically unstable and kinetically stable
Methods of preparation	High- and low-energy methods	Low-energy methods	High- and low-energy methods
Polydispersity	Often high (>40%)	Typically, low (<10%)	Typically, low (<10–20%)

9.12.1 Methods of preparation of nanoemulsions

The preparation of nanoemulsions typically involves the use of so-called "high-energy" methods and specialized equipment, such as high-pressure homogenizers or ultrasound generators, to provide sufficient energy to enhance the oil/water interfacial area to produce submicronic droplets. Low-energy methods make use of the intrinsic physicochemical features of the constituents to manufacture submicronic droplets, and they also enable the formulation of nanoemulsions, but through spontaneous emulsification without the need for any apparatus or energy. In short, the process only consists of mixing two liquid phases at room temperature, one containing a lipophilic phase into which a hydrophilic surfactant is solubilized to form a homogeneous liquid (plus potentially a solvent, polymer, or drug) and the other an aqueous phase, which can be pure water. Once these two liquids are brought into contact, the hydrophilic species contained in the oily phase (i.e., surfactants) are rapidly solubilized into the aqueous one, inducing the demixtion of the oil in the form of nanodroplets, immediately stabilized by the amphiphiles. The droplet size of nanoemulsions is easily controllable as a function of the oil/ surfactant weight ratio. Low-energy spontaneous emulsification is an effective approach for producing stable and concentrated emulsion droplets. Size ranges from 10 to 300 nm. The simple formulation method of mixing two liquid phases leads to confusion between nanoemulsions and microemulsions. The EIP and PIT are the two most commonly utilized low-energy methods. EIP is often referred to as PIC [58].

9.13 Gel and magma

Gels are defined as a semisolid system consisting of a dispersion made up of either small inorganic particles (like minerals) or organic (like polymer) microparticles. The gel is a semisolid preparation dispersion of small and large molecules in aqueous liquid carriers. Gels are semisolid systems in which interactions (physical or covalent) occur between colloidal particles within a liquid carrier. They add a gelling agent, which gives them a gelatinous consistency. The gels are formed using synthetic polymers, such as Carbomer 934 and cellulose (such as hydroxypropyl cellulose and hydroxypropyl methyl cellulose). Tragacanth gum, pectin, and natural agar gum are used in the gel formula.

9.13.1 Single-phase gels

Gels in which macromolecules are distributed through the liquid in such a way that no apparent boundaries exist between them and the liquid are called single-phase gels [59].

9.13.2 Two-phase system

If the gel system consists of floccules of small, distinct particles, the gel is called a two-phase gel or magma, or milk. Gels and magmas are both colloidal dispersions as they consist of particles of colloidal dimension [59].

Terminologies related to gel
Most commonly used terms for characteristics of gels include imbibition, swelling, syneresis, thixotropy, and xerogel.

Imbibition: It is the absorption of a specific amount of liquid without a noticeable increase in volume.

Swelling: It is the process by which a gel absorbs liquid and increases its volume. Only liquids that solvate gels can produce swelling. The pH and presence of electrolytes influence the swelling of protein gel.

Syneresis: It occurs when the contact between the particles of the dispersed phase is sufficiently strong that, when standing, the dispersing medium is forced out in droplets and the gel contracts. Unpredictability occurs in both aqueous and non-aqueous gels.

Thixotropy: It is the process in which a semisolid form is converted into a liquid form of the substance, and then the liquid form is converted into semisolid form by variation of temperature or agitation.

Sol: It is a common name used for the dispersion medium in which solid/liquid is dispersed in:

    Water; then called hydrosol
    Alcohol; then called alcosol
    Air; then called aerosol

9.13.3 Preparation of gel and magma

Gels and magmas can be prepared by the following methods:

Precipitation
Most of the magmas and gels are prepared by precipitation. The dispersed phase is freshly prepared by precipitation in order to achieve a fine degree of subdivision of the particles and gelatinous character to these particles. When the solutions of inorganic substances react to form insoluble compounds having high attraction for water, they attract water to form a gel.

Hydration
Some magmas and gels can be prepared by direct hydration of the inorganic chemicals in water.

9.13.4 Bentonite gel

It is chemically 5% w/v colloidal hydrated aluminum silicate. It can be prepared by two methods. In the first method, gel is prepared in a blender when bentonite is added directly to the purified water while the machine is running. In this method, bentonite is sprinkled on hot water, in portions. When first portions are wetted, then add another portion. Leave it for 24 h so that it swells up to 12 times. In the second method, no stirring is used. This is used as an emulsifying agent. Bentonite is a thixotropic gel. The thixotropy occurs only when the bentonite concentration is somewhat above 41%.

9.13.5 Aluminum hydroxide gel

It is an aqueous suspension of a gelatinous precipitate composed of insoluble aluminum hydroxide and hydrated aluminum oxide, 4%. Here chemical reaction is performed to get the required product. Aluminum is usually obtained from $AlCl_3$ or alum. Some sweeteners and flavorings may be added. It is a white viscous suspension used to counteract the hyper acidity of the stomach and is also helpful in healing peptic wounds by covering the inflamed and untreated area. Its disadvantages are that it is an anterior heal drug, i.e., it has constipating effects. Another is that it interferes with the bioavailability of tetracycline by chelating with it in gastrointestinal tract (GIT). It is stored in air-tight containers, and freezing should be avoided.

9.13.6 Milk of magnesia

It is composed of 7–8.5% of magnesium hydroxide.
    It is usually prepared by two types of reaction:
– Hydration of magnesium oxide:
    – $MgO + H_2O \rightarrow Mg(OH)_2$
    – It is the most commonly used method.
– Reaction of magnesium sulfate and sodium hydroxide: Precipitates obtained are very fine when we use a diluted solution.

Then hydration of precipitates results in product formation. The precipitates are washed with water to remove sodium sulfate. It is a white, opaque, and viscous preparation. It should be shaken well because water is usually separated from it on standing. It has a pH of 10, so it has a bitter taste. To solve this problem, add 0.1% citric acid and 0.051 flavoring oil. It is also an antacid but causes diarrhea. So, a combination of this and $Al(OH)_3$ gel is used, which acts as both antidiarrheal and antacid.

References

[1]   Khan BA, Akhtar N, Khan HMS, Waseem K, Mahmood T, Rasul A, et al. Basics of pharmaceutical emulsions: A review. African Journal of Pharmacy and Pharmacology. 2011; 5(25): 2715–25.
[2]   Fatima M, Sheraz MA, Ahmed S, Kazi SH, Ahmad I. Emulsion separation, classification and stability assessment. RADS Journal of Pharmacy and Pharmaceutical Sciences. 2014; 2(2): 56–62.
[3]   Sarathchandraprakash N, Mahendra C, Prashanth S, Manral K, Babu U, Gowda D. Emulsions and emulsifiers. The Asian Journal of Experimental Chemistry. 2013; 8: 30–45.
[4]   Benichou A, Aserin A, Garti N. Double emulsions stabilized with hybrids of natural polymers for entrapment and slow release of active matters. Advances in Colloid and Interface Science. 2004; 108: 29–41.

[5]   Tadros TF. Emulsion science and technology: A general introduction. Emulsion Science and Technology. 2009; 1(1): 1–55.

[6]   Sinko PJ. Martin's physical pharmacy and pharmaceutical sciences: Lippincott Williams & Wilkins; Philadelphia, USA, 2023.

[7]   AboulFotouh K, Allam AA, El-Badry M, El-Sayed AM. Role of self-emulsifying drug delivery systems in optimizing the oral delivery of hydrophilic macromolecules and reducing interindividual variability. Colloids and Surfaces B: Biointerfaces. 2018; 167: 82–92.

[8]   Garg T, Rath G, Goyal AK. Comprehensive review on additives of topical dosage forms for drug delivery. Drug Delivery. 2015; 22(8): 969–87.

[9]   Ali MY, Tariq I, Ali S, Amin MU, Engelhardt K, Pinnapireddy SR, et al. Targeted ErbB3 cancer therapy: A synergistic approach to effectively combat cancer. International Journal of Pharmaceutics. 2020; 575: 118961.

[10]   Ali MY, Tariq I, Sohail MF, Amin MU, Ali S, Pinnapireddy SR, et al. Selective anti-ErbB3 aptamer modified sorafenib microparticles: In vitro and in vivo toxicity assessment. European Journal of Pharmaceutics and Biopharmaceutics. 2019; 145: 42–53.

[11]   Tariq I, Pinnapireddy SR, Duse L, Ali MY, Ali S, Amin MU, et al. Lipodendriplexes: A promising nanocarrier for enhanced gene delivery with minimal cytotoxicity. European Journal of Pharmaceutics and Biopharmaceutics. 2019; 135: 72–82.

[12]   Idrees M, Rahman N, Ahmad S, Ali M, Ahmad I. Enhance transdermal delivery of flurbiprofen via microemulsions: Effects of different types of surfactants and cosurfactants. DARU Journal of Pharmaceutical Sciences. 2011; 19(6): 433.

[13]   Gupta A, Bajaj S. Introduction to pharmaceutics-II: CBS; New Delhi, India, 2008.

[14]   Garti N. Hydrocolloids as emulsifying agents for oil-in-water emulsions. Journal of Dispersion Science and Technology. 1999; 20(1–2): 327–55.

[15]   Pasquali RC, Taurozzi MP, Bregni C. Some considerations about the hydrophilic–lipophilic balance system. International Journal of Pharmaceutics. 2008; 356(1–2): 44–51.

[16]   Porter CJ, Trevaskis NL, Charman WN. Lipids and lipid-based formulations: Optimizing the oral delivery of lipophilic drugs. Nature Reviews Drug Discovery. 2007; 6(3): 231–48.

[17]   Sreenathkumar S. Current updates on global phytoceuticals and novel phyto drug delivery system in herbal medicine. In: Natural drugs from plants: IntechOpen; London, England, 2021.

[18]   Jones DS. FASTtrack Pharmaceutics dosage form and design: Pharmaceutical press; London, England, 2016.

[19]   Tackie-Otoo BN, Mohammed MAA, Yekeen N, Negash BM. Alternative chemical agents for alkalis, surfactants and polymers for enhanced oil recovery: Research trend and prospects. Journal of Petroleum Science and Engineering. 2020; 187: 106828.

[20]   Navare B, Thakur S, Nakle S. A review on surfactants: Role in skin, irritation, SC damage, and effect of mild cleansing over damaged skin. International Journal of Advanced Research, Ideas and Innovations in Technology. 2019; 5: 1077–81.

[21]   Mukherjee A, Mullick A, Vadthya P, Moulik S, Roy A. Surfactant degradation using hydrodynamic cavitation based hybrid advanced oxidation technology: A techno economic feasibility study. Chemical Engineering Journal. 2020; 398: 125599.

[22]   Al-Adham I, Haddadin R, Collier P. Types of microbicidal and microbistatic agents. Russell, Hugo & Ayliffe's: Principles and Practice of Disinfection, Preservation and Sterilization. Wiley-Blackwell; New Jersey, USA, 2013; pp. 5–70.

[23]   Salager J-L. Surfactants types and uses. FIRP Booklet. Mérida, Venezuela, 2002; 300.

[24]   Sakač N, Madunić-Čačić D, Karnaš M, Đurin B, Kovač I, Jozanović M. The influence of plasticizers on the response characteristics of the surfactant sensor for cationic surfactant determination in disinfectants and antiseptics. Sensors. 2021; 21(10): 3535.

[25] Obłąk E, Futoma-Kołoch B, Wieczyńska A. Biological activity of quaternary ammonium salts and resistance of microorganisms to these compounds. World Journal of Microbiology and Biotechnology. 2021; 37(2): 22.

[26] Ríos F, Fernández-Arteaga A, Lechuga M, Fernández-Serrano M. Ecotoxicological characterization of polyoxyethylene glycerol ester non-ionic surfactants and their mixtures with anionic and non-ionic surfactants. Environmental Science and Pollution Research. 2017; 24: 10121–30.

[27] Schick MJ. Nonionic surfactants: Physical chemistry: CRC Press; Florida, USA, 1987.

[28] Schott H. Hydrophile-lipophile balance and cloud points of nonionic surfactants. Journal of Pharmaceutical Sciences. 1969; 58(12): 1443–49.

[29] Brooks BW, Richmond HN. Phase inversion in non-ionic surfactant – Oil – Water systems – I. The effect of transitional inversion on emulsion drop sizes. Chemical Engineering Science. 1994; 49(7): 1053–64.

[30] Fine R. A review of ethylene oxide condensation with relation to surface-active agents. Journal of the American Oil Chemists' Society. 1958; 35(10): 542–47.

[31] Quinquenet S, Ollivon M, Grabielle-Madelmont C, Serpelloni M. Polymorphism of hydrated sorbitol. Thermochimica Acta. 1988; 125: 125–40.

[32] Mittal N, Athony RL, Bansal R, Kumar CR. Study of performance and emission characteristics of a partially coated LHR SI engine blended with n-butanol and gasoline. Alexandria Engineering Journal. 2013; 52(3): 285–93.

[33] Snelling JR, Scarff CA, Scrivens JH. Characterization of complex polysorbate formulations by means of shape-selective mass spectrometry. Analytical Chemistry. 2012; 84(15): 6521–29.

[34] Iwata H, Shimada K, Iwata H, Shimada K. Practice of designing cosmetics formulations. Formulas, Ingredients and Production of Cosmetics: Technology of Skin-and Hair-Care Products in Japan. Springer; London, England, 2013; 113–217.

[35] De Villiers M. Surfactants and emulsifying agents.In: A practical guide to contemporary pharmacy practice: Thompson JE, Editor. Philadelphia: Lippincott Williams and Wilkins; 2009. pp. 251.

[36] Fruijtier-Pölloth C. Safety assessment on polyethylene glycols (PEGs) and their derivatives as used in cosmetic products. Toxicology. 2005; 214(1–2): 1–38.

[37] Kothekar SC, Ware AM, Waghmare JT, Momin S. Comparative analysis of the properties of Tween-20, Tween-60, Tween-80, Arlacel-60, and Arlacel-80. Journal of Dispersion Science and Technology. 2007; 28(3): 477–84.

[38] Depraetere P, Florence A, Puisieux F, Seiller M. Some properties of oil-in-water emulsions stabilized with mixed non-ionic surfactants (Brij 92 and Brij 96). International Journal of Pharmaceutics. 1980; 5(4): 291–304.

[39] Casiraghi A, Selmin F, Minghetti P, Cilurzo F, Montanari L. Nonionic surfactants: Polyethylene glycol (peg) ethers and fatty acid esters as penetration enhancers. Percutaneous Penetration Enhancers Chemical Methods in Penetration Enhancement: Modification of the Stratum Corneum. Springer; London, England, 2015; 251–71.

[40] Shah J, Arslan E, Cirucci J, O'Brien J, Moss D. Comparison of oleo-vs petro-sourcing of fatty alcohols via cradle-to-gate life cycle assessment. Journal of Surfactants and Detergents. 2016; 19: 1333–51.

[41] Atta AM, Allohedan HA, El-Mahdy GA. Dewatering of petroleum crude oil emulsions using modified Schiff base polymeric surfactants. Journal of Petroleum Science and Engineering. 2014; 122: 719–28.

[42] Mansour FR, Arrua RD, Desire CT, Hilder EF. Non-ionic surface active agents as additives toward a universal porogen system for porous polymer monoliths. Analytical Chemistry. 2021; 93(5): 2802–10.

[43] Barry B, Saunders G. Rheology of systems containing cetomacrogol 1000 – Cetostearyl alcohol. I. Self-bodying action. Journal of Colloid and Interface Science. 1972; 38(3): 616–25.

[44] Kazmi SJA. Distribution and antimicrobial activity of preservatives in solubilized and emulsified systems: University of British Columbia; Vancouver, Canada, 1974.

[45] Wenande E, Garvey L. Immediate-type hypersensitivity to polyethylene glycols: A review. Clinical and Experimental Allergy. 2016; 46(7): 907–22.

[46] Casiraghi A, Di Grigoli M, Cilurzo F, Gennari CGM, Rossoni G, Minghetti P. The influence of the polar head and the hydrophobic chain on the skin penetration enhancement effect of poly (ethylene glycol) derivatives. American Association of Pharmaceutical Scientists PharmSciTech. 2012; 13: 247–53.

[47] Stjerndahl M, Lundberg D, Holmberg K. Cleavable surfactants. Surfactant Science Series. 2003; 317–46.

[48] Bharath S. Pharmaceutics: Formulations and dispensing pharmacy: Pearson Education India; Noida, 2013.

[49] Gadhave A. Determination of hydrophilic-lipophilic balance value. International Journal of Scientific Research. 2014; 3(4): 573–75.

[50] Schmidts T, Dobler D, Guldan A-C, Paulus N, Runkel F. Multiple W/O/W emulsions – Using the required HLB for emulsifier evaluation. Colloids and Surfaces A: Physicochemical and Engineering Aspects. 2010; 372(1–3): 48–54.

[51] Pant A, Jha K, Singh M. Role of excipient's HLB values in microemulsion system. Journal of Pharmaceutical and Biological Sciences. 2019; 14: 1–6.

[52] Mehta S, Kaur G. Microemulsions: Thermodynamic and dynamic properties. Thermodynamics. Intechopen; London, England, 2011; 381–406.

[53] Paul BK, Moulik SP. Uses and applications of microemulsions. Current Science. 2001; 80(8): 990–1001.

[54] Tariq I, Ali MY, Janga H, Ali S, Amin MU, Ambreen G, et al. Downregulation of MDR 1 gene contributes to tyrosine kinase inhibitor induce apoptosis and reduction in tumor metastasis: A gravity to space investigation. International Journal of Pharmaceutics. 2020; 591: 119993.

[55] de Oca-ávalos JMM, Candal RJ, Herrera ML. Nanoemulsions: Stability and physical properties. Current Opinion in Food Science. 2017; 16: 1–6.

[56] Amin MU, Ali S, Ali MY, Tariq I, Nasrullah U, Pinnapreddy SR, et al. Enhanced efficacy and drug delivery with lipid coated mesoporous silica nanoparticles in cancer therapy. European Journal of Pharmaceutics and Biopharmaceutics. 2021; 165: 31–40.

[57] Ali S, Amin MU, Tariq I, Sohail MF, Ali MY, Preis E, et al. Lipoparticles for synergistic chemo-photodynamic therapy to ovarian carcinoma cells: In vitro and in vivo assessments. International Journal of Nanomedicine. 2021; 16: 951–76.

[58] Gupta A, Eral HB, Hatton TA, Doyle PS. Nanoemulsions: Formation, properties and applications. Soft Matter. 2016; 12(11): 2826–41.

[59] Farjami T, Madadlou A. An overview on preparation of emulsion-filled gels and emulsion particulate gels. Trends in Food Science & Technology. 2019; 86: 85–94.

Shams ul Hassan, Muhammad Yasir Ali, Tanzeela Awan, Humaira Gul,
Rabia Munir, Arshad Mahmood

10 Suspension

10.1 Introduction

It is a heterogeneous and unstable dispersed system of two components [1] defined as
"it is an unstable dispersed system of two components in which one is insoluble solid
(dispersed phase) and the other is liquid or semi-solid (dispersion medium)." The particles of the dispersed phase are in the range of 0.5–5 μm [2]. These formulations are
extensively employed within the pharmaceutical industry for numerous therapeutic
and practical objectives. Some suspensions are available in a ready-to-use form that
has already been distributed through a liquid vehicle, with or without stabilizers and
other pharmaceutical additives. Some other suspensions, however, are provided as a
dry powder combination containing the drug plus a suitable suspending agent that,
when diluted and agitated with a precise amount of vehicle, forms a suspension.

10.2 Reasons for the formulation of pharmaceutical suspension

The formulation of pharmaceutical suspensions is mainly required for drug stability,
to boost patient compliance, provide flexibility in dosing, and improve the bioavailability of specific poorly soluble drugs.

Enhancing the stability of the active pharmaceutical ingredient (API) is one of the
main reasons for the development of pharmaceutical suspensions. Some medicines
can be chemically unstable in solution forms, primarily due to hydrolysis or oxidation
in water. These medications can be manufactured in a suspended or undissolved
state, which improves stability, since the level of surface area that contacts the solvent
is minimized. Antibiotics and corticosteroids, for example, are more stable in suspension than in solution [3, 4].

Pharmaceutical suspensions provide a versatile strategy for dosing. This is especially helpful for children, whose weight and health can have a significant impact on
the dosage needed. Instead of being restricted to tablets with preset dosages, care-
givers and healthcare professionals can easily modify the dosage with suspensions by
measuring precise volumes using syringes or spoons.

Another important aspect is the improvement of patient acceptance and compliance, especially in pediatric and geriatric groups; the addition of flavoring agents,
sweeteners, and colorants can enhance the taste and appearance of the suspension.

https://doi.org/10.1515/9783111438108-010

On the contrary, the development of drugs is otherwise unappealing in flavor or hard to swallow in solid formats.

Suspensions enable the adjustment of drug release profiles. Through the manipulation of particle size and the application of appropriate suspending agents or coatings, a sustained or controlled drug release effect can be realized. This can lead to a reduction in dosing frequency and the maintenance of therapeutic drug levels over an extended timeframe, which proves beneficial in chronic treatment strategies. Furthermore, suspensions play an essential role in the solubility of insoluble API in water, facilitating precise dosing and improving bioavailability by changing the surface area available for dissolution once they enter the body.

When solid dosage forms are not suitable, suspensions are sometimes used for topical, parenteral, or rectal administration. To provide prolonged therapeutic benefits from a single dose, depot injections and long-acting intramuscular formulations, for example, are usually created as sterile suspensions.

10.3 Advantages of pharmaceutical suspension

The following are a few advantages of suspension [5]:
- Suspensions serve to improve the stability of pharmaceuticals that are chemically unstable in a solution, notably those that are at risk of hydrolysis or oxidation, for example, Procaine penicillin G.
- Suspensions are notably appropriate for the administration of drugs that are poorly soluble in water, facilitating their delivery in a bioavailable and dispersed manner.
- Suspensions facilitate flexible dosing, which is particularly advantageous in pediatric, geriatric, and veterinary contexts.
- Insoluble medications in suspension are less prone to activate taste receptors, thus facilitating the masking of bitter or unwanted tastes, for example, chloramphenicol.
- Liquid suspensions are typically simpler to ingest than tablets or capsules, making them more suitable for both children and elderly individuals.
- By adjusting the size of particles or the components of the formulation, it is possible to achieve controlled or sustained release of drugs, for example, protamine zinc-insulin suspension.

10.4 Disadvantages of pharmaceutical suspension

The following are a few disadvantages of suspension:
- Over time, suspensions are likely to experience sedimentation, Ostwald ripening, and caking, which can result in irregular dosing if they are not shaken properly.

– Chemical destabilization of suspensions may occur as a consequence of changes in pH and their inherent chemical degradation.
– This is in contrast to solid dosage forms like tablets, where suspensions generally exhibit a shorter expiration period once they are reconstituted or opened.
– To achieve a stable and homogeneous suspension, it is essential to carefully select the appropriate suspending agents, surfactants, and preservatives.
– The suspension could yield an imprecise dose due to the irregular distribution of particles and the method of dosing with a spoon.
– Water-based suspensions can encourage the development of microorganisms, which consequently necessitates the use of preservatives that could result in allergic responses in some people.
– Some suspensions might present a grittiness in the mouth, which can negatively impact patient acceptance.

10.5 Properties of a good suspension

The following are a few properties of a good suspension:
– Dispersed particles should not settle down readily, and the settled particles should disperse easily upon shaking.
– The particle should be controlled enough not to form a cake or compact mass on sedimentation.
– The suspension must possess an appropriate color, scent, and flavor to encourage patient adherence, particularly in oral formulations.
– It should have a good flow property so be easily removed from the container, leaving less content stuck to the container.
– It should be chemically stable under the influence of pH change and temperature.
– Suspension of internal use must be palatable.
– It should be prepared with suitable preservatives (e.g., parabens and benzoates) to prevent microbial contamination.
– It should be free from large particles or particles having a gritty feeling if it is for external use.

10.6 Applications of pharmaceutical suspension

– Suspension is typically used for drugs that are insoluble or poorly soluble, for example, prednisolone suspension.
– Improve drug stability or limit deterioration, for example, oxytetracycline suspension.

- To conceal the bitter taste of an undesirable drug. As an example, consider chloramphenicol palmitate suspension.
- Drug suspensions can be prepared for topical administration, such as Calamine lotion.
- Formulating a suspension for parenteral administration can control medication absorption.
- Vaccines are usually available in suspension form, for example, cholera vaccination.
- X-ray contrast agents are also available as suspensions. For example, barium sulfate is used to examine the alimentary canal.

10.7 Types of suspensions

10.7.1 According to the route of administration

Oral suspensions
These suspensions are administered orally; thus, they contain flavoring and sweetening agents to mask the unpleasant taste of the drug.

Topical suspensions
These are applied to the exterior surface of the body and must be devoid of any gritty particles that could cause skin irritation.

Parenteral suspensions
These suspensions are delivered by a parenteral route, such as intravenously or intramuscularly; thus, they should be sterile and free of extraneous particles.

Ophthalmic suspensions
Ophthalmic suspensions are utilized for the treatment of eye disorders; thus, they should be sterile and devoid of any foreign particles.

10.7.2 On the proportion of solid particles

Dilute suspension
The concentration of solid particles in dilute suspensions ranges between 2% and 10% w/v.

Concentrated suspension
In concentrated suspension, the concentration of solid particles is around 50% w/v.

10.7.3 Based on the particle size of solids

Nanosuspensions
These suspensions contain particles with sizes ranging from 1 nm to 1 μm. They have incredibly fine particles that are hard to settle. They usually have a translucent or transparent appearance and are commonly used in specialized formulations, like injectable or ophthalmic preparations.

Fine suspensions
These suspensions contain particles with sizes ranging from 1 μm to 10 μm. These are the most commonly encountered pharmaceutical suspensions. The particles in these suspensions settle gradually and can be readily redispersed with a little shaking.

Coarse suspensions
These suspensions contain particles larger than 10 μm. They have a propensity to settle faster and may create sediment that is difficult to redistribute. To keep them stable, strong suspending agents may be required.

10.7.4 Based on the kinetic nature of solids

Suspensions can be divided into two categories depending on the electrokinetic properties of solid particles:
- Flocculated system
- Deflocculated system

Flocculated system
In a flocculated system, the particles of the dispersed phase settle to form loose aggregates that form a network-like structure [6]. In this form of suspension, particles settle down easily and can be easily redistributed when shaken [7]. The upper clear layer created during sedimentation is known as the supernatant layer [2]. The flocculated suspension intended for oral, parenteral, ophthalmic, and external application may not be elegant because it is difficult to remove from the bottle, and the floccules adhere to the container (Fig. 10.1).

Deflocculated system
In this type of system, the particles from the dispersed phase settle down and form the compact mass. In this type of suspension, the particles of the dispersed system are not easily settled down, and also not dispersed easily [5]. The compact mass formed is called a cake and is formed due to the reactions of the particles with one another [8]. In a flocculated system, the separated layer is turbid, so this is a more elegant system than a flocculated system (Fig. 10.1). After some time, no change is observed, but a supernate turbid layer is present, which contains a cake at the bottom (Tab. 10.1).

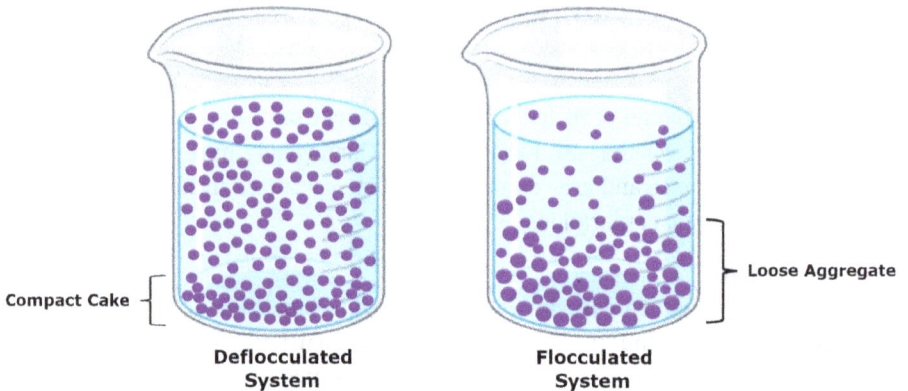

Fig. 10.1: Deflocculated and flocculated suspension showing compact cake and loose aggregates.

Tab. 10.1: Comparison between flocculated and deflocculated suspensions.

S. no.	Flocculated	Deflocculated
1.	Particles generate loose aggregate.	Particles develop compact mass.
2.	The sedimentation rate is high.	The sedimentation rate is low (2).
3.	They are easily redistributed.	They are not easily redistributed.
4.	Floccules adhere to the walls of the bottle.	Floccules don't adhere to the walls of the bottle.
5.	The suspension does not have a pleasing appearance.	Suspension has a pleasant appearance.
6.	The particle size is greater.	The particle size is smaller.
7.	Attractive forces are operative among the particles.	Repulsive forces are operative among the particles.

10.8 Suspensoid

Suspended solids are a type of colloidal system in which there is little attractive force between the colloidal particles and the dispersion medium. They are part of a larger group of colloids, which are mixtures of microscopic dispersed insoluble particles of one substance suspended in another. The dispersed phase in the case of suspension is also called a suspensoid. Now question about suspension that arises is why we make suspensions of these substances. One reason may be the chemical stability of these suspensoids [9]. Some drugs are very unstable when present in the solution form, so for these drugs, suspension is the liquid dosage form. In most patients, liquid dosage forms are preferred over solid dosage forms as they are easily swallowable. One other purpose for which we make a liquid preparation in the suspension form is that some drugs have a bitter taste in solution form [10]. To cover the bitter taste, we can make suspensions to mask the bad taste of the drug [11]. Initially, we have to make so derivatives that are insoluble in water, then we are able to form suspensions, e.g., chloramphenicol, which is water soluble, but Chloramphenicol palmitate is water insoluble, so we can easily make its suspensions [12].

10.8.1 Properties of suspended solids

Suspensions are characterized by particle size, surface tension, and viscosity almost identical to the solvent. Suspended particles carry a specific charge that determines the stability of the suspended solid. These can be characterized by techniques such as suspension characterization using a rheometer [13].

10.8.2 Advantages of suspensoid

- Suspensions containing suspensoids are useful drug delivery systems for poorly soluble therapies.
- It can be used in paints, where the paint molecules remain suspended in the mixture, causing it to be applied uniformly.
- Suspensions improve the taste and smell of medicines and make them easier to swallow, especially for children.
- It can prevent drug degradation caused by microbial, oxidative, or hydrolytic activity.

10.8.3 Antacids

Here are some examples of suspensoids, antacids, and their uses.

Aluminum hydroxide
This is a common antacid and is often combined with magnesium hydroxide. It is not exactly aluminum hydroxide ($Al(OH)_3$), but rather an aluminum "hydroxycarbonate."

Aluminum hydroxide + magnesium carbonate
This combination is used to neutralize the stomach acid [14].

Aluminum hydroxide + magnesium hydroxide
This combination is also used to neutralize the stomach acid [15].

Magnesium hydroxide (milk of magnesia)
Magnesium hydroxide is used to neutralize the stomach acid.

10.8.4 Antibiotics

Beta-lactam antibiotics
These include penicillin derivatives (penams), cephalosporins (cephems), monobactams, and carbapenems. They work by inhibiting cell wall synthesis [16].

Vancomycin
This antibiotic is used to treat serious bacterial infections and works by inhibiting cell wall synthesis.

Aminoglycosides
These are usually considered bactericidal, although they may be bacteriostatic in some organisms.

Tetracyclines, sulfonamides, spectinomycin, trimethoprim, chloramphenicol, macrolides, and lincosamides
These are examples of bacteriostatic antibiotics that limit bacterial growth by interfering with bacterial protein production, DNA replication, or other aspects of bacterial cellular metabolism.

10.8.5 Antifungal

Azoles
Azoles are some of the most commonly used antifungals. They interfere with an enzyme that is important for creating the cell membrane of the fungus. There are two subgroups of azole antifungals: imidazoles and triazoles [17].

Examples of imidazole antifungals include:
Ketoconazole: Used for skin and hair infections, candida infections of the skin and mucous membranes, blastomycosis, and histoplasmosis.

Clotrimazole: Used for skin and mucous membrane infections.

Miconazole: Used for skin and mucous membrane infections.
Examples of triazole antifungals include:

Fluconazole: Used for *Candida* infections, including mucosal, systemic, and invasive infections; cryptococcosis.

Itraconazole: Used to treat aspergillosis, blastomycosis, histoplasmosis, *Candida* mucosal infections, coccidioidomycosis (off-label), and onychomycosis.

Posaconazole: Used to treat aspergillosis (unapproved treatment) and mucosal and invasive *Candida* infections.

Voriconazole: Used to treat aspergillosis, mucosal or invasive *Candida* infections, infections caused by *Fusarium* species.

Isavuconazole: Used for aspergillosis and mucormycosis.

Polyenes
Polyenes kill fungal cells by making the fungal cell wall more porous, making the fungal cell prone to bursting.

Nystatin: A topical and oral antifungal that treats candida infections involving the mouth or skin.

Amphotericin B: Treats a wide variety of fungal diseases, including invasive aspergillosis, blastomycosis, candidiasis, coccidioidomycosis, cryptococcal meningitis, cryptococcosis, histoplasmosis, mucormycosis, sporotrichosis, and others.

10.8.6 NSAIDs

Below are some examples of suspension NSAIDs.
Ibuprofen: Ibuprofen is used to treat mild to moderate pain from a wide variety of conditions, including headaches, menstruation, migraines, osteoarthritis or rheumatoid arthritis, sprains and strains, and toothaches [18].

Naproxen: Naproxen is also used to treat mild to moderate pain from a wide range of conditions.

Indomethacin: The COX-2 selectivity of Indomethacin is less than 50 times that of COX-1.
Diclofenac: Diclofenac is used for arthritis and mild to moderate pain in adults.

10.8.7 Bronchodilators

Below are some examples of suspension bronchodilators.
Anticholinergic drugs: used to treat chronic obstructive pulmonary disease (COPD) and asthma. An example is ipratropium.

Theophylline: This is a bronchodilator drug used to open the airways in the lungs and reduce asthma symptoms such as wheezing, coughing, and shortness of breath [19].

10.9 Features of dispersed phase

10.9.1 Wetting

The presence of hydrophilic agents facilitates the wetting of hydrophilic materials. One hydrophilic material that can successfully encourage the formation of a suspension of a hydrophilic component is sodium carboxymethyl cellulose, or Na-CMC. Bentonite and aluminum magnesium silicate, which aid in dispersing the dispersed phase, further increase the viscosity. They can also be categorized as hydrophilic or lipophilic. Aluminum hydroxide and hectorite make it easier to wet materials like light liquid paraffin that are difficult to dissolve in water [20].

10.9.2 Wet point

The amount of a liquid required for wetting a substance is called the wet point, which is taken in mL/100 g.

10.9.3 Flow point

The volume or amount of liquid used for pourability is called the flow point. It is taken in mL/100 g.

10.10 Preparation of suspensions

We have two types of methods for the preparation of suspension.

10.10.1 Dispersion method

A pharmaceutical suspension can be prepared using the dispersion method (general method).

Dissolve the soluble substances. The soluble materials should be dissolved in the proper volume of diluent (vehicle). Before volume correction, the solid therapeutic agent is mixed and distributed throughout the vehicle. Development of vehicles wetting and dispersion of the solid phase are facilitated by a vehicle formulation that is as wet and dispersible as feasible. Wetting and suspending agents can be used to accomplish this.

The speed at which the solid therapeutic agent is mixed is a crucial aspect of the production of pharmaceutical products. The coagulated suspension can function at high mixing speeds because of this system's quasi-plasticity (shear thinning). However, mixing at a high speed can produce a sticky product (also known as a spreading flow) if the formulation design is flawed and the adhesion properties are insufficient [21].

The complexity of various pharmaceuticals has increased with the advent of international pharmaceutical regulations. Thanks to the creation of innovative forms and packaging materials, pharmacists now have access to a wide range of packaging materials that are directly related to the acceptability and durability of the dose to be recognized. For instance, in order to maximize shelf life, industrial pharmacists must understand the relationships between constituent qualities, while retail pharmacists must not cut corners on dosage form storage. As a result, labeling and preservation requirements are essential for both patients and pharmacists. Suspensions of oral medications are frequently packaged in wide-mouth containers with sufficient headspace to ensure uniform mixing. There should be enough air space above the liquid to enable simple pouring and shaking. A 5-mL medicinal spoon or oral syringe should be used when taking the suspension orally.

The most crucial extra indication for suspensions is "shake well before use." Usually, some precipitation is expected from the drug. Shaking the bottle redistributes the medication and allows the patient to accurately measure the dose. "Store in a cool place." Fluctuations in temperature might have an impact on suspension stability. It might be necessary to refrigerate particular suspensions, such as one made by reconditioning dry powders. Instantly prepared and reconstituted suspensions have a limited shelf life. It should usually be fresh or freshly prepared, and the best time before that is between 1 and 4 weeks. Some official formulas have an expiration date, but many do not. Pharmacists may need to determine the expiration date of a particular drug product on the basis of its ingredients and anticipated storage conditions. Remanufactured items come with specified storage conditions in the manufacturer's datasheet. Remanufactured items come with specified storage conditions in the manufacturer's datasheet.

Examples of available commercial suspensions are:

Amoxicillin oral suspension: This is an antibiotic used to treat many bacterial infections.

Mebendazole oral suspension: This is an anthelmintic drug used to treat various parasitic infestations.

Albendazole oral suspension: This is also an anthelmintic drug and is used to treat some infections caused by parasites such as pig tapeworm and dog worm.

Sephrazine suspension (25 mg/mL): This is an antibiotic used to treat many bacterial infections.

Sucralfate suspension: This drug is used to treat and prevent intestinal ulcers.

Natamycin eye drops: This is an antifungal medication used to treat fungal eye infections.

Suspension for diffusible solids
Some insoluble powders are light and easily wettable; hence, they are readily homogenized upon shaking with water, giving each dosage an appropriate distribution by distributing it equally throughout the liquid for an adequate duration of time. Such substances are known as diffusible solids [18]. The following steps are observed during this preparation [19].

When making a mixture, begin by making sure that every ingredient is finely powdered, separating any that are still coarse. Subsequently, combine the insoluble powders in a mortar, starting with the ingredient with the smallest bulk and progressively diluting it with those with larger bulk. After the powders are well combined, add enough vehicle to make a paste. Next, add the volatile oil, which needs to dissolve in the vehicle before being added to the paste. Transfer the mixture into a graduated cylinder after adding more vehicle until it is pourable. Lastly, add all the other ingredients, like volatile oil, and adjust the volume with the external phase if necessary.

By this method, we make suspension of diffusible solids, e.g., rhubarb powder, calcium carbonate, magnesium trisilicate, light kaolin, and light magnesium carbonate.

Suspension for in-diffusible solids
These are those solids that don't remain evenly distributed for long enough to ensure the uniform dose [20]. The simplest way to solve the problem is to increase the viscosity of the vehicle by adding a thickening agent [22], e.g., in diffusible solids are aspirin, chalk, calamine, phenobarbitone, and aluminum hydroxide compressed gel (Tab. 10.2).

Tab. 10.2: Composition of aluminum hydroxide compressed gel suspension.

S. no.	Ingredients	Concentration
1.	Aluminum hydroxide compressed gel	326.8 g
2.	Sorbitol syrup	282.0 mL
3.	Simple syrup	93.0 mL
4.	Glycerin	25.0 mL
5.	Methyl paraben	0.9 g
6.	Propyl paraben	0.3 g
7.	Flavorant	q.s.
8.	Purified water	1,000.0 mL

According to the USP monograph, suspension of amorphous aluminum hydroxide is known as aluminum hydroxide gel, in which the hydroxide is partially replaced by carbonate. The pH of the gel is 5.5–8.0 as measured by potentiometry. The gel is stored in a closed container. And freezing should be avoided. Microbial limits meet the test requirements for total aerobic microorganisms not exceeding 100 CFU/mL and absence of *E. coli*. The acid neutralizing capacity is greater than 65.0% of the expected mEq value, calculated from the results of tests [23, 24]:

- The expected acid neutralizing capacity per mg of $Al(OH)_3$ is 0.0385 mEq.
- Wet aluminum hydroxide compressed gel with glycerin thoroughly.
- Add sorbitol syrup, simple syrup, and flavorant to this.
- Take methyl paraben and propyl paraben in a separate beaker and dissolve them in water. Propyl paraben is dissolved in hot water.
- Add these two solutions to the first solution.
- Finally, make up the volume.

10.10.2 Precipitation method

This method is used for those substances dissolved in water. In this method, precipitates of these materials are formed by the pH-based precipitation method and the anti-solvent precipitation method [25].

Examples

Estradiol (hormone) is soluble in an alkaline medium, and if acid is added, then precipitation occurs [26].

Protamine Zn solution is also prepared by precipitation of pH change type.

10.11 Pharmaceutical application of suspension

10.11.1 Oral suspension

The amount of medicine absorbed by the GIT determines an oral suspension's bioavailability. The makeup of oral suspensions varies. The buffer capacity, pH, and viscosity of the vehicle differ. To put it briefly, choosing the right drug particle sizes, where to optimize absorption, particle densities, and vehicle viscosities, will optimize the bioavailability of the oral solution [27]. Although aluminum hydroxide also exists in tablets, its higher rate of absorption causes it to get suspended. The flavor of chloramphenical palmitate is not as bitter as it would be in solution [28].

10.11.2 Dry powder or granules

These contain antibiotics, e.g., ampicillin, amoxicillin, cefaclor, cephalexin, and erythromycin. These powders contain all other components of the suspension, e.g., preservatives, antibiotics, flavorants, colorant, and dispersing agent [29].

10.11.3 Suspension for external use

Skin lotion may be present in solution, suspension, and emulsion forms [30, 31]. Semisolid or solid pastes have a greater amount of dispersed phase. The outer phase is liquid paraffin [32].

10.11.4 Parental suspensions

Powders with a particle size of 5–10 μm are a suitable candidate for the suspension intended for intramuscular and subcutaneous [33, 34]. Water, organic solvents (glycerine, propylene glycol, and polyethylene glycol), and non-hazardous oils (sesame, peanut, and olive) are suitable suspension vehicles for subcutaneous and intramuscular delivery. Drugs that have been dissolved in water quickly permeate into bodily tissue, leaving an accumulation of remaining drug at where it was injected. When it came to parenteral suspensions, the medication's capacity to dissolve at the injection site governed how quickly the drug entered the bloodstream and how bioavailable it became.

10.11.5 Ophthalmic and ear suspension

The bioavailability of the ocular solution is affected by the vehicle's viscosity and the suspended medication particles' sizes. Polymers such as polyvinyl alcohol, polyvinyl pyrrolidone, and cellulose derivatives are utilized to provide the necessary viscosity, hence delaying the process of particle settling. Particles either dissolve or are expelled from the eye at the lid margin. Particle size has an impact on both retention and dissolution rates inside the sac of the conjunctiva. For the purpose of particle retention, its dissolution and corneal absorption times must be less than the drug's residence duration in the conjunctival sac. The retained particles sustain the reaction as the medication is absorbed and the particles disintegrate, whereas the saturated state of a suspension taken by the cornea produces the first response. When an elevated particle content suspension is used, more drug mass is left in the cul-de-sac after the applied volume is drained [35, 36]. The extra particles then dissolve in the tear fluids, releasing an additional drug, such as an antibiotic or corticosteroid called hydrocortisone, which is transported across the cornea into the aqueous humor [37].

10.11.6 Suspensions for immunity

Vaccines are usually prepared in the form of a suspension. The cholera vaccine is a dead bacterium of cholera absorbed on the surface of certain substances [38], e.g., aluminum hydroxide $Al(OH)_3$; similarly, we have toxicities for diphtheria [39, 40].

10.11.7 Suspension for diagnosis

$BaSO_4$ is used for alimentary canal examination [41]. The enema device was used to provide the medicine solution through the rectum. Enemas have a big capacity (50–100 mL) and a restricted patient pool. Rectal blood flow and absorption from rectal tissues determine the rate of absorption of rectal solution.

10.12 Physical stability of suspension

For the stability of the suspension, we should ensure that
- the particle settles as slowly as possible and
- when sediment is formed, it should be redistributed on shaking [42, 43].

In a suspension concentrate, it is possible to identify three different physical processes that may lead to instability in a particulate system, namely a change in particle size due to Ostwald ripening, particle aggregation, and particle sedimentation. An experimental investigation of the sedimentation of monodisperse colloidal silica spheres with grafted octadecyl chains with three different interaction potentials is presented. Small particles (0.27 µm) behaved as hard spheres in cyclohexane, but larger ones (0.60 and 0.94 µm) are weakly flocculated by van der Waals attractions. The smallest particles (0.08 µm) in hexadecane are strongly flocculated by attractions between the octadecyl layers.

10.12.1 Cake formation

This size decrease is extremely tiny, costly, and time-consuming. Increased surface area often results in uncontrollably clumped particles due to the high surface free energy created. More physical stability would be in keeping with this size decrease. Reduced particle size, however, can be a clear drawback if the product is used for IM use and is intended for long-term effects. The deflocculation due to the existence of a shared surface charge is an additional issue. Re-dispersion is frequently almost impossible due to the gradual merging of the distinct particles that initially made up the tightly packed sediments. We refer to this phenomenon as caking. The existence of repelling or attractive forces between the particles determines whether a cake forms. There is a slight force of attraction between particles when they are a specific distance apart, but they are unable to develop a strong attraction and go closer because of a strong aversion. This energy barrier is breached, though, and a strong attraction between the particles that make up the cake is formed. As a result, the system flocculates. Since particles settle on the bottom to create high-density sediment, it is hypothesized that this large energy barrier is surpassed in the sediments of the deflocculated system [44]. Increasing the viscosity of the dispersion medium was another way to slow down sedimentation, in addition to the first one. But this made it difficult to shake the substance toward the bottle and redistribute the settling particles.

Cake formation refers to the process where particles in suspension are separated by forming a filter cake on the filter media. This is a common phenomenon in filtration processes, especially in the separation of solids and liquids [23]. The resistance of the filter cake and the filter medium causes a specific pressure drop, which subsequently defines the energy demand of the process. The micromechanics of filter cake formation, including interactions between particles, fluid, other particles, and the filter medium, must be considered to properly describe pore plugging, filter cake growth, and consolidation. The cake can have elastic-plastic properties, and the filtration characteristics can change significantly depending on the model parameters characterizing these properties [24]. The parameters of plasticity in terms of particle concentration and permeability mainly affect the corresponding indicators, i.e., the distribution of particle concentration and the relative permeability of the cake. The cake formation process can be studied using a variety of methods, including coupled discrete element method coupled with computational fluid dynamics (CFD-DEM) allows for the precise prediction of filter cake development during separation between solids and The created model considers particle-fluid and particle-particle interactions in a four-way connection. The filtering characteristics of suspensoids can be strongly impacted by the cake-building process [24]. According to the Stokes equation,

$$dx/dt = d^2(pi-pe)g/18\ \eta$$

where dx/dt is the rate of sedimentation, d is the diameter of the particle, pi is the density of the particle, pe is the density of the medium, g is the force of gravity, and η is the viscosity of the medium.

So, reduction in particle size of the suspensoid is beneficial in the stability of the suspension in that sense that if the sedimentation rate is less, the suspensoid will remain in suspended form, but size reduction should not be such that particles tend to form a compact cake. The result may be that the cake resists breakup upon shaking and forms rigid aggregates which are less suspensible. Particle shape also affects the formation of cake, e.g., symmetrically barrel-shaped particles of calcium carbonate produce a more stable suspension than do asymmetric particles of the compound. Needle-shaped crystals form a compact cake on sedimentation, but barrel-shaped ones don't form such a cake and are then easily re-distributed on shaking. One common method of preventing the rigid cake formation from small particles is the intentional formation of a less rigid or loose aggregation of the particles held together by weak particle binding forces. These aggregates are called floccules and form a lattice-type structure that resists the compact mass formation. However, these are easily sedimented and then easily redistributed.

10.12.2 Controlled flocculation

Most suspending agents belong to the class of negatively charged hydrophilic colloids [45]. When added to a suspension containing positively charged flocculating agents, they form a massive structure of suspending agents. But phosphate ion agents, which are used to flocculate positively charged particles, are compatible with the commonly used suspending agents. The difficulty with negatively charged drugs (mentioned above) may be overcome by absorbing onto the drug particle certain agents that will reverse the surface charge from negative to positive. This may be accomplished by fatty acid amines.

When individual suspended particles come together to form loose, fluffy clusters called flocs, a phenomenon known as flocculation takes place. These flocs are cohesive because of weak van der Waals forces or bridging mechanisms that prevent the particles from becoming too close to one another and forming a stiff cake. A flocculated suspension settles more quickly and produces a soft sediment that is easily re-dispersed, as opposed to a deflocculated suspension, where particles settle more slowly and create a compact sediment.

A pharmaceutical suspension's objective is to keep the drug particles evenly distributed over the course of the product's shelf life so that precise and reliable dosing is possible. The particles in a deflocculated system, on the other hand, stay apart and settle slowly due to their small size. At first, this might seem beneficial, but eventually, the particles tend to pack tightly, creating a hard cake that is challenging to spread out. A flocculated system, on the other hand, settles more quickly but produces a porous sediment that is easily shaken back into a uniform mixture, increasing physical stability and decreasing the formation of cakes.

10.12.3 Deliberate flocculation

Deliberate flocculation involves altering the formulation environment or adding flocculating agents to encourage controlled floc formation. To maximize the overall effectiveness of the suspension, the primary goal is to strike a balance between the rate of sedimentation and the ease of redispersion.

10.12.3.1 Hydrophilic polymers

Because of their mesh-like interactions and connections, certain polymers can form loose aggregates called flocs. Hydrophilic polymers, such as xanthan gum, sodium alginate, and gelatin, can function as flocculants in this way. As loose sediments are produced, this tactic increases viscosity, which is associated with less sedimentation and a faster rate of redistribution. However, this method may not produce a satisfactory suspension when a much higher concentration of polymer is used.

10.12.3.2 Electrolyte addition

Electrolytes, including sodium, ammonium, calcium, and magnesium ions, are commonly employed to reduce the zeta potential of suspended particles. By neutralizing surface charges, these ions enable particles to come closer to one another and create flocs. Because of their increased charge density, multivalent cations – like calcium or magnesium ions (such as calcium nitrate, calcium chloride, magnesium hydroxide, and magnesium sulfate) – are especially effective. However, by reducing interparticle repulsion, monovalent electrolytes – which contain sodium and ammonium ions – such as sodium citrate, sodium chloride, sodium phosphate, and ammonium chloride, can also encourage flocculation. However, too many of these electrolytes might cause permanent aggregation and hard cake formation.

10.12.3.3 pH

Since pH levels frequently affect the surface charge on particles, flocculation can be improved by adjusting the pH to the isoelectric point, where the net surface charge is neutral. Particles are more likely to aggregate loosely at this point because the zeta potential is at its lowest. By adjusting the pH between 5 and 6, where its zeta potential is almost zero, aluminum hydroxide, a common component of antacids, can be flocculated.

10.12.3.4 Zeta potential

Most of the suspended particles have an electrical charge on their surfaces, which causes an ion double layer to form around them. As a result, the particles are kept apart and in a deflocculated state by a repulsive force. The strength of this repulsive force is measured by the zeta potential. A deflocculated system is created when a positive zeta potential produces strong repulsion. On the other hand, a flocculated system results from a negative zeta potential, which causes weaker repulsive forces.

Advantages of deliberate flocculation

- Consistent dosage is ensured by flocculated particles, which form loose, porous aggregates that are easily redispersed with light shaking.
- Over time, a flocculated suspension shows a more stable physical structure, which reduces the possibility of irreversible settling.
- The formation of a hard, compact sediment (cake), which is frequently impossible or very difficult to redistribute, is prevented by intentional flocculation.
- More user convenience is provided by easily redispersible suspensions, especially for elderly and pediatric patients who need a consistent dosage.
- Flocculation guarantees that each dose contains the appropriate amount of active ingredient by promoting uniform particle distribution upon redispersion.
- Because flocculated suspensions don't have to worry about forming permanent sediment over time; they are easier to process, fill, and store.

10.12.4 Flocculating agents

Particles in suspension that have acquired a surface charge will tend to repel one another, resulting in the formation of de deflocculated system. Solutions and suspensions of drugs are used widely in the pharmaceutical industry for the production of dosage forms for different routes of administration, for example, oral, parenteral, and inhalation. Pharmaceutical solutions and suspensions might appear to be simple formulations, but they can present many technical problems both for the manufacturing industry and for the individual pharmacist. Substances can be chemically unstable, insoluble in water, distasteful, etc. Suspensions are often used as a dosage form when the drug is insoluble in water and when the use of solubilizing agents is not possible. Based on the method of preparation, suspensions can be divided into two categories: flocculated and deflocculated systems. This can be explained as follows. Electrolytes are frequently used for deliberate flocculation of the particles. These ions reduce the electrical barrier between the particles and also form a bridge between the particles so as to link them. Thus, the articles are held in a loosely arranged structure in suspension. The use of electrolytes may be illustrated by the addition of monobasic potassium phosphate, a negative flocculating agent, to a suspension of bismuth subnitrate, the particles of which are positively charged. When monobasic potassium phosphate is added, the positive charge on particles is decreased to the point at which flocculation is increased to maximum, but if more amount is added, then again a deflocculated system is obtained because again a similar charge, but now a negative charge, is developed on the surface of the particles [46]. Deflocculated suspensions of coarse powders tend to cake as the individual particles settle out and form compact, cohesive sediments. Limited flocculation results in looser sediments because the settled-out flocs incorporate large amounts of the liquid-suspending medium. Tiny amounts of bentonite were added to the bismuth subnitrate solutions to obtain controlled flocculation. The interaction of the coarse, positively charged bismuth subnitrate particles in aqueous suspension with negatively charged, colloidally dispersed bentonite was investigated by measuring electrophoretic mobility, sedimentation volume, and viscosity. The gradual addition of bentonite dispersion to bismuth subnitrate suspensions first reduced the zeta potential of the bismuth subnitrate particles from +28 mV to zero, then inverted it, and finally caused it to level off at −20 mV for bismuth subnitrate-bentonite weight ratios below 200. Both anionic and nonionic detergents also cause the floccules to be formed. Non-ionic surfactants at appropriate concentrations can produce floccules [47].

10.13 In-process quality control tests for suspensions

In-process quality control tests ensure product stability, safety, and quality. They are as follows.

10.13.1 Phase test for appearance

Appearance tests are typically done on both the dispersed phase and the dispersion medium. The suspension is normally prepared using purified water. The syrup's purity, solid particle distribution, gum dispersion consistency, and water quality are often examined during this test. Rheological tests are carried out to guarantee that the medium has the required viscosity, enabling the development of a stable and re-dispersible solution. The dispersion medium's viscosity is determined before mixing the dispersed phase. The test results are compared to a standard reference, and action is taken if an issue arises [48].

10.13.2 Particle size of dispersed phase test

The drug particle size has a considerable impact on the stability of the final product. This test examines the particle size under a microscope. The drug particle size is compared to the desired particle size, and immediate action is taken if there is a difference [48–50].

10.13.3 Pourability test

This test is performed to make sure that the final composition is pourable and does not cause any issues when filling the container or when handled by patients.

10.13.4 pH test

The formulation's pH is critical to its stability. As a result, a variety of vehicles and suspension phases are monitored prior to and following mixing. Records are also kept regularly to ensure that the appropriate pH is maintained.

10.13.5 Final product assay test

This test determines if active components are distributed equally throughout the formulation. In this test, the sample is extracted and an assay is performed to determine the extent of homogeneity. When a fault is detected, it is corrected by closely monitoring the formulation procedures.

10.13.6 Zeta potential

The zeta potential gives information about the suspension's future stability. To detect zeta potential, utilize microelectrophoresis or a zeta meter [51–53].

10.13.7 Centrifugation test

The physical stability of the suspension is determined by performing a centrifugation test. Before packaging, a consistent color distribution and the lack of air globules are checked [45].

References

[1] Patel N, Kennon L, Levinson R. Pharmaceutical suspensions. The Theory and Practice of Industrial Pharmacy. 1986; 3: 479–501.
[2] Kumar RS, Yagnesh TNS. Pharmaceutical suspensions: Patient compliance oral dosage forms. World Journal of Pharmacy and Pharmaceutical Sciences. 2016; 7(12): 1471–537.

[3]   Ali MY, Tariq I, Ali S, Amin MU, Engelhardt K, Pinnapireddy SR, et al. Targeted ErbB3 cancer therapy: A synergistic approach to effectively combat cancer. International Journal of Pharmaceutics. 2020; 575: 118961.

[4]   Ali MY, Tariq I, Sohail MF, Amin MU, Ali S, Pinnapireddy SR, et al. Selective anti-ErbB3 aptamer modified sorafenib microparticles: In vitro and in vivo toxicity assessment. European Journal of Pharmaceutics and Biopharmaceutics. 2019; 145: 42–53.

[5]   Doye P, Mena T, Das N. Formulation and bio-availability parameters of pharmaceutical suspension. International Journal of Current Pharmaceutical Research. 2017; 9(3): 8–14.

[6]   Matthews B, Rhodes C. Some studies of flocculation phenomena in pharmaceutical suspensions. Journal of Pharmaceutical Sciences. 1968; 57(4): 569–73.

[7]   Sartin RD The measurement of caking tendencies in sulfa drug suspensions 1968.

[8]   Machkovech SM, Foster TP. Aqueous suspensions. Long acting injections and implants: Springer; London, England, 2011. pp. 137–51.

[9]   Peng F, Johnson DL, Effler SW. Suspensoids in New York city's drinking water reservoirs: Turbidity apportionment 1. JAWRA Journal of the American Water Resources Association. 2002; 38(5): 1453–65.

[10]   Rowell H. Principles of vehicle suspension. Proceedings of the Institution of Automobile Engineers. 1922; 17(1): 455–541.

[11]   Kumar KS, Bhowmik D, Srivastava S, Paswan S, Dutta AS. Taste masked suspension. The Pharma Innovation. 2012; 1(2): 1–7.

[12]   Banerjee S, Chakrabarti K, Haldar AK. Determination of chloramphenicol palmitate in pharmaceutical suspensions. Journal of Pharmaceutical Sciences. 1973; 62(11): 1841–44.

[13]   Howard-Williams C, editor. Suspensoids and turbidity: Chairman's summary. In: Perspectives in southern hemisphere limnology: Proceedings of a symposium, held in wilderness: South Africa: Springer; 1985. July 3–13, 1984.

[14]   Bonn D, Eggers J, Indekeu J, Meunier J, Rolley E. Wetting and spreading. Reviews of Modern Physics. 2009; 81(2): 739.

[15]   Zhang J, Zhou CH, Petit S, Zhang H. Hectorite: Synthesis, modification, assembly and applications. Applied Clay Science. 2019; 177: 114–38.

[16]   Chen Y-S, Hsiau -S-S. Cake formation and growth in cake filtration. Powder Technology. 2009; 192(2): 217–24.

[17]   Hennart S, Wildeboer W, Van Hee P, Meesters G. Stability of particle suspensions after fine grinding. Powder Technology. 2010; 199(3): 226–31.

[18]   Party P, Klement ML, Szabó-Révész P, Ambrus R. Preparation and Characterization of Ibuprofen Containing Nano-Embedded-Microparticles for Pulmonary Delivery. Pharmaceutics. 2023 Feb 6; 15(2): 545.

[19]   Arshady R, Ledwith A. Suspension polymerisation and its application to the preparation of polymer supports. Reactive Polymers, Ion Exchangers, Sorbents. 1983; 1(3): 159–74.

[20]   Okubo T. Amorphous solid-like distribution of polydisperse particles of colloidal clay and microcrystalline cellulose in deionized suspension. Colloid and Polymer Science. 1990; 268: 1159–66.

[21]   Ali S, Amin MU, Tariq I, Sohail MF, Ali MY, Preis E, et al. Lipoparticles for synergistic chemo-photodynamic therapy to ovarian carcinoma cells: In vitro and in vivo assessments. International Journal of Nanomedicine. 2021; 2021: 951–76.

[22]   Taylor N, Bagley E. Dispersions or solutions? A mechanism for certain thickening agents. Journal of Applied Polymer Science. 1974; 18(9): 2747–61.

[23]   Ajmal M, Demirci S, Siddiq M, Aktas N, Sahiner N. Betaine microgel preparation from 2-(methacryloyloxy) ethyl] dimethyl (3-sulfopropyl) ammonium hydroxide and its use as a catalyst system. Colloids and Surfaces A: Physicochemical and Engineering Aspects. 2015; 486: 29–37.

[24] Nail SL, White JL, Hem SL. Structure of aluminum hydroxide gel I: Initial precipitate. Journal of Pharmaceutical Sciences. 1976; 65(8): 1188–91.

[25] RUbin AJ, Johnson JD. Effect of pH on ion and precipitate flotation systems. Analytical Chemistry. 1967; 39(3): 298–302.

[26] Shareef A, Angove MJ, Wells JD, Johnson BB. Aqueous solubilities of estrone, 17β-estradiol, 17α-ethynylestradiol, and bisphenol A. Journal of Chemical & Engineering Data. 2006; 51(3): 879–81.

[27] Ajala TO, Silva BO. The effect of pharmaceutical properties on the acid neutralizing capacity of antacid oral suspensions. Journal of Pharmaceutical Investigation. 2015; 45: 433–39.

[28] LK P. Clinical pharmacology of two chloramphenicol preparations in children: Sodium succinate (IV) and palmitate (oral) eaters. Journal of Pediatrics. 1980; 96: 757.

[29] Hempenstall J, Irwin W, Po ALW, Andrews A. Antibiotic granules for reconstitution as syrups: Product uniformity and stability dependent upon reconstitution procedure. International Journal of Pharmaceutics. 1985; 23(2): 131–46.

[30] Buhse L, Kolinski R, Westenberger B, Wokovich A, Spencer J, Chen CW, et al. Topical drug classification. International Journal of Pharmaceutics. 2005; 295(1–2): 101–12.

[31] Lu GW, Gao P. Emulsions and microemulsions for topical and transdermal drug delivery. In: Handbook of non-invasive drug delivery systems: Elsevier; Edinburgh, England, 2010. pp. 59–94.

[32] Juch R, Rufli T, Surber C. Pastes: What do they contain? How do they work?. Dermatology. 1994; 189(4): 373–77.

[33] Brazeau G, Sauberan SL, Gatlin L, Wisniecki P, Shah J. Effect of particle size of parenteral suspensions on in vitro muscle damage. Pharmaceutical Development and Technology. 2011; 16(6): 591–98.

[34] Gulati N, Gupta H. Parenteral drug delivery: A review. Recent Patents on Drug Delivery & Formulation. 2011; 5(2): 133–45.

[35] Baer RL, Litt JZ. Treatment of otitis externa with hydrocortisone suspension. Journal of the American Medical Association. 1954; 155(11): 973–74.

[36] Laval J. Use of hydrocortisone (hydrocortone) acetate in ophthalmology. AMA Archives of Ophthalmology. 1953; 50(3): 299–302.

[37] Brummett RE, Harris RF, Lindgren JA. Detection of ototoxicity from drugs applied topically to the middle ear space. The Laryngoscope. 1976; 86(8): 1177–87.

[38] Kabir S. Critical analysis of compositions and protective efficacies of oral killed cholera vaccines. Clinical and Vaccine Immunology. 2014; 21(9): 1195–205.

[39] He P, Zou Y, Hu Z. Advances in aluminum hydroxide-based adjuvant research and its mechanism. Human Vaccines & Immunotherapeutics. 2015; 11(2): 477–88.

[40] Goto N, Kato H, Maeyama J-I, Eto K, Yoshihara S. Studies on the toxicities of aluminium hydroxide and calcium phosphate as immunological adjuvants for vaccines. Vaccine. 1993; 11(9): 914–18.

[41] Miller RE. Barium sulfate suspensions. Radiology. 1965; 84(2): 241–51.

[42] Luckham PF. The physical stability of suspension concentrates with particular reference to pharmaceutical and pesticide formulations. Pesticide Science. 1989; 25(1): 25–34.

[43] Auzerais FM, Jackson R, Russel W, Murphy W. The transient settling of stable and flocculated dispersions. Journal of Fluid Mechanics. 1990; 221: 613–39.

[44] Shojaei A, Arefinia R. Analysis of the sedimentation process in reactive polymeric suspensions. Chemical Engineering Science. 2006; 61(23): 7565–78.

[45] Edman P. Pharmaceutical formulations – Suspensions and solutions. Journal of Aerosol Medicine. 1994; 7(s1): S–3–S–6.

[46] Funck JA, Schnaare RL, Schwartz JB, Sugita ET. Some observations on conductivity and flocculation in bismuth subnitrate suspensions. Drug Development and Industrial Pharmacy. 1991; 17(14): 1957–70.

[47] Mundhada D, Chandewar A. An overview on cationic surfactant. Research Journal of Pharmaceutical Dosage Forms and Technology. 2015; 7(4): 294–300.

[48] Amin MU, Ali S, Ali MY, Tariq I, Nasrullah U, Pinnapreddy SR, et al. Enhanced efficacy and drug delivery with lipid coated mesoporous silica nanoparticles in cancer therapy. European Journal of Pharmaceutics and Biopharmaceutics. 2021; 165: 31–40.

[49] Tariq I, Hassan H, Ali S, Raza SA, Shah PA, Ali MY, et al. Ameliorative delivery of docetaxel and curcumin using PEG decorated lipomers: A cutting-edge in-vitro/in-vivo appraisal. Journal of Drug Delivery Science and Technology. 2024; 97: 105814.

[50] Ali S, Amin MU, Ali MY, Tariq I, Pinnapireddy SR, Duse L, et al. Wavelength dependent photo-cytotoxicity to ovarian carcinoma cells using temoporfin loaded tetraether liposomes as efficient drug delivery system. European Journal of Pharmaceutics and Biopharmaceutics. 2020; 150: 50–65.

[51] Tariq I, Pinnapireddy SR, Duse L, Ali MY, Ali S, Amin MU, et al. Lipodendriplexes: A promising nanocarrier for enhanced gene delivery with minimal cytotoxicity. European Journal of Pharmaceutics and Biopharmaceutics. 2019; 135: 72–82.

[52] Arshad S, Asim MH, Mahmood A, Ijaz M, Irfan HM, Anwar F, et al. Calycosin-loaded nanostructured lipid carriers: In-vitro and in-vivo evaluation for enhanced anti-cancer potential. Journal of Drug Delivery Science and Technology. 2022; 67: 102957.

[53] Amin MU, Ali S, Ali MY, Fuhrmann DC, Tariq I, Seitz BS, et al. Co-delivery of carbonic anhydrase IX inhibitor and doxorubicin as a promising approach to address hypoxia-induced chemoresistance. Drug Delivery. 2022; 29(1): 2072–85.

Mulazim Hussain Asim, Muhammad Yasir Ali, Rabia Jabeen,
Sadia Hakim, Uzma Saher, Muhammad Irfan, Nisar ur Rahman,
Saeed Ahmad, Udo Bakowsky

11 Aerosols

11.1 Introduction

Aerosols are a collection of particles that can be of any size, which may be suspended in air or a gaseous medium. They may contain one or more active ingredients which, upon actuation, emit a fine dispersion of liquid or gaseous medium that is emulsified or suspended in propellants (liquefied or compressed gas). The particle size of the released product is usually below 50 μm, 2 μm reaches the alveolar sac, and 6 μm reaches the bronchioles. A 1-s burst from this type of aerosol produces 120 million particles, which may be suspended in the air for 1 h. In aerosols, compressed or liquefied gas exerts a force upon the internal surfaces of the container in which it is enclosed. This force per unit area is expressed in the form of pounds per square inch gauge (psig). Aerosol's gaseous mixture's vapor pressure obeys Raoult's law. Valve discharge rate, particle size distribution are measured for aerosols. Aerosols nowadays find their novel applications in many diseases cancer, gene therapy, and diabetes.

Aerosols are specialized pressurized dosage forms that contain one or more active pharmaceutical ingredients. When actuated, they emit a fine dispersion of liquid or solid materials in a gaseous medium. The active pharmaceutical ingredient or drug is either suspended or emulsified in propellants. These propellants serve to propel the contents outside. Aerosols can be administered orally, topically, or to body cavities. They are designed to expel their contents in various ways, such as a fine mist, a coarse or dry spray, a steady stream, or stable foam that breaks quickly.

When aerosols are used to provide an airborne mist, they are referred to as "space spray". Examples of this class of aerosols include room disinfectants, room deodorizers, and space insecticides. The particle size of the released product is typically below 50 μm, with 2 μm particles reaching the alveolar sac and 6 μm particles reaching the bronchioles. A 1-s burst from this type of aerosol can produce 120 million particles that remain suspended in the air for up to 1 h. On the other hand, aerosols used for surface applications are called surface sprays or surface coatings. This category includes dermatologic aerosols, cosmetic hair lacquers, perfume sprays, shaving lathers, toothpaste, and surface pesticide sprays [1].

Aerosols are classified into two types based on their origin mechanisms: primary aerosols and secondary aerosols. Primary aerosol particles are formed as a result of processes such as fragmentation or combustion and appear in the carrier gas as well-formed objects. Their shape can, of course, alter according to a variety of physicochemical phenomena such as humidification, gas-particle interactions, coagulation,

https://doi.org/10.1515/9783111438108-011

and so on. The size, shape, and chemical content of aerosol particles distinguish primary from secondary aerosols. In terms of shape, one would generally assume that the particles are spheres. This assumption is an idealization required for the simplification of mathematical issues relating to aerosol particle behavior. Numerous aerosols contain irregularly shaped particles. The non-sphericity of particles causes numerous issues, like agglomerate formation.

11.2 Background

The use of inhaled chemicals for medical purposes stretches back thousands of years, with the first treatments including the burning and inhalation of fragrant leaves such as tobacco. A definition of boiling to produce vapor or used topically, i.e., nasal snuff. Atomization, first powered by hand in the 1800s, then by steam, and finally by electricity by the 1930s, considerably advanced the discipline. Because of the qualities of these chemicals, early users were able to harness naturally aromatic plants and oils and create powders that could be used.

The use of ultrasonic nebulizers dates back to 1949, and the first pressurized metered-dose inhaler (pMDI), the Medihaler, was introduced in 1956, closely followed by the creation of the current jet nebulizers. The technology underlying these devices laid the groundwork for modern medical aerosol therapy, and with the addition of the dry powder inhaler (DPI) in the 1970s, these devices remained the mainstays of aerosol therapy into the twenty-first century.

The term "aerosol" initially originated around 1920 to describe the suspension of solid or liquid particles in a gaseous medium with little settling velocity. These equate to particles smaller than 100 μm in air and under typical conditions. The dimensions of some of the contaminants commonly found in air are given, along with some comparison components. By definition, an aerosol includes both the suspended particles and the gas in which they are suspended.

11.3 Advantages

There are advantages to the aerosol dosage form over other forms of medication.
- The drug can be easily drawn from the package without being contaminated.
- Aerosol protects active ingredients against atmospheric oxygen and moisture.
- In a thin layer, it can be applied to the skin without contacting the affected area.
- The particle size is nearly constant, which contributes to the efficacy of the drug.
- Aerosol application is a process that is both clean and effortless.
- The dosage form is elegant.
- Calibrating dosage is possible with the use of a metered dose inhaler.

11.4 Pulmonary drug delivery system

The shortcomings of traditional methods for treating chronic conditions have led to a growing focus on creating targeted drug delivery systems. Recent studies have focused a lot of attention on the pulmonary route of drug administration because it allows drugs to be delivered directly to the lungs for both local and systemic therapy. Over the past 20 years, research on both humans and animals has shown that a wide range of medicinal substances can be absorbed throughout the body after being given through the lungs.

Since the lung can absorb drugs for local deposits or systemic delivery, the pulmonary delivery of drugs is an attractive target for research into health care and has enormous scientific and biomedical interests. The respiratory epithelial cells play an important role in regulating the tone of the lungs and in the production of the lining airways. In this respect, more and more attention has been paid to the potential of the lung route as a noninvasive systemic and local therapeutic agent to be administered because of the high permeability and large absorption surface area of the lungs in adult humans with very thin absorption mucosa membranes and good blood supply. The alveolar epithelium of the lateral lungs is a point of absorption for most therapeutics and different macromolecules. Local respiratory diseases and some systemic diseases have already been reported to be well-treated by delivering drugs through the lungs. These include the current treatment of asthma, local infectious diseases, pulmonary hypertension, the systemic use of insulin, human growth hormones, and oxytocin; this is currently the case with many biotherapeutic drugs that are currently injected into the human body, such as growth hormones, glucose, or insulin, each of which can be administered to humans by inhalation if the efficiency of inhalation therapy is high [2].

11.4.1 Anatomy and morphology of the lungs

The intricate organs called the lungs are in charge of the body's exchange of carbon dioxide and oxygen. Their complex shape is essential for respiratory function and allows for effective gas exchange. We will examine the structure, function, and physiological adaptations of the lungs as well as their anatomy and morphology in this in-depth conversation.

11.4.2 Anatomy of lungs

The lungs are two conical organs that are situated on either side of the mediastinum in the thoracic cavity. The rib cage encloses them, and the pleural membranes shield them. While the left lung only has two lobes, upper and lower, the right lung has three lobes: upper, middle, and lower. The tertiary bronchi and their branches supply the bronchopulmonary segments, which are functionally separate units of lung tissue that make up each lung.

11.4.3 Bronchopulmonary segments

Tertiary bronchi and their branches supply the lungs' anatomical and functional units known as bronchopulmonary segments. The pulmonary artery, pulmonary vein, lymphatic vessels, and segmental bronchus are all found within each bronchopulmonary segment, which is encased in connective tissue. Specific tertiary bronchi supply air to bronchopulmonary segments, which are functionally separate units of lung tissue that take part in ventilation and gas exchange.

11.4.4 Microscopic anatomy of the lungs

At the microscopic level, the lungs are composed of specialized structures that facilitate efficient gas exchange. These structures include the respiratory bronchioles, alveolar ducts, and alveoli.

11.4.5 Respiratory bronchioles

The terminal bronchioles give rise to respiratory bronchioles, which are tiny airways with alveoli in their walls. Smooth muscle lines the walls of respiratory bronchioles, which are lined with simple cuboidal epithelium. They split off to form alveolar ducts, which eventually lead to alveolar sacs. By permitting air to pass through their walls into the alveoli, where carbon dioxide is expelled and oxygen is absorbed, respiratory bronchioles take part in gas exchange.

11.4.6 Alveolar ducts and alveoli

The respiratory bronchioles and the alveolar sacs are connected by alveolar ducts, which are narrow passageways. At the end of the alveolar ducts are tiny, grape-like structures called alveoli. A system of pulmonary capillaries envelops the alveolar ducts and alveoli, which are lined with simple squamous epithelium. In the lungs, they serve as the main locations for gas exchange. Alveoli help the bloodstream and the air in the lungs exchange carbon dioxide and oxygen. While carbon dioxide diffuses from the capillaries into the alveoli, oxygen diffuses from the alveoli into the capillaries.

11.4.7 Pulmonary vasculature

By moving blood from the lungs to the heart, the pulmonary vasculature is essential to gas exchange. It is made up of the pulmonary capillaries, pulmonary veins, and pulmonary arteries. The heart's right ventricle sends deoxygenated blood to the lungs via pulmonary arteries so that it can be oxygenated. Following the bronchi and bronchioles into the lungs, pulmonary arteries emerge from the pulmonary trunk. Eventually, they create an arteriole network that supplies the capillaries that encircle the alveoli.

The left atrium of the heart receives oxygenated blood from the lungs via pulmonary veins before it is distributed throughout the body. After emerging from the pulmonary capillaries, pulmonary veins unite to form larger veins that leave the lungs and enter the bloodstream. By permitting carbon dioxide to diffuse from the bloodstream into the alveoli and oxygen to diffuse from the alveoli into the bloodstream, pulmonary capillaries help to facilitate gas exchange. These capillaries form a dense network of small blood vessels surrounding the alveoli, ensuring proximity between air and blood for efficient gas exchange [3].

11.5 Defensive mechanisms of the pulmonary system and their impact

The complex network of organs known as the pulmonary system is in charge of the body's gas exchange with the outside world. The pulmonary system's primary job is to facilitate breathing, but it also acts as the body's first line of defense against pathogens, foreign substances, and particulate matter that could cause harm. In order to maintain respiratory health, this defense mechanism consists of a number of physiological processes intended to capture, eliminate, and neutralize dangerous substances.

11.5.1 Defense mechanisms of the pulmonary system

The pulmonary system uses a number of defense mechanisms to keep pathogens and dangerous substances out. These mechanisms fall into four general categories: immune responses, mucociliary clearance, reflex actions, and physical barriers. Physical barriers that stop foreign substances from entering the lower airways and alveoli are the pulmonary system's first line of defense. These barriers include the mucous membranes lining the airways, which secrete mucus to trap pathogens and particulate matter, and the nasal hairs, which trap larger particles. Furthermore, the airways' branching structure serves as a mechanical barrier, with smaller alveoli and airways offering greater resistance to foreign object penetration.

A crucial defense mechanism, mucociliary clearance depends on the cooperation of cilia lining the respiratory epithelium and goblet cells, i.e., mucus-producing cells, which produce mucus. The cilia beat in unison to push the trapped material upward toward the pharynx, where it can be expectorated or swallowed. The mucus layer traps inhaled particles, pathogens, and foreign substances. This system efficiently clears the airways of potentially dangerous materials before they can enter the deeper parts of the lungs.

The pulmonary system is equipped with a robust immune system that can detect and neutralize invading pathogens and foreign substances. Immune cells such as macrophages, neutrophils, and lymphocytes patrol the airways and alveoli, scavenging for and engulfing any foreign material they encounter. Macrophages, in particular, play a crucial role in phagocytosing and clearing bacteria, viruses, and other debris from the lungs. In addition to the aforementioned mechanisms, the pulmonary system also employs reflex actions to protect against potentially harmful stimuli [4].

11.5.2 Impact of defense mechanisms on drug delivery and deposition

The fate of inhaled substances, such as medicinal medications meant for local or systemic action, is largely determined by the pulmonary system's defense mechanisms. Particle size, solubility, surface characteristics, and the interplay with mucociliary clearance and immune responses are some of the variables that affect how well drugs are delivered and deposited in the lungs. One important factor in determining where inhaled substances will deposit in the respiratory tract is particle size. The rate at which inhaled substances are cleared from the respiratory tract can be greatly impacted by the presence of mucus and the activity of cilia. Mucociliary transport, which can vary based on factors like mucus viscosity, ciliary beat frequency, and the density of ciliated cells along the respiratory epithelium, is responsible for clearing particles trapped in the mucus layer. While substances that avoid mucociliary clearance may have a longer residence time and a higher potential for systemic absorption, those that are quickly removed from the airways may have less time to produce their therapeutic effects.

The fate of substances inhaled within the lungs can also be influenced by the immune system's reaction to those substances. For instance, phagocytic cells like neutrophils and macrophages may target and eliminate particles that the immune system identifies as foreign. The duration of action and therapeutic effectiveness of inhaled medications may be restricted by this clearance mechanism, especially if the drugs are quickly absorbed and expelled from the lungs [5].

11.6 Advantages of pulmonary drug delivery systems

- Targeted delivery: Pulmonary drug delivery enables targeted delivery of medications directly to the site of action in the lungs, minimizing systemic side effects.
- Improved bioavailability: The large surface area and extensive vascularization of the alveoli facilitate efficient absorption of drugs, resulting in higher bioavailability and lower doses required for therapeutic effect.
- Inhalation therapy offers a noninvasive route of drug administration, eliminating the need for injections and reducing patient discomfort and risk of infection.
- Patient convenience: Pulmonary drug delivery systems are often portable and easy to use, allowing patients to self-administer medications at home or on the go, improving treatment adherence and overall quality of life.
- Reduced systemic toxicity: By delivering medications directly to the lungs, pulmonary drug delivery systems minimize systemic exposure and toxicity, particularly beneficial for drugs with narrow therapeutic windows [6].
- Tailored formulations show advances in formulation technology to enable the development of tailored drug formulations optimized for pulmonary delivery, including dry powder inhalers, metered-dose inhalers, and liposomal formulations, enhancing drug stability and efficacy [7].

11.7 Aerosol components

The mechanism of an aerosol product is quite simple compressed or liquefied gas exerts a force on the internal surfaces of the container in which it is enclosed. The force per unit area is represented as pounds per square inch gauge (psig). An aerosol product is made up of four components that include propellant, container, valve, and product concentrate.

11.7.1 Raoult's law

When two substances whose molecules are very similar form a liquid solution, the vapor pressure of the mixture is very simply related to the vapor pressures of the pure substances. Suppose, for example, we mix 1 mol benzene with 1 mol toluene. It is easy to explain this behavior if we assume that because benzene and toluene molecules are so nearly alike, they behave the same way in solution as they do in the pure liquids. Since there are only half as many benzene molecules in the mixture as in pure benzene, the rate at which benzene molecules escape from the surface of the solution will be half the rate at which they would escape from the pure liquid. In consequence, the partial vapor pressure of benzene above the mixture will be one-half the vapor pressure of pure benzene. By a similar argument, the partial vapor pressure of the toluene above the solution is also one-half that of pure toluene (Fig. 11.1).

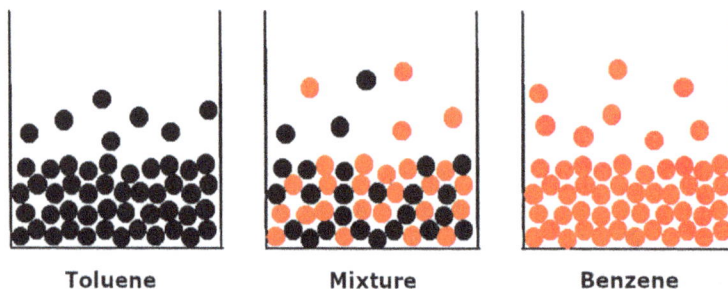

| Toluene | Mixture | Benzene |

Fig. 11.1: Vapor-liquid equilibrium.

In the mixture, only half of the molecules are benzene molecules, and so the concentration of benzene molecules in the vapor phase is only half as great as above pure benzene. Note also that although the initial amounts of benzene and toluene in the solution were equal, more benzene than toluene escapes to the gas phase because of benzene's higher vapor pressure.

11.7.1.1 Ideal solutions

An ideal solution obeys Raoult's law. In a pure liquid, some of the more energetic molecules have enough energy to overcome the intermolecular attractions and escape from the surface to form a vapor.

The smaller the intermolecular forces, the more molecules will be able to escape at any particular temperature. If we have a second liquid, the same thing is true. At any particular temperature, a certain proportion of the molecules will have enough energy to leave the surface. In an ideal mixture of these two liquids, the tendency of the two different sorts of molecules to escape is unchanged (Fig. 11.2).

One might think that the diagram shows only half as many of each molecule escaping, but the *proportion* of each escaping is still the same. The diagram is for a 50/50 mixture of the two liquids (as mentioned earlier, in case of a 50/50 ratio of benzene and toluene). That means that there are only half as many of each sort of molecule on the surface as in the pure liquids. The total vapor pressure of the system will be

Total vapor pressure = $P_b + P_t$

Common examples discussed under the heading of ideal solution are

– hexane and heptane
– benzene and methylbenzene
– propan-1-ol and propan-2-ol

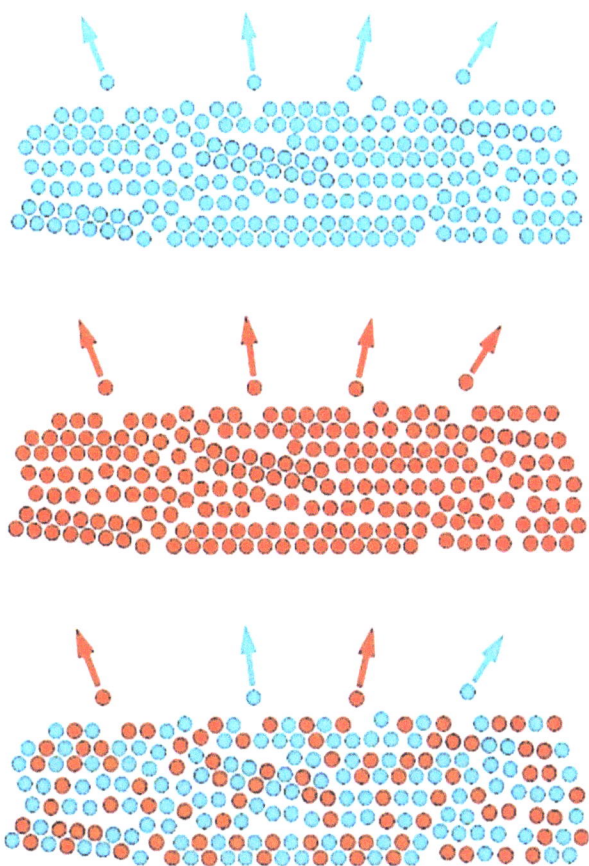

Fig. 11.2: Effect on evaporation of one solvent in the presence of a second solvent.

11.7.1.2 Real solution

These are those solutions that don't obey Raoult's law. Real solutions show two kinds of deviation from ideal behavior, explained by Raoul's law. These are discussed as follows:

Negative deviation: The vapor pressure is lower than would be expected from Raoult's law, i.e., the intermolecular forces increase when the liquids are mixed, e.g., chloroform and acetone.

Positive deviation: For a positive deviation, the vapor pressure for a given mixture is greater than would be expected (and therefore the boiling point is lower). Because the vapor pressure is higher, the liquid is evaporating more easily than would be expected. This means that some of the intermolecular bonds in the liquid must have been broken when the liquids were mixed, e.g.,

- benzene and ethyl alcohol
- carbon disulfide and acetone

11.7.1.3 Vapor pressure calculation using Raoult's law

What is the vapor pressure of a 60:40 mixture of propane and isobutene (Tab. 11.1)?

Tab. 11.1: Properties of solvents.

S. no.	Property	Propane	Isobutane
1.	Molecular formula	C_3H_8	C_4H_{10}
2.	Molecular weight	44.1	58.1
3.	Vapor pressure	110	30.4

1. Number of moles

   $n_{propane} = 60/44.1 = 1.36$

   $n_{isobutane} = 40/58.1 = 0.69$

2. Partial pressure (by Raoult's law)

   $P_{propane} = [n_p/n_p + n_i] * v.p$

   $= [1.36/1.36 + 0.69] * 110$

   $= 72.98$ psi

   $P_{isobutane} = [n_i/n_p + n_i] * v.p$

   $= [0.69/1.36 + 0.69] * 30.4$

   $= 10.23$ psi

3. Vapor pressure exerted by the system ideally

   $P_{total} = 72.98 + 10.23$

   $= 83.21$ psi

11.7.2 Propellants

Two types of propellants are used in aerosol products, including liquefied propellants and compressed gas propellants.

11.7.2.1 Liquefied propellants

Frequently, propellants that are liquid gases or mixtures of liquefied gases serve as both a propellant and solvent or vehicle for the concentrated product. These propellants are of two types: fluorocarbons and hydrocarbons.

Fluorocarbons: At room temperature, fluorocarbons are gaseous substances. They may be liquefied by cooling below their boiling points or by compressing the gas at room temperature, e.g., dichlorodifluoromethane gas will form a liquid when cooled to −22 °F or when compressed to 70 psig at 70 °F. Fluorocarbons are used in aerosols for inhalation.

Once the container is compressed, the vaporized state in the upper portion and the liquid state in the container of these gases will quickly establish equilibrium. The wall assembly and the surface of the liquid phase, which is made up of liquefied gas and the product concentrate, are all subjected to pressure in all directions by the vapor phase. The liquid phase is pushed up the dip tube and out of the aerosol valve's orifice into the atmosphere upon activation due to this pressure.

They reduce the amount of ozone in the upper atmosphere, which increases the amount of UV radiation reaching the Earth. In case of abuse, these compounds can harm epithelial tissues, raise the reactivity of the mammalian heart to endogenous epinephrine, and lead to fatal cardiac arrhythmias. The ability of fluorocarbon gases to freeze tissue and cause necrosis is what causes the topical toxicity effects, as they have a low boiling point. During skin planning, fluorocarbons can be used alone or in combination with ethyl chloride to freeze tissue, and this is a unique application of their tissue chilling effects. Examples include dichlorodifluoromethane (p.12), trichloromonofluoromethane (p.11), monochlorodifluoroethane (p.142), and dichlorotetrafluoroethane (p.114 and 114a), where "p" stands for propellant [8].

There is a numeric system for these fluorocarbons that consists of numbers in which the extreme right number is the number of fluorine atoms, second number from the right is one number greater than hydrogen atoms, and third from the right is one number less than carbon atoms. Carbon valence is adjusted finally by chlorine atoms after following the above rules.

However, the FDA only allows the use of fluorocarbons when the manufacturer ensures them. There is no alternative solution available, and the product provides substantial health benefits by using fluorocarbons that are not released during the operation and processing.

Hydrocarbons: These propellants are mostly used in topical aerosols and those used in cosmetics. These are explosive in nature and hence are not used in oral or inhalation aerosols. These are also low-cost propellants and are more compatible with the newer water-based formulations. Examples include propane, butane, pentane, and hexane.

Fluorocarbons and hydrocarbons have some pressure in the containers. If the pressure is high in the container, then the container wall cannot withstand the pressure, so use the propellants having low vapor pressure. Fluorocarbons and hydrocarbons may be used in combined form to get the required pressure in two- or three-phase systems.

A two-phase system is mostly used for inhalation or nasal application. This sort of system contains a solution or suspension of product concentrate with propellants that are miscible with one another, forming a solution form and a vapor state. Propellants used are p.12, p.12/11, and p.12/14. Three-phase systems are mostly used for topical preparation. These systems contain propellant, which is immiscible with the solution or emulsified product concentrate. These different phases are product concentrate, liquefied propellant, and vaporized propellant. Mostly difluoroethane (p.152) is used because it is superior to hydrocarbons due to less toxicity and less flammability.

11.7.2.2 Compressed propellants

The compressed gases that have been used in aerosol products include nitrous oxide, carbon dioxide, and nitrogen. But both nitrous oxide and carbon dioxide are capable of liquefaction at room temperature; their vapor pressure at the point is about 720 and 835 psig, which are beyond those that available containers can withstand. They cannot provide constant pressure as they expand with increasing volume.

11.7.3 Containers

We usually use glass, metal, and plastic-made containers.

11.7.3.1 Glass containers

Glass aerosol containers, especially those that are plastic-coated, are being used for pressurized medicines. They are usually inert materials (although they have some alkalinity), but if we use high-quality materials, then they are compatible with the product. They are mostly used because they are nonreactive, free from corrosion and deterioration, patient acceptance, and can be viewed if we want to see the product. Mostly, coating the glass from the inner side is done if the product is not compatible. The outer side of the glass is coated to protect it from light or to avoid breakage of glass into very small pieces if broken.

11.7.3.2 Metal containers

We have usually tin-plated, aluminum, and stainless-steel containers. All these containers are produced by the extrusion process. After preparing, we check these containers to see whether they have sieves or not in their body. They are used for some perfumes and inhalation pharmaceutical products. Stainless steel vials permit the use of increased pressures more safely than glass.

11.7.3.3 Plastic containers

These containers are the most widely used containers because they have both the safety of metal plate containers and the inertness of glass. But there may also be problems, including leaching (a process in which the product comes out of the container) and sorption (a process in which some plastic material is added to the product, just like hydroxyl ions of glass) [9].

11.7.4 Valve assembly

It is the most critical component of the aerosol package because it permits the expulsion of the contents of the can in the desired form and at the desired rate. One of the major parts of the assembly is the actuator. Then we have a stem, a gasket, a spring, a mounting cap, a dip tube, and housing.

11.7.4.1 Actuator

It is often referred to as the button or spout. It permits the easy opening and closing of the valve to emit or expel the product. An actuator is of four types that are spray actuator used to have a product in spray form for topical use, a foam actuator that is in liquid form but due to the big orifice and the big chamber it forms foam, a solid actuator discharged in semisolid form, and a solid actuator that is discharged in solid form.

11.7.4.2 Stem, gasket, and spring

The stem provides support to the actuator and brings the product forth from the container. A gasket is made up of plastic and is used to prevent leakage from the container. The spring is made up of stainless steel and brings the actuator to its place when the pressure is released. The gasket is also present in its proper place because of this spring.

11.7.4.3 Mounting cup

A mounting cup or ferrule fixes all the parts of the valve assembly and is self-fixed in the container after filling it. As its inner side is also exposed to the product, like a container, so may be cooled by some inert material.

11.7.4.4 Housing and dip tube

It serves to support the stem and spring, and the lower part of the housing also contains the upper part of the dip tube that serves to bring the product from the container to the housing.

Metered dose valves are employed when the formulation is a potent medication. The metering valve is constructed to deliver a prearranged measured quantity of product each time the valve is actuated. This is achieved by a valve design that isolates a given volume of product in a separate chamber and releases it when the valve is actuated, while simultaneously shutting off the flow of remaining contents. Most metering valves employ a molded plastic or machined stainless-steel chamber whose internal diameter determines the metered quantity.

A trans-lingual aerosol formulation of nitroglycerine has been developed for the relief of an acute attack or prophylactically for angina pectoris. At the onset of the attack, two metered spray emissions, each containing 0.4 mg of nitroglycerine, are administered. The product contains a total of 200 doses [10].

11.7.5 Product concentrate

Aerosol product concentrate is composed of two parts: propellants and aerosol formulations that can be in the form of a solution, suspension, or emulsion.

11.7.5.1 Aerosol solution

In aerosol solutions, we first dissolve the active ingredients in pure propellants or in a mixture of propellants and solvents. The solvent is used to dissolve the active ingredients or to retard the evaporation of propellants.

Aerosol solutions are easy to formulate, but propellants are nonpolar and are, in most cases, poor solvents for commonly used aerosol ingredients. There is no limit to the number of solvents used, but toxicity should be considered. Ethyl alcohol is a commonly used solvent. Remember that the particle size should be very small, even less than 2 μm. Propellants used are mostly 11, 12, and 114. These are used in high concentrations because the active ingredients are used in very small concentrations. Examples include isoproterenol oral inhalant.

11.7.5.2 Aerosol suspension

Aerosol suspension is quite difficult to stabilize because crystal growth or caking may occur, or agglomeration may take place along with the blockage of the orifice. Aerosol suspension contains the active drug, propellants, suspending agent, and moisture content (200–300 ppm). The drug should have fine particles. These should be uniformly distributed. For inhalant purposes, particles should be 1–5 μm and should not be more than 10 μm. But, for topical use, it may be 50 μm.

The system is useful for insoluble drugs, e.g., antibiotics and steroids. But these should be soluble in body fluids. One drug is epinephrine bitartrate, which is insoluble in the media but is soluble in body fluids. The density of the drug and propellant should be the same; otherwise, stability is affected. If this is not the same, then it may be equalized by adjusting propellants (using alone or in combination).

11.7.5.3 Aerosols emulsion

Emulsions can be dispersed from an aerosol container as a spray, stable foam, or quick-breaking foam depending upon the valve assembly and formulation. W/O are in the form of spray, but stable foams are formed when it is O/W. Quick-breaking foam allows for the application of medication efficiently and conveniently because we need no mechanical force, and they are converted into a liquid without rubbing.

11.8 Filling of aerosols

Filling of aerosols is performed by cold filling and pressure filling.

11.8.1 Cold filling

As fluorocarbons and hydrocarbons may be liquefied by cooling, this property is used in the filling of aerosols. In this method, both the propellants and the product concentrate must be cooled to −30° to −40 F°. Now the container is also cooled equally, and then the product is added. The vapors of still cold propellant displace the air of the container, and then immediately the valve is inserted. Freon (p.12) has a boiling point of −30 °C or 21 °C at 70 psi.

When we fill the container, some air may be entrapped, which causes the loss of propellant when it comes to normal temperature, and liquid products cannot be filled by this method.

11.8.2 Pressure filling

Here we have filling machines. The product concentrate is placed in the container, and the valve assembly is fixed. Then propellant is passed from a container to the aerosol container up to the required capacity. This may also be done when the pressure of the container having propellant initially is adjusted. During filling, when the pressure becomes equal, the filling is automatically stopped, and shaking is required.

11.9 Testing

After filling, the aerosol container is tested under various environmental conditions for leakage or weakness in the valve assembly. Proper functioning of the assembly is also tested. Weigh the container, discharge a portion of the contents, then again weigh. Finally, the valve discharge rate is determined per the given time. Particle size distribution is also tested [11].

11.10 Packaging, labeling, and storage

Most aerosol products have a cap or cover that fits snugly over the valve and the mounting cup. This protects the valve against contamination with dust and dirt. The cap is generally made up of plastic or metal and also serves a decorative function.

Some instructions are labeled by manufacturers, which include specific instructions that include do not puncture the container, do not expose to heat or place at high temperature, and avoid inhalation (if it is not for inhalation). Keep away from the eyes and mucous membranes if it is for topical use because it contains hydrocarbons, which are toxic and flammable. The product should be stored at 15–30 °C because if it is a metered dose inhaler, then due to lower temperature or cold containers, the amount of spray is affected [12].

11.11 Principle of aerosol generation

Aerosol generation involves the production of a suspension of fine solid particles or liquid droplets in a gaseous medium. This section discusses the basic principles behind aerosol generation, including atomization and aerosolization techniques, the role of surfactants and co-solvents, factors affecting droplet size and distribution, and formulation considerations.

11.11.1 Atomization and aerosolization techniques

Atomization and aerosolization techniques are crucial procedures utilized for the generation and delivery of aerosols. Atomization techniques involve breaking up a liquid into small droplets or particles, while aerosolization refers to the process of converting a liquid or solid into an aerosol – a colloidal suspension of fine solid particles or liquid droplets in a gas. Commonly used techniques for atomization and aerosolization of aerosols are air-jet, ultrasonic, and pressure atomization. Air-jet atomization utilizes compressed air or gas to break up the liquid into droplets by passing compressed air or gas through a narrow nozzle, creating a high-velocity jet. This jet then shears the liquid into small droplets, forming an aerosol. Ultrasonic atomization uses high-frequency sound waves to break the liquid into small droplets that generate aerosols. Pressure atomization involves forcing the liquid through a small orifice at high pressure, which results in the creation of a fine mist or aerosol [13].

11.11.2 Role of surfactants and co-solvents

Surfactants and cosolvents are essential to generate aerosols. Surfactants are compounds with both hydrophilic (water-loving) and lipophilic (oil-loving) properties. They are used to reduce the surface tension between the liquid phase and propellant phase of aerosol formulations, which helps to disperse and stabilize active ingredients. Examples of surfactants are Span, Tween, and lecithin.

Co-solvents are substances that improve the solubility of active ingredients in the propellant phase. They help dissolve active ingredients and ensure their uniform distribution in the aerosol product. Examples of co-solvents are ethanol and propylene glycol.

Surfactants play an important role in enhancing the dispersion of active ingredients in aerosol formulations, promoting uniform distribution and preventing clumping or aggregation. Co-solvents play an important role in enhancing the solubility of active ingredients, improving the stability and shelf life of aerosol formulations. By using the right combination of surfactants and co-solvents, aerosol formulations can be optimized for better performance and efficacy [14].

11.11.3 Factors affecting droplet size and distribution

The size and distribution of droplets during aerosol generation can be influenced by several factors.

11.11.3.1 Physicochemical properties

The surface tension of the aerosol formulation can determine the resistance to droplet formation and droplet size, with lower surface tension resulting in smaller droplets. Similarly, the viscosity of the formulation can affect the flow properties and atomization process, with higher viscosity leading to larger droplets.

11.11.3.2 Formulation parameters

The size of droplets can be impacted by the concentration of the active drug in the formulation. Larger droplets can result from higher drug concentrations. Droplet size and distribution can be affected by the choice of excipients in the formulation. Different excipients can alter the viscosity and surface tension, influencing droplet formation.

11.11.3.3 Device characteristics

The size of the nozzle or orifice through which the aerosol is emitted is crucial in determining droplet size. Smaller nozzle sizes generally produce smaller droplets. Droplet size can be affected by the pressure applied to the formulation during atomization. Higher air pressure can result in smaller droplets.

11.11.3.4 Environmental conditions

The temperature of the environment can impact the rate at which the solvent in the aerosol formulation evaporates, which can affect droplet size and distribution. The level of humidity can influence the rate of evaporation and condensation of water vapor on the aerosol droplets, which can potentially alter droplet size and distribution [15].

11.11.4 Formulation considerations for aerosol generation

Formulation considerations are crucial for the successful generation of aerosols.

11.11.4.1 Excipient selection

The selection of excipients is crucial; they should be compatible with the active ingredient. The type of propellant used is crucial in aerosol formulations as it affects the delivery mechanism and stability of the aerosol. Commonly used propellants include hydrofluoroalkanes (HFAS) and chlorofluorocarbons (CFCS). To maintain the stability and uniform dispersion of the active pharmaceutical ingredient (API) in the aerosol formulation, a suitable suspending agent, such as polymers, surfactants, or viscosity enhancers, must be chosen.

11.11.4.2 Particle size control

The size of the API particles can impact their performance and deposition in the lungs. To improve aerosol performance, micronization techniques such as jet milling or spray drying can be used to reduce the API particle size. It is important to control the particle size distribution to ensure consistent and targeted delivery of the API. Narrow particle size distributions are preferred to minimize variability in drug delivery.

11.11.4.3 Stability considerations

It is crucial to maintain the therapeutic efficacy of the API in the aerosol formulation by ensuring its chemical stability. During the development of the formulation, various factors such as pH, temperature, and exposure to light should be considered. Physical stability refers to the ability of the aerosol formulation to maintain its integrity over time. This includes preventing particle aggregation, sedimentation, or creaming. Proper selection of excipients and optimization of the formulation can help enhance physical stability.

11.11.5 Patient factors

Age-related factors such as lung development and function can have an impact on aerosol deposition and drug delivery. Thus, formulation considerations should be adjusted to account for age-related differences in lung capacity and inhalation pattern. It is important to understand the patient's lung capacity, as it can affect the deposition and distribution of aerosol particles. By doing so, the formulation and delivery system can be optimized to ensure effective drug delivery [15].

11.12 Pharmaceutical aerosol formulation

11.12.1 Drug selection and compatibility

The careful selection of the appropriate drug is essential for aerosol formulation. The drugs should be chosen based on factors such as solubility, stability, and compatibility with the propellant and excipients. Compatibility studies are crucial to ensure the drug's stability and effectiveness within the aerosol formulation throughout the formulation process and during storage. These studies evaluate the physical and chemical interactions between the drug and other components of the formulation.

11.12.2 Excipients and formulation ingredients

Excipients play a significant role in the aerosol formulation and are added to enhance drug solubility and stability, improve aerosol performance, and ensure patient acceptability. Common excipients include propellants, e.g., hydrofluoroalkanes, 1,1,1,2-tetrafluoromethane (propellant 134a), co-solvents, surfactants, and antioxidants to optimize drug delivery and minimize adverse effects. Consider the use of other formulating agents like viscosity modifiers, pH adjusters, antioxidants, and tonicity agents.

11.12.3 Particle engineering and size control

Particle engineering is crucial for ensuring efficient drug delivery to the target site. It is imperative to control the particle size within the respirable range, typically ranging from 1 to 5 μm, for optimal drug deposition in the lungs. The techniques such as spray drying, micronization, and nanotechnology are used to achieve the desired particle characteristics, including aerodynamic properties and deposition patterns [16].

11.12.4 Stability and shelf-life considerations

Stability studies are conducted to evaluate the physical, chemical, and microbiological stability of aerosol formulations over their intended shelf life. These studies are crucial for determining the shelf-life of the product and ensuring both efficacy and safety. Various factors such as temperature, humidity, light exposure, and container closure systems are considered during stability testing to simulate real-time storage conditions. Accelerated stability testing is frequently conducted to predict the long-term stability of the product under various storage conditions. Adjustments to the formulation are made to enhance stability and extend shelf life, such as optimizing packaging materials, adjusting pH, or incorporating stabilizing agents.

11.13 Inhalation devices and techniques

11.13.1 Nebulizer

It involves the generation of aerosols from liquid formulations using nebulizers. These devices convert liquid medication into a fine mist or aerosol by breaking it up into small droplets using atomizing techniques like compressed air or ultrasonic vibrations.

11.13.2 Dry powder inhalers (DPIs)

DPIs are medical devices that deliver medication through a dry powder aerosol. When patients inhale through the device, the airflow creates a pressure drop that disperses the medication into the air. DPIs are an effective way to treat respiratory conditions and offer a convenient way to deliver medication.

11.13.3 Metered dose inhalers (MDIs)

Handheld devices deliver a specific amount of medication in the form of a pressurized aerosol. The medication is released when the patient activates the device by pressing down on the canister [17].

11.14 Marketed products of the aerosol dosage form

11.14.1 Inhalers for respiratory diseases

Albuterol inhalers deliver a bronchodilator medication called albuterol, which helps relieve symptoms of asthma by relaxing the airway muscles. Fluticasone/salmeterol inhalers. These combination inhalers contain fluticasone (a corticosteroid) and salmeterol (a long-acting bronchodilator) and are used for the long-term management of asthma and COPD.

11.14.2 Nasal sprays for allergic rhinitis

Fluticasone nasal sprays work as anti-inflammatory agents and are effective in reducing the symptoms of allergic rhinitis, including nasal congestion, sneezing, and itching. Azelastine nasal sprays are antihistamine medications that provide relief from seasonal and perennial allergic rhinitis symptoms.

11.14.3 Topical aerosols for dermatological conditions

Corticosteroid aerosols, such as hydrocortisone or betamethasone, are used to treat inflammatory skin conditions like eczema, psoriasis, and dermatitis. Antibacterial aerosols, like mupirocin or fusidic acid, deliver antibacterial agents and are used to treat localized skin infections, such as impetigo.

11.14.4 Oral inhalation aerosols for pulmonary hypertension

Iloprost inhalation aerosol contains iloprost, a prostacyclin analog, and is delivered via inhalation aerosol and used for the treatment of pulmonary arterial hypertension, a condition characterized by high blood pressure in the arteries of the lungs.

11.15 Future trends and innovations in pharmaceutical aerosols new medicines as aerosol dosage

Pharmaceutical aerosols have proven to be an effective and convenient method of drug delivery, offering advantages such as targeted lung deposition, rapid onset of action, and reduced systemic side effects. As technology continues to evolve, the future of pharmaceutical aerosols holds great promise for innovative advancements. One notable trend is the development of personalized medicine delivered through aerosol dosage. This approach tailors treatments to individual patient needs, optimizing efficacy and minimizing side effects.

Furthermore, advancements in nanotechnology have the potential to revolutionize pharmaceutical aerosols. Nanoparticle-based drug delivery systems can improve the solubility and stability of drugs, enhancing their effectiveness when administered as aerosols. Additionally, the integration of smart inhaler technology with pharmaceutical aerosols is a significant innovation on the horizon. This enables real-time monitoring of medication usage and patient adherence, leading to better treatment outcomes and disease management. Aerosol dosage forms have become increasingly popular for the delivery of medications and therapeutic agents. These products offer a unique way of administering drugs directly to the targeted sites within the body, such as the lungs or nasal passages.

11.16 Products under trial

11.16.1 Aerosolized liposomes and vaccines

Aerosolized liposomal formulations of the chemotherapy drug doxorubicin are being developed for the treatment of lung cancer. The liposomes encapsulate the drug, improving its solubility and allowing for targeted delivery to tumor cells in the lungs. Aerosolized liposomal formulations are being explored as carriers for vaccines. Liposomes can encapsulate antigens and adjuvants, protecting them from degradation and facilitating targeted lung delivery. This approach holds promise for respiratory infections and diseases.

11.16.2 Aerosolized nanoparticles

Nanoparticle-based vaccines: Researchers are exploring the use of aerosolized nanoparticles as carriers for vaccines. Nanoparticles can be engineered to encapsulate antigens and adjuvants, enhancing their stability and immunogenicity. Aerosol delivery of nanoparticle-based vaccines may offer advantages such as improved mucosal immunity and needle-free administration.

11.16.3 Aerosolized imaging agents

Aerosolized nanoparticle imaging agents aerosolized nanoparticles are being explored as imaging agents for lung imaging and the diagnosis of respiratory conditions. Nanoparticles can be labeled with imaging probes or contrast agents, enabling their detection in specific tissues or organs. This approach may have applications in lung cancer diagnosis and monitoring disease progression [18].

References

[1]   Wu T, Boor BE. Urban aerosol size distributions: A global perspective. Atmospheric Chemistry and Physics. 2021; 21(11): 8883–914.
[2]   Leach C. The CFC to HFA transition and its impact on pulmonary drug development. Respiratory Care. 2005; 50: 1201–08.
[3]   Chaudhry R, Anatomy BB, Thorax L. StatPearls: Treasure Island (FL): StatPearls PublishingCopyright © 2024, StatPearls Publishing LLC; 2024.
[4]   Nicod LP. Lung defences: An overview. European Respiratory Review. 2005; 14(95): 45.
[5]   Ruge CA, Kirch J, Lehr C-M. Pulmonary drug delivery: From generating aerosols to overcoming biological barriers—therapeutic possibilities and technological challenges. The Lancet Respiratory Medicine. 2013; 1(5): 402–13.

[6]   Huchon G. Advantages and difficulties of aerosol therapy. La Revue du praticien. 1992; 42(13): 1657–61.

[7]   Chow MYT, Pan HW, Lam JKW. Chapter Ten – Delivery technology of inhaled therapy for asthma and COPD. In: Wong WSF, editor. Advances in Pharmacology (Vol. 98): Academic Press; Republic of Singapore, 2023. pp. 273–311.

[8]   Kulkarni VS, Shaw C. Chapter 6 – Aerosols and Nasal Sprays. In: Kulkarni VS, Shaw C, editors. Essential chemistry for formulators of semisolid and liquid dosages: Boston: Academic Press; San Antonio, Texas, USA, 2016. pp. 71–97.

[9]   Nasa P. A review on pharmaceutical packaging material. World Journal of Pharmaceutical Research. 2014; 3(5): 344–68.

[10]   Boden WE, Padala SK, Cabral KP, Buschmann IR, Sidhu MS. Role of short-acting nitroglycerin in the management of ischemic heart disease. Drug Design, Development and Therapy. 2015; 9: 4793–805.

[11]   Hickey AJ. Chapter 3 – Quality and Performance Tests. In: Hickey AJ, editor. Inhaled pharmaceutical product development perspectives: Elsevier; NC, USA, 2018. pp. 33–51.

[12]   Niemiec K, Fitrzyk A, Grabowik C. Technological solutions and innovations within aerosol packaging. In: IOP conference series: Materials science and engineering (Vol. 400 (2)): IOP Publishing Ltd; USA, 2018. pp. 022040.

[13]   Baghdan E, Pinnapireddy SR, Vögeling H, Schäfer J, Eckert AW, Bakowsky U. Nano spray drying: A novel technique to prepare well-defined surface coatings for medical implants. Journal of Drug Delivery Science and Technology. 2018; 48: 145–51.

[14]   Suhail M, Kumar A, Khan A, Naeem A, Badshah S. Surfactants and their role in pharmaceutical product development: An overview. International Journal of Pharmaceutical Sciences. 2019; 6: 72–82.

[15]   Bourlon M, Feng Y, Garcia-Contreras L. Designing aerosol therapies based on the integrated evaluation of in vitro, in vivo, and in silico data. Pharmaceutics. 2023; 15(6): 1695.

[16]   Darquenne C. Deposition mechanisms. Journal of Aerosol Medicine and Pulmonary Drug Delivery. 2020; 33(4): 181–85.

[17]   Cripps A, Riebe M, Schulze M, Woodhouse R. Pharmaceutical transition to non-CFC pressurized metered dose inhalers. Respiratory Medicine. 2000; 94(Suppl B): S3–9.

[18]   Rahimpour Y, Hamishehkar H, Nokhodchi A. Chapter: 6 Lipidic micro- and nano-carriers for pulmonary drug delivery – a state-of-the-art review. Pulmonary Drug Delivery (1st edition). John Wiley & Sons, Ltd.,; Chichester, West Sussex, UK, 2015; 123–42.

Part 3: **Solid dosage form**

Ghazala Ambreen, Muhammad Yasir Ali, Sufyan Iqbal, Laraib Zeeshan,
Abida Sabir, Udo Bakowsky, Saeed Ahmad

12 Powders and granules

12.1 Powders

Powders are dry, finely separated drugs or substances that can be used internally
(oral) or externally (topically). Powders spread and dissolve faster than compacted
dosage forms due to their larger specific surface area. Powders may be more palatable to children and adults who struggle to swallow tablets or capsules. Drugs that are
too large to fit into tablets or capsules of an appropriate size can be delivered as powders. Oral powders should be added to a beverage or apple sauce before use. Powdered dosage forms frequently avoid stability issues that are common in liquid dosage
forms. Pharmacopoeias distinguish two types of powders: oral and topical. Powders
can be coarse powder, moderately coarse powder, fine powder, or very fine powder.
In United States Pharmacopoeia, there are two types of powders: (a) animal or plant
crude powder and (b) chemical crude powders.

Remember that plant crude powders are of a fine powder type, but animal crude
powders are not fine [1, 2].

12.2 Advantages

- Dosage can be easily adjusted.
- Easy to be absorbed.
- These are easier to carry than liquid.
- Incompatibility is less in the case of powders than in liquids.
- More stable than liquid dosage forms because of resistance to hydrolysis and oxi-
  dation.

12.3 Disadvantages

- Cannot mask the bad taste of the drug.
- Bulk powders are inconvenient for self-administration.
- Difficult to preserve deliquescent and hygroscopic drugs because they undergo
  hydrolysis.
- Bulk powders are less convenient to carry than a small container of tablets or
  capsules.

https://doi.org/10.1515/9783111438108-012

- Bulk powders are not suitable for the administration of more potent drugs in smaller doses.
- Powders are not suitable for the drugs that are degraded in the stomach; for these drugs, gastro-resistant tablets are formulated.

12.4 Classification of powders based on route of administration

12.4.1 Oral powder

Oral powders are preparations that consist of solid, loose dry particles of varying degrees of particle size. They may contain one or more active pharmaceutical ingredients (APIs), with or without excipients. Flavoring agents can be added to enhance palatability, and colorants to make it more appealing to the patient. They are generally administered with water or with another suitable liquid, like juice or milk, before administration. They may also be swallowed directly. This dosage form is suitable for the patient who has difficulty swallowing tablets or capsules. They may provide faster dissolution than tablets. To ensure dose uniformity, powders are usually measured with a spoon or in a sachet. Antibiotics, electrolyte solutions, and nutritional supplements are usually available in the form of oral powders. These powders may be available as single- or multi-dose containers.

12.4.1.1 Single-dose powder

In single-dose powders, each dose is enclosed in an individual container. Traditionally, single-dose powders were wrapped in suitable material like paper, which was not suitable for hygroscopic, deliquescent, and volatile products. Now, modern packaging materials of plastic laminate and foil are being used; they offer better protection and can tolerate high-speed packaging machines. Paper wrapping is still in use for certain over-the-counter products.

12.4.1.2 Multidose powders

These are packed in a suitable bulk container such as a wide-mouthed glass or plastic jar. A suitable measuring device should be provided for accurate measurement of the dose. Owing to difficulty in precise dosing, generally non-toxic medicaments are packed in bulk containers, which require large doses. Many food and dietary products are packed in this way, while a few proprietary examples are present.

12.4.1.3 Preparation requirements

Oral powders are used to yield oral solutions and oral suspensions using a suitable vehicle. This may be done at the time of dispensing or at the time of administration by the patients. The selection of a vehicle for oral use is dependent on the nature of the active substance, and it should maintain the organoleptic characteristics required for intended use.

Powders for oral solution and suspensions: Active compounds with high aqueous solubility can be prepared as a powder for oral solution and packed in a bottle for multiple-use products and a stick pack, packet, or sachet for single-use products. Active compounds with limited aqueous solubility can be made into a powder for oral suspension and packed in a container for multiple-use products and a stick pack or sachet for single use. These powders may contain excipients, particularly to facilitate the dissolution or dispersion and to prevent cracking. The product should comply with the requirements of oral solutions or oral suspensions after dissolution. A label should be attached that explains the method of preparation of the solution or suspension from powder or granules. Conditions of storage, along with duration of storage, should also be mentioned on the label [3].

Powders and granules for syrups: Syrups are aqueous preparations characterized by a sweet taste and a viscous consistency. Sucrose may be present in a concentration of at least 45%. Sweetening agents and polyols can be added to sweeten the taste. They may also contain aromatic or flavoring agents. The necessary ingredients of the syrups can be prepared and stored in powder form and reconstituted (using water alone or other suitable liquid) at the time of use or dispensing. The resulting syrup should fulfill the Pharmacopoeia requirements of syrups.

Antibiotic syrups are prepared for patients who resist taking capsules or tablets, so these are suitable alternatives. Unfortunately, antibiotics are unstable physically and chemically when formulated as a solution or suspension. To counter this instability problem, prepare the required dry ingredients in powder or granule form. It can be constituted with water to make a solution or suspension at the time of dispensing or administration. After the constitution, the patient should be informed about the shelf life of this product. A shelf life of 1–2 weeks for the reconstituted product will not be a problem. Amoxicillin oral suspension and erythromycin ethyl succinate oral suspension are examples.

Powders for oral drops: Oral drops are solutions, emulsions, or suspensions that are administered in small volumes, such as in drops, by means of a suitable device. These powders shall conform to all the requirements of all other oral powders. The excipients may be present to facilitate dissolution or prevent caking. The label should inform about the number of drops per milliliter or per gram of preparation.

Powders for injections: Injections of the active substances that are not stable in aqueous medium should be prepared immediately at the time of administration. The ingredients are filled in ampoules or vials and are sterile. Sufficient diluent (e.g., sterile water for injection) is added from a separate container at the time of administration. Some excipients (e.g., additives to produce the required isotonic solution on reconstitution) may also be present in addition to the drug in the formulation.

Freeze drying method is commonly used to manufacture powder for injection. The label for powder for injection should include the amount of the API present in the sealed container, directions regarding the preparation of injections or intravenous infusion for the period, and preparation is intended for parenteral use upon reconstitution [4].

12.4.2 Effervescent powders

Effervescent powders are also available as single- and multi-dose preparations, containing APIs in addition to acid substances, carbonates, or hydrogen carbonates, which react rapidly as they are added to water, producing effervescence and may cause a drought. Sodium bicarbonate plus citric acid is a common example that produces effervescence. The drug is dissolved or dispersed in water before ingestion. Preferably, they are packed as single-dose preparations in air-tight containers (laminated sachets are ideal). It is necessary to prevent the product from moisture while manufacturing and during subsequent storage to avoid the premature occurrence of a reaction [5].

Key ingredient
Commonly utilized acids include citric and tartaric acids. These acids react with the carbonate component, producing carbon dioxide gas. The commonly used carbonate component is sodium bicarbonate (baking soda). When it combines with acid, it emits carbon dioxide, resulting in effervescence.

Other ingredients
–   Flavoring agents are used to improve taste.
–   Sweeteners are used, like sucrose and artificial sweeteners.
–   Coloring agents are used for aesthetic purposes.
–   Active ingredients are combined with an effervescent base for therapeutic benefits.

Applications
–   Effervescent powders are utilized in a variety of applications.
–   Vitamins and supplements: Primarily utilized for vitamin and mineral supplements.
–   Quick relief from pain and antacids with an effervescent formula.

–   Rehydration solutions: Prepare oral rehydration solutions to quickly restore electrolytes.
–   Oral medications offer an alternative to traditional tablets and liquids, particularly for patients who have trouble swallowing.

Advantages
–   Effervescent response enhances API absorption and commencement of action.
–   Ease of use: Ideal for those who struggle to swallow medications.
–   Improved taste: Effervescence helps disguise the unpleasant taste of certain drugs [6].

12.4.3 Insufflated powders

Insufflated powders are finely divided powders that are intended to be applied to body cavities, such as the nose, ears, vagina, throat, or tooth socket. While using an insufflator (puffer), the patient simply puffs the required quantity of powder onto the affected area or body cavity. This device is appropriate for anti-infectives. A moisture-activated adherent, such as polyvox, can also be added to these powders, which is an ethylene oxide polymer having a large molecular weight that forms a viscous, mucoadhesive gel when it comes in contact with moisture. The gel serves as a depot for long-term delivery of the drug.

12.4.3.1 Composition and function

These powders typically include a moisture-actuated adhesive, e.g., polyvox, an ethylene oxide polymer, which is a high-molecular-weight one added thereto. When exposed to moisture, it undergoes polyvox to produce a viscous, mucoadhesive gel, which then works as a depot for its sustained drug release in the presence of fluids. It is a kind of gel that enhances the residence time and, so, the therapeutic effect for a long time [4].

12.4.4 Powders for other routes of administration

12.4.4.1 Powders for inhalation

Powders are extensively used for pulmonary drug delivery nowadays. Some medicated powders can be administered with the aid of dry powder inhalers (DPIs), for delivery of micronized particles of API in metered quantities. This dosage form is one of the most effective methods for delivering APIs to the lungs for the treatment of chronic obstructive pulmonary disease and asthma. To accomplish this, the medication should have a particle size in the range of 1–6 μm in diameter.

Inhalation powders are made of a mixture of APIs, the carrier, and all the parameters of formulation packed in a suitable closure system. The formulation may be packaged in a device with metered units. Pre-metered DPIs consist of a measured amount of individual units (e.g., capsules and blisters) of formulation. They may also contain measured unit doses as ordered multi-dose assemblies in the delivery system. In this delivery system, a mechanism of piercing the capsules or opening the unit-dose container allows mobilization and aerosolization of the powder. Nowadays, dose counter or dose indicator system is being employed in DPIs to facilitate dosing compliance. The following are a few examples of powder for inhalation:

–   Advair Diskus 100/50, 250/50, and 500/50 contain fluticasone propionate 100, 250, and 500 mg, respectively, along with salmeterol 50 mg in a powder for inhalation.
–   The Foradil Aerolizer is a capsule dosage form for oral inhalation only in the Aerolizer inhaler. The capsule contains a dry powder formulation of 12 mg of formoterol fumarate and 25 mg of lactose as a carrier [2, 7].

12.4.5 Powders for external use

12.4.5.1 Powders for cutaneous application (topical powders)

Medicated or non-medicated powder that can be applied easily on the skin for external use is called a topical powder.

Examples of different powders are as follows:

Nystatin topical powder: It is used for the treatment of mycotic infection.

Neomycin and polymyxin sulfate and bacitracin zinc topical powder: It is used as an antibiotic and anti-infective.

Tolnaflate powder: It is used for the treatment of fungal infection.

Methyl benzethonium chloride: It is used for the treatment of diaper rashes and chafing on babies.

Miconazole nitrate powder: It is used for the treatment of athlete's foot.

Compound liquinol: It is used for the treatment of vaginal trichomoniasis. It is in the form of insufflations.

Microcrystalline cellulose: It is helpful in the preparation of different dosage forms.

Depending upon the intended use of powder, it may be supplied to the patient in a perforated or sifter-type can or container for external dusting, in an aerosol container for spraying on the skin, or a wide-mouthed glass or plastic jar which permits the entrance of a spoon and the easy removal of a spoonful of powder. All the powders should be stored in a tightly closed container.

Powders containing potent substances and those that should be administered in controlled doses are supplied to the patient in divided amounts. Powders supplied to patients in either bulk or divided portions that are intended for external use should bear an external use only label.

For divided powders after it has been properly mixed, they may be divided into individual units based upon the dose at a single time. Each divided portion of powder may be placed on a small piece of paper, which is then folded to close the medication. Today, only headache powder, laxatives, and douche powder are available.

Topical powders should have a small particle size and uniformity to prevent irritation of the skin upon application. They should be free-flowing, impalpable, and they should easily adhere to skin and should pass through at least a No.100-mesh sieve to minimize irritation of skin. Highly sorptive powders should not be used for topical powders that are applied to an oozing wound, because a hard crust may form. For this type of wound, a hydrophobic, water-repellent powder should be used, as it will prevent water loss from skin and will not cake. Talc, or any other naturally occurring preparation that is to be applied to an open wound, should be sterilized first to avoid infection.

Topical powders usually contain a base or vehicle (cornstarch or talc), an adherent (magnesium stearate, calcium stearate, or zinc stearate), and possibly an active ingredient, along with an aromatic substance. The powder should provide a large surface area, flow easily, and spread uniformly [4].

12.4.5.2 Dusting powders

These are powders for cutaneous application having a suitable fineness. For example, talc dusting powder, which is a mixture of 10% starch and 90% purified talc, with a controlled particle size using a 250 µm sieve. Dusting powders contain ingredients that are used for therapeutic, lubricant, or prophylactic purposes, intended for external use only. For open wounds, sterile dusting powders can only be used. These powders are usually dispensed in a glass or metal container having a perforation in the lid, to be dusted in the affected area. Therefore, the active ingredients should be mixed with a diluent having good flow ability [8].

12.4.6 Classification based on packaging of powders

Powders are packed in two ways: as prepared by a pharmacist and dispensed to the patient by a pharmacist. Some powders are packaged by manufacturers, whereas others are prepared and packaged by the pharmacist.

12.4.6.1 Bulk powders

Some bulk powders available in pre-packaged form are (a) laxatives (e.g., psyllium), antacids (sodium bicarbonate), which are taken by patients either by dissolving in water or other beverages; (b) douche powders (e.g., Messengill powder), for vaginal use after dissolving in warm water, (c) brewer's yeast powders, consisting of vitamin B complex and other supplements, and (d) medicated powders for external use, usually anti-infectives (e.g., polymyxin B sulfate and bacitracin zinc). For measurement of an accurate dose, a small measuring scoop, spoon, or other device can be supplied in some cases.

Usually non-potent medicament powders are prepared or dispensed in bulk quantities. Powders containing active substances whose doses should be given in controlled dosage are given to patients in folded packets or papers. Instructions should be given to the patients regarding appropriate handling, measurement, storage, and preparation of prescription and non-prescription bulk powders. Counselling should also be provided to the patients about the use, measurement of dose, and type of vehicle to be used. These products are stored at room temperature in a clean, dry place. They should be kept out of the reach of children and animals.

12.4.6.2 Divided powders

After proper blending of powder (geometric dilution for potent substances), it may be divided into individual dose units depending on the amount to be taken for single use. Each dose of powder may be placed in folded packets or papers. A number of commercially available products include headache powders (e.g., BC powders), douche powders (e.g., Messengill powder), and powdered laxatives (e.g., psyllium mucilloid and cholestyramine resin). Pharmacists can prepare divided powders as follows. The pharmacist made a decision whether to weigh each portion of powder separately before enfolding in a paper or to approximate each portion by using the block-and-divide method, depending on the potency of the medicament [5].

12.4.7 Classification of powders based on particle size

Powders of vegetable and animal origin drugs are officially defined as follows:

Very coarse (no. 8)
All particles pass through a no. 8 sieve, and not more than 20% pass through a no. 60 sieve.

Coarse (no. 20)
All particles pass through a no. 20 sieve, and not more than 40% pass through a no. 60 sieve.

Moderately coarse (no. 40)
All particles pass through a no. 40 sieve, and not more than 40% pass through a no. 80 sieve.

Fine (no. 60)
All particles pass through a no. 60 sieve, and not more than 40% pass through a no. 100 sieve.

Very fine (no. 80)
All particles pass through a no. 80 sieve. There is no limit to greater fineness [9].

The sieve number, along with the standard sieve opening and respective type of powders, is mentioned in Tab. 12.1.

Tab. 12.1: Opening of standard sieves.

S. no.	Sieve no.	Sieve opening	Powder
1	2.0	9.5 mm	
2	3.5	5.6 mm	
3	4.0	4.75 mm	
4	8.0	2.36 mm	
5	10.0	2.0 mm	
6	20.0	850.0 μm	
7	30.0	600.0 μm	Coarse
8	40.0	425.0 μm	Moderately coarse
9	50.0	300.0 μm	
10	60.0	250.0 μm	Fine
11	70.0	212.0 μm	
12	80.0	180.0 μm	Very fine
13	100.0	150.0 μm	
14	120.0	125.0 μm	
15	200.0	75.0 μm	
16	230.0	63.0 μm	
17	270.0	53.0 μm	
18	325.0	45.0 μm	
19	400.0	35.0 μm	

In a sieve are the holes are distributed in an inch. As sieve no. 10 has 10 holes in an inch and each hole has a size 2 mm, the total area of holes is $10 \times 2 = 20$ mm or 2 cm, but in 1 inch, we have 2.54 cm, so the remaining 0.54 cm is the distance between all the holes:

1 inch = 2.54 cm

= 2.00 + 0.54

= total area of holes + total area of distance between the holes.

12.5 Comminution of powders

Size reduction procedures for medicinal powders are evolving, with new equipment becoming accessible. Different approaches are classified based on the milling process used to separate powder particles. It's important to note that size reduction is always proportional to milling duration [10].

12.5.1 Manual comminution

In extemporaneous compounding, there are three methods of comminution.

12.5.1.1 Trituration

The process of grinding a drug in a mortar with a pestle to reduce its particle size is termed trituration. Trituration is best accompanied by rotating the firmly held pestle in a circular movement with downward pressure, starting at the center of the mortar and gradually moving in successive circles until the side of the mortar is touched and then back again at the center. This method is used when working on hard and fracturable powders [11]. It is usually used in the manufacturing of bulk powders and tablets, and also ensures the mono-sized particles. Trituration also eases the combination of powders thoroughly by adding the inert diluent. Crystalline substances are handled in porcelain mortars, while potent or colored drugs are handled in glass mortars to avoid the wastage of material. It is an ancient and dependable method in extemporaneous compounding.

12.5.1.2 Pulverization

This method is used for hard crystalline powders that are unable to crush or triturate easily or gummy-type substances. It is the process of reducing a substance to a fine powder utilizing a solvent that can be removed easily. The first step is to dissolve the drug in the intervening solvent, such as alcohol or acetone. Then, dissolve the powder and properly mix it in the mortar or spread it on the slab to speed up the evaporation process [9]. The quick evaporation of solvents results in minute particles with decreased size to an extent. In case of sticky and volatile substances, such as camphor and iodine, which triturate when dry, pulverization is highly convenient. It also avoids caking and agglomeration of powders in mortar, in addition to minimizing particle size. The process provides a sleek and smooth end product, which is further applied in the compounding process.

12.5.1.3 Levigation

In this method, particle sizes are reduced by triturating in a mortar or spatulating on an ointment slab with a small amount of solvent in which the solid is not soluble. The solvent used should be viscous, such as mineral oil or glycerin. In calamine lotion, we use glycerin for this purpose. This solvent is called a levigating agent, and the process is also termed levigation. While carrying out the process, in some cases levigating agent is removed by evaporation, e.g., camphor is levigated in the presence of a few drugs or alcohol. It makes a much more elegant, non-gritty product in a considerably shorter time than simple incorporation [12]. It is also commonly employed for the preparation of creams, suspensions, and ointments in order to obtain smooth products. It also helps in dispersing the insoluble products in semisolid bases uniformly. Suitable selection of levigating agent is crucial to ensure compatibility with the drug and vehicle. Due to the absence of an irritating effect caused by rough particles of topical preparations, this technique enhances the patient's compliance.

12.5.2 Mechanical size reduction methods

12.5.2.1 Cutting method

Cutter mill: A cutter mill uses knives mounted to a horizontal rotor to cut against fixed knives on the mill casing. Milling reduces particle size by fracturing them between two blades with a few millimeters of space. A screen in the mill casing retains material until it is sufficiently reduced in size, allowing for self-classification. Cutter mills may reduce dry granulations to a coarse size before tableting due to their high shear rates. The size reduction range by this mill is approximately 100–100,000 μm.

Particles of varying sizes and size distributions demonstrate distinct behaviors due to differences in surface area, bulk density, porosity, flow characteristics, and solubility between smaller and larger particles. These properties play a critical role in the formulation, packaging, and processing of pharmaceutical dosage forms, underscoring the importance of comminution as a vital step in manufacturing. This chapter delves into the concept of size reduction, emphasizing its essential role in pharmaceutical processes, the mechanisms involved, factors that influence size reduction, and various techniques for particle size analysis. Furthermore, it offers comprehensive insights into the design features and operational principles of equipment utilized for the reduction of solids, dispersions, and semisolids [11].

12.5.2.2 Compression method

Roller mill: The roller mill is a type of compression mill consisting of two horizontally placed cylindrical rollers that spin around their long axes. In roller mills, one roller is operated directly, while the second is turned by friction as material flows through the space between the rollers [13].

Chitin has been investigated as a potential excipient for direct compression through roller compaction, with a simultaneous comparison to ball milling techniques. A variety of chitin powder forms, raw, processed, dried, and humidified, were subjected to comprehensive analysis, including assessments of morphology, X-ray diffraction patterns, densities, Fourier Transform Infrared (FT-IR) spectra, flowability, compressibility, and compactibility. The roller compaction process effectively converted the fluffy raw chitin powder into denser granules with enhanced flow properties. X-ray diffraction analysis revealed only a slight reduction in crystallinity, in contrast to the significant decreases noted with ball milling, indicating minimal lattice deformation during roller compaction. The compacted chitin exhibited a high resistance to compression, attributed to improved granule compactibility resulting from substantial plastic deformation. In contrast, neither drying nor humidification improved the compressibility or compactibility of the directly compressed chitin. Additionally, formulations containing metronidazole demonstrated comparable drug release profiles between compacted chitin and microcrystalline cellulose, highlighting chitin's potential as a direct compression excipient in pharmaceutical formulations [14].

12.5.2.3 Impact method

Hammer mill: A hammer mill is effective for size reduction by impact. Hammer mills are made up of four or more hammers connected by a central shaft and encased in a sturdy metal casing. During milling, the hammers swing out radially from the center shaft. Hammers with high angular velocity can cause strain rates of up to $80\ s^{-1}$, leading to brittle fracture in most particles. Hammer mills create powders with narrow size distributions due to lower particle mass and inertia. The mill's screen prevents particles from passing through unless they are sufficiently comminuted. Particles going through a particular mesh can be considerably finer than the mesh apertures because the hammers carry particles around the mill and approach the mesh tangentially. Square, rectangular, or herringbone slots are commonly employed. Hammers can be squarely faced, tapered to a cutting edge, or stepped on to suit the job.

Typically, a hammer mill is not ideal as a mixing device due to its limited capacity for holding materials. It is generally more effective to grind individual pure components rather than mixtures, as there is a risk of losing active ingredients during the milling process. After comminution in the hammer mill, it is essential to remix the components. This sequence of grinding followed by remixing significantly improves the homogeneity of the mixture. The comminution process produces a diverse range of particle sizes, which can lead to the segregation of ordered units, ultimately resulting in a final mixture characterized as a randomized ordered mixture [15].

Vibration mill: Vibration milling is a viable alternative to hammer milling for size reduction. Vibration mills typically include around 80% porcelain or stainless steel balls. Milling involves vibrating the whole mill body and reducing the size through repeated impacts. The mill's base screen filters out comminuted particles. The vibratory ball mill is particularly effective in combining milling and mixing operations, quickly producing highly homogeneous fine powder mixtures.

12.5.2.4 Attrition method

Roller mills employ attrition to reduce the size of solids in suspensions, pastes, and ointments. Two or three porcelain or metal rollers are positioned horizontally with an adjustable spacing (as tiny as 20 μm). The rollers revolve at different speeds, shearing the material as it passes through the gap. The material is then transferred from the slower roller to the faster roller and removed using a scraper [16]. Comparatively, attrition consumes less energy than other crushing techniques, so an insignificant amount of thermal damage is done. They are mostly employed in cream, ointment, and suspension preparation, where uniformity of particles is vital for the stability and therapeutic effectiveness. Particle size is also controllable with the respective method by the exact change in pressure and the separation between rollers.

12.5.2.5 Combined impact and attrition methods

Ball mill: A ball mill uses impact and attrition to reduce particle size. Ball mills are made from a hollow cylinder that rotates horizontally along its longitudinal axis. The cylinder includes balls that make up 30–50% of the overall volume, with ball size determined by feed and mill size. Mills use balls of various sizes to optimize the operation. Large balls break down coarse feed materials, while smaller balls make fine products by minimizing the vacuum spaces. The ball mill's rotational speed is crucial for its operation. At low angular velocities, the balls travel with the drum until gravity's pull surpasses the frictional force of the bed, at which point they slide back to the drum's base.

Repeating ball mill operation over time results in negligible ball movement and minimal size decrease. At high angular velocities, balls are flung out to the mill wall and remain there due to centrifugal force, resulting in no size decrease. At around two-thirds of the essential angular velocity for centrifugation, a cascade effect begins. Lift the balls on the drum's rising side until they reach their maximum dynamic angle of repose. The particles cascade over the mill's diameter before returning to the drum's base. The most efficient size reduction happens by particle-ball impact and attrition. The optimal rate of rotation depends on the mill's diameter and is typically about 0.5 revolutions per second [17–19].

Chi Kwan and colleagues [20] provided a detailed exploration of the ball milling process, introducing a novel method to evaluate the grinding behavior of pharmaceutical powder blends. This method emphasizes the assessment of their chemical and physical properties, determined through impact testing of individual particles. The study investigated the grinding behavior of two widely used pharmaceutical excipients: microcrystalline cellulose and α-lactose monohydrate, utilizing a rotating single ball mill. The results indicated that the grinding characteristics of these components could be compared to a first-order rate reaction, except for α-lactose monohydrate at milling frequencies of 18 Hz. Additionally, impact testing was utilized to derive the physical properties related to the potential for powder disintegration. The milling rate for these powders demonstrated a strong correlation with specific physical properties, including particle diameter, density, hardness, and critical stress intensity. This finding provides a new perspective on understanding the grinding behavior of materials through a method that is both straightforward and reliable [20].

Fluid energy mill: Fluid energy milling is a size-reduction process that involves particle impaction and attrition. A sort of fluid energy, jet mill, or micronizer. Circular and oval route designs are provided. The round design is currently the most frequent. This comprises a hollow toroid with a diameter ranging from 20 to 200 mm. Typically, a high-pressure jet of air is pumped through nozzles at the bottom of a loop. The air's high velocity creates turbulence zones that attract solid particles. The air's high kinetic energy causes particles to strike one another and the mill sides, resulting in fracture. Turbulence causes high collision rates between particles, resulting in size reduction and attrition. The design includes a particle size classifier that retains particles in the toroid until they are fine enough to be entrained in the mill's exhaust stream [21]. The surface energetics of different samples, when evaluated using vacuum microbalance and microcalorimetry to monitor water adsorption, reveal that milling significantly affected surface energetics depending on the energy characteristics of the milling process. Powders subjected to two milling methods retain the surface properties established during the first milling.

Pin mill: In addition to ball mills and fluid energy mills, other comminution processes use particle impact and attrition. Pin mills use two discs with closely spaced pins to revolve at high speeds. Particle size is reduced by pin impaction and attrition as the centrifugal force moves particles outside.

12.6 Blending powders

Blending depends on the nature of the powder ingredients, the amount of powder, and the equipment used for blending. Mixing of powders is carried out by following the methods.

12.6.1 Spatulation

Spatulation involves combining small portions of powders with a spatula on paper or an ointment tile. Homogeneous blending is not reliable for large quantities or powders with powerful substances. Other procedures are better suited. Spatulation is ideal for mixing solid components that create eutectic mixtures (liquefaction) when in close contact. It results in little particle compression or compacting. So, therapeutic materials (materials that are liquefied on compression), e.g., phenol, camphor, menthol, thymol, and aspirin, are difficult to prepare by this method. To reduce interaction, a powder made from these compounds is combined with an innocuous diluent, such as magnesium oxide or magnesium carbonate, to physically separate the harmful agents.

Particle size distributions are typically measured for systems with single, well-dispersed (primary) particles. Free-flowing powders disperse easily, while stickier powders may require air dispersion or liquid dispersion to achieve primary particle separation. Liquid dispersion, enhanced by wetting agents, surfactants, stirrers, or ultrasonic baths, reduces inter-particle forces and facilitates dispersion. Stabilizing chemicals can further improve dispersion quality, which is crucial for accurate evaluation. Inadequate dispersion results in incomplete de-agglomeration and larger particle sizes. Spatulation is important in this context as it aids in the initial manual mixing and dispersion of powders before more intensive methods are applied. Various techniques are available to evaluate dispersion quality and stability over time, addressing issues like swelling, dissolution, or breakage.

12.6.2 Trituration

Trituration can be used to finely grind or combine particles. If a simple admixture is wanted without any specific requirements. For comminution, glass mortars are typically preferred. The geometric dilution method evenly distributes a potent material when combined with a significant volume of diluent. This strategy is recommended when the potent chemical and other substances are similar in color and there is no obvious trace of mixing. To use this procedure, combine the potent drugs with an equal volume of diluent in a mortar and thoroughly triturate them.

This procedure includes the following steps. Take a potent drug and mix an equal amount of diluents. Take the mixture in a mortar and then mix an amount of diluents which is equal to the first mixture. In this way, all the diluents are mixed. This process is continued by adding an equal volume of diluents to the powder present in the mortar and repeating the mixing until all diluents are used. This is called geometric mixing. It is used for a small amount of a potent drug.

12.6.3 Sifting

Powders can be combined by running them through sifters, similar to how flour is sifted. Sifting produces a light and fluffy product. This technique is not suitable for incorporating strong medicines into diluent powders.

With the implementation of good manufacturing practice, sifting equipment is required to adhere to stringent sanitation standards, prompting the development of technologies aimed at minimizing contamination. Recent advancements, such as ultrasonic sifters, facilitate the sifting of materials with very fine meshes below 100 μm, addressing previous challenges and representing a significant advancement in sifting technology. This report provides an overview of the evolution of sifting equipment and technologies designed to prevent contamination [22].

12.6.4 Tumbling

Powder can also be mixed by tumbling it in a spinning chamber. Motorized powder blenders, both small and large-scale, use tumbling to combine powders. This method involves thorough mixing yet takes time. Blenders and mixers with motorized blades are commonly used in industry to combine materials in large vessels. In this method, we have a hollow tube. It has no blades, and powder is allowed to fall freely. This process can be performed simply with a polyethylene bag by adding the drug to it, closing the neck, and then mixing it by shaking. Experimental validation indicates that cohesion affects the spatial distribution of particles over time; specifically, higher cohesion initially enhances the mixing rate in uniform mixtures, but this effect diminishes as mixing progresses. In non-uniform systems, the mixing patterns differ based on the interactions between particles.

12.7 Evaluation of pharmaceutical powders

The flowability of pharmaceutical powders is one of the key elements to be determined during tablet processing and capsule filling. The reason is that before the compression, powders or granules must be flown from the hopper shoe into the die cavity of the tablet punching machine, i.e., rotary multi-station tablet press. This also applies during capsule filling as the beads or powders must be flown from the hopper into capsule shells. This depends on the smoothness of the process since the powder flow from the hopper into the dies often determines the weight, hardness, and content uniformity of tablets. Understanding of powder flow is also crucial during mixing, packaging, and transportation. And thus, it becomes essential to measure the flow properties of these materials before tableting or capsule filling.

Among all the methods, cohesivity determination, avalanching determination, shear cell, dielectric imaging, atomic force microscopy [23], and penetrometry are new tests designed after technological advancements. There are various limitations of these methods, such as reproducibility, performance conditions, and predictability. For example, the angle of repose and avalanching determination are not applicable for cohesive powders, since they don't flow through the hopper or funnel during tableting or capsule filling. Other characterization techniques, such as bulk density and tapped density, are semiquantitative and unable to give completely accurate information. So far, no single test has been accepted as a standard for the measurement of powder flow [24].

12.7.1 Angle of repose

The angle of repose is the angle formed by the horizontal base of the bench surface and the edge of a cone-like pile of granules. Here is the formula to determine the angle of repose:

$$\theta = \tan^{-1}\left(\frac{h}{r}\right)$$

To measure the angle of repose, allow the powder to fall to the flat surface from the funnel that is positioned at a certain height. The funnel is slowly lifted upward to maintain a constant distance between the powder tip and the funnel bottom. The angle that is produced by the powder over the surface is known as the angle of repose. The alternate approach is to allow powder to fall freely from a centrally driven hole in a flat-bottom container. In this case, the slope is the angle of repose [25]. The angle of repose, whose tangent equals the coefficient of friction between the two bodies, is the angle that the plane of contact between two bodies forms with the horizontal when the upper body is almost at the point of sliding [26].

A conical pile is created when large granular materials are placed over a horizontal surface. The angle of repose, which is the internal angle formed between the surface of the pile and the horizontal surface, is influenced by the material's coefficient of friction, density, surface area, and particle morphologies. Compared to material having a high angle of repose, material with a low angle of repose generates flatter piles [27]. The angle of repose may be measured using a variety of techniques, and the outcomes vary slightly. The particular approach used by the experimenter has an impact on the results as well. Data from several laboratories are therefore not always comparable. The direct shear test and the triaxial shear test are two techniques. The following function may be used to approximate the angle of repose with a fair degree of accuracy if the material's coefficient of static friction is known. For stacks containing tiny individual objects arranged in an arbitrary sequence, this function is fairly accurate, where θ is the angle of repose and μs is the coefficient of static friction.

12.7.1.1 Tilting box technique

This technique is suitable for non-cohesive, fine-grained materials with individual particle sizes less than 10 mm. To view the granular test material, the material is placed within a transparent-sided box. At first, it ought to be parallel to the box's base and level. The tilt angle is determined when the box is gradually tilted until the material starts to move in bulk.

12.7.1.2 Fixed funnel method

To create a cone, the material is poured through a funnel. To reduce the effect of falling particles, the funnel tip should be held near the developing cone and lifted gradually as the pile rises. When the base reaches a certain width or the pile reaches a predetermined height, stop pouring the material. Divide the height by half the base width of the resultant cone instead of trying to estimate the angle directly. The angle of repose is the inverse tangent of this ratio.

12.7.1.3 Method of rotating cylinder

The substance is put within a cylinder that has at least one transparent end. The material within the revolving cylinder is observed by the observer while the cylinder rotates at a set speed. It's like seeing clothes tumble over each other in a clothes dryer that rotates slowly. As the granular material passes through the revolving cylinder, it will take on a certain angle. The dynamic angle of repose, which might differ from the static angle of repose determined by other techniques on different materials, should be obtained using this method. The different materials and their angles of repose are listed below. Every measurement is an estimate [28].

12.7.2 Bulk density

Bulk density is the weight of a volume unit of powder and is usually expressed in g/cm³, kg/m³, or g/100 mL:

$$\rho = \frac{m}{V}$$

The mass of an untapped powder sample divided by its volume, which includes the interparticulate void volume contribution, is the powder's bulk density. Therefore, the density of the powder particles and their spatial arrangement in the powder bed both affect the bulk density. Because the measurements are conducted using cylinders, the bulk density is represented in grams per milliliter (g/mL), even though the international unit is kilogram per cubic meter (1 g/mL = 1,000 kg/m³).

Another way to describe it is as grams per cubic centimeter or g/cm³. A powder's bulking qualities rely on how the sample was treated during production, storage, and treatment. Particles may be arranged to have a variety of bulk densities, and even the smallest disruption to the powder bed can cause the bulk density to vary. Because of this, it can be challenging to measure a powder's bulk density with excellent repeatability. Therefore, it is crucial to describe the methodology used in the determination when presenting the results. The bulk density of a powder can be found by measuring the mass of a known volume of powder that has been passed through a volumeter into a cup (Method B), a measuring vessel (Method A), or the volume of a known mass of powder sample that may have been passed through a sieve into a graduated cylinder (Method A).

12.7.2.1 Method A measurement in a graduated cylinder process

If required, gently break up any agglomerates that may have developed during storage by passing through a sieve with openings larger than or equal to 1.0 mm with an amount of powder needed to finish the test. Be careful not to alter the material's composition in any way. Gently add about 100 g of the test sample (m), which has been precisely weighed to within 0.1% of its capacity, into a dry graduated cylinder measuring 250 mL (readable to 2 mL), without compacting. When reading the unsettled apparent volume (V_0), take care not to compress the powder and round it up to the closest graded unit. Use the formula m/V_0 to determine the bulk density in units of (g/mL). For the determination of this attribute, duplicate determinations are generally preferred. It is not feasible to utilize 100 g of powder sample if the density of the powder is too high or too low, resulting in an untapped apparent volume of either more than 250 mL or less than 150 mL in the test sample. As a result, a different quantity of powder must be chosen for the test sample, with the mass of the sample being indicated in the expression of findings, and its unblocked apparent volume having to be between 150 and 250 mL (apparent volume more than or equal to 60% of the cylinder's total capacity).

12.7.2.2 Method B measurement in a volumeter

Use a minimum of 25 cm^3 of powder with the cubical cup and 35 cm^3 of powder with the cylindrical cup, and allow an excess of powder to flow through the device and into the sample receiving cup until it overflows. Using a spatula blade that is smoothly moved perpendicular to the cup's top surface and in touch with it, carefully scrape off any extra powder from the top of the cup. Be careful to maintain the spatula perpendicular to avoid packing or removing powder from the cup. To the closest 0.1%, find the mass (M) of the powder after removing any material from the cup's side [28].

12.7.3 Tapped density

The tapped density of a powder is the ratio of the mass of the powder to the volume occupied by the powder after it has been tapped for a defined period. The following is the formula to calculate the tapped density:

$$\rho_t = \frac{m}{V_t}$$

To measure tapped density, gently pour the specified amount of powder into the measuring cylinder without any mechanical jerks to the cylinder. Adjust the level of powder and now mechanically tap the cylinder by raising it in the air and allowing it to drop under its own weight using a suitable mechanical tapped density tester that provides a suitable fixed drop distance and the nominal drop rate.

12.7.4 Carr's compressibility index and Hausner's ratio

The most usual applied indirect methods to predict the powder flowability are Carr's index (CI) and Hausner's ratio (HR). CP < 15% and high HR (approaching 1) are desirable because they imply excellent flow properties, which are critical in achieving consistent die filling. A greater CI (>25%) and HR (>1.34) show that it has a poor flow because of inner particle friction, irregular shape, or cohesive nature of the powder. They are also applied in the formulation stage, where the excipients are screened to optimize powder blends. Examples include micronized drugs, which tend to be cohesive and have proper flow properties, where spray-dried lactose exemplifies granular excipients that have good flow. In industrial practice, CI and HR are used as quality control tools in small-scale production before the development of large tablets or capsules. Furthermore, the flow improvers (e.g., glidants) could be added to minimize CI and HR values and improve processing.

The bulk and tapped densities are used to determine CI and HR as follows (Tab. 12.2):

$$CI = \frac{\rho_t - \rho_b}{\rho_t} \times 100$$

$$HR = \frac{\rho_t}{\rho_b}$$

Tab. 12.2: Flow properties of powders [29].

S. no.	Carr's index (%)	Flow property	Hausner's ratio
1	≤10	Excellent	1.00–1.11
2	11–15	Good	1.12–1.18
3	16–20	Fair	1.19–1.25
4	21–25	Passable	1.26–1.34
5	26–31	Poor	1.34–1.45
6	32–37	Very poor	1.46–1.59
7	>38	Very very poor	>1.60

12.8 Granules

Granules are huge agglomerates formed during the granulation process from tiny fine or coarse particles [30]. They are generally irregularly shaped but may be prepared as spherical. The size of these varies from 4 to 12 mesh range. However, the granules of different sizes can be prepared depending on the application. Pharmaceutical granules can vary in size from 0.2 to 4.0 mm, depending on their intended application. Granules can be packaged as a dosage form or blended with other excipients before being compacted or filled into a capsule [31]. Granulation is needed to prevent segregation, enhance flow characteristics, and improve compaction [32].

12.8.1 Types of granules

There are several categories of orally administered granules:

12.8.1.1 Effervescent granules

Effervescent granules are uncoated granules that typically contain acidic compounds, carbonates, or hydrogen carbonates, which react quickly with water to release carbon dioxide, in addition to the medication. Sodium bicarbonate and citric acid are a common combination (Tab. 12.3). Before being administered, effervescent granules are meant to dissolve or distribute in water. In 5 min, the granules should have completely disintegrated; at this point, the substances should be completely dissolved or distributed throughout the water. It is recommended to keep effervescent granules in an airtight container [33].

Tab. 12.3: Sodium phosphate effervescent granules.

S. no.	Ingredients	Concentration
1	Dibasic Na phosphate	200 g
2	Na bicarbonate	478 g
3	Citric acid	162 g
4	Tartaric acid	252 g

12.8.1.2 Reactions

For citric acid:

$$3NaHCO_3 + C_6H_8O_7 \cdot H_2O \longrightarrow 4H_2O + 3CO_2 + Na_3C_6H_5O_7$$

$$162 \text{ (citric acid)}/210 \text{ (mol. weight)} = x(NaHCO_3)/84 \times 3$$

$$x(NaHCO_3) = 162 \times 84 \times 3/210$$
$$= 196 \text{ g}$$

For tartaric acid:

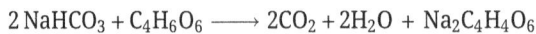

$$2\,NaHCO_3 + C_4H_6O_6 \longrightarrow 2CO_2 + 2H_2O + Na_2C_4H_4O_6$$

$$252 \text{ (tartaric acid)}/150 \text{ mol. weight} = x(NaHCO_3)/84 \times 2$$

$$x(\text{NaHCO}_3) = 252 \times 84 \times 2/150$$
$$= 282 \text{ g}$$

Total amount of NaHCO_3 used = 196 + 282 = 478

12.8.1.3 Coated granules

Granules coated in one or more layers of different excipient blends form coated granules. The materials that are employed as coatings (which are often polymers) are typically applied as a suspension or solution in a situation where the vehicle evaporates and leaves a coating film behind. The coating is used to fulfil various functions like masking the taste of a material, increasing the resistance to water and light, and also avoiding the interaction with other ingredients present. Moreover, coatings can be functional, enabling the modified release or protecting against the gastric acid. The aesthetic attraction can also be improved by the use of coated granules, which also allows for combination therapy with different drugs coated separately and missing in a single formulation. To achieve a uniform or near uniform coating, developments of fluidized-bed technology have enabled coating of very fine particles.

12.8.1.4 Modified-release granules

Modified-release granules can be coated or uncoated, consist of specific excipients, are manufactured using specific methods, or both, and are designed to change the place, time, or rate at which the active ingredient or substances are released. Extended-release or delayed-release properties are possible for modified-release granules and achieve long-term drug release, thereby avoiding high dose frequency and also improving adherence. These preparations are particularly employed when drugs have a brief half-life, when drugs cause gastric irritation, or when drugs are absorbed by a localized part of the intestine. It is also possible to adjust the dose of modified release granule as they can be added in capsules, packets, or processed into tablets.

12.8.1.5 Gastro-resistant granules

Enteric-coated granules, formerly known as gastro-resistant granules, are delayed-release granules designed to withstand stomach fluid and release the active ingredient(s) into the intestinal fluid. This is generally achieved by coating the granules with a gastro-resistant polymer. These granules are most importantly significant in the case where medications are unstable in gastric acid, or any medication that tends to irritate the interior lining of the stomach, like NSAIDs or medicines that can act locally in the intestines. Capsule filling and tablet compression are likely to be used with gastro-resistant granules to ease the pharmaceutical administration. Additionally, they have an advantage in pediatric and geriatric patients, as the granules can be sprinkled over food without affecting the protection offered by the enteric coating.

12.8.2 Methods of preparation

Granules can be prepared by two different methods.

12.8.2.1 Wet method

Using a granulating liquid, a mixture of dry primary powder particles is assembled in the process of wet granulation. The solvent in the granulating fluid needs to be non-toxic and volatile in order to be eliminated by drying. Water, ethanol, and 2-propanol are common appropriate liquids that can be used individually or in combination. The granulation liquid can be employed either alone or, more often, in combination with a solvent to ensure particle adherence after the granule has dried. This solvent is known as a binder, or binding agent. Using the wet granulation process, wet granules are formed by passing the wet mass through a screen and then drying them. The fine material, which may be recycled, is removed in a second screening stage that breaks apart granule agglomerates (Tab. 12.4).

Tab. 12.4: Merits and demerits of wet granulation.

S. no.	Merits	Demerits
1	The cohesion and compressibility of powders are improved	Because of the large number of processing steps, it requires a large area with temperature and humidity control
2	Good distribution of content due to uniform mixing	It requires a number of pieces of equipment and is time-consuming; hence, it is more expensive
3	A wide variety of powders can be possessed together in a single batch	There is a possibility of material loss during processing due to the transfer of materials from one unit to another, and there is a possibility of cross-contamination
4	The hydrophilicity of powders is increased	
5	A controlled-release dosage form can be accomplished by the selection of a suitable binder and solvent	

Lixia Cai et al. [34] proposed and validated an innovative methodology for efficiently developing high drug load (>85%) formulations using high shear wet granulation. This approach correlates API properties, binder types, and granulation fluid levels to optimize product attributes, emphasizing the addition of excipients only when necessary to enhance formulation processability or performance. Simvastatin, etoricoxib, and metformin hydrochloride were selected as model drugs due to their varied particle characteristics. The granulation process involved varying fluid levels and employing polymeric binders, with evaluations on granule size distribution, strength, flowability, dissolution, and compactibility. Some formulations were compressed into tablets and further assessed with extragranular excipients. One formulation underwent in vivo pharmacokinetic studies in dogs, demonstrating the successful development of high drug load formulations with desirable attributes. This study underscores the importance of wet granulation in achieving efficient drug formulation development and optimizing pharmaceutical performance [34].

12.8.2.2 Dry method

Dry granulation includes the formation of granules without involving fluid arrangement, as the item may be delicate to moisture and heat. In this process, dry powder particles may be united precisely by pressure to form slugs or by roller pressure to form flakes. In both situations, an appropriate milling process is used to break the intermediate product into granular material, which is then typically sieved to separate the required size fraction. To avoid waste, the fine material that is not utilized may be reprocessed. Granules made by this technique are porous and profoundly compressible. This process is excellent for the drugs that are sensitive to moisture [35].

Oscar-Rupert Arndt et al. (2016) investigated 18 different dry binders in roll compaction/dry granulation to address challenges such as lower tablet tensile strength compared to direct compression methods. They evaluated three types of microcrystalline cellulose, five types of hydroxypropyl cellulose (HPC), three types of povidone, three types of copovidone, and four types of crospovidone at a 10% fraction to optimize granule size and minimize fines. Consistent dry granulation and tableting methods were applied across all batches. The results highlighted significant variations in granule size, fines fraction, tablet tensile strength, friability, and disintegration time depending on the binder type. Linear regressions within chemical binder groups elucidated binding behaviors influencing granule and tablet properties. Fine grades of HPC and copovidone emerged as the most effective binders, with HPC producing larger, less friable granules, and copovidone facilitating faster disintegration, likely due to viscosity effects. This study underscores the importance of dry granulation in optimizing binder selection to achieve desired tablet characteristics and effectively address formulation challenges [36].

12.9 Granulation mechanisms

Particle adhesion happens in the dry method due to applied pressure, and a sheet is formed which is bigger than the required granule size; so, desired size of granules is achieved by milling and sieving.

The wet granulation method uses the mechanical agitation produced in the granulator to disperse added liquid throughout the powder. Liquid films hold the particles to one another, and more agitation and/or liquid addition increases the adhesion of the particles. Each type of granulation apparatus has a different exact mechanism for converting a dry powder into a bed of granules. The suggested mechanism for granulation can be segmented into three phases: transition, nucleation, and ball growth.

12.9.1 Nucleation

Particle-to-particle contact and adhesion that result from liquid bridges initiate granulation. The pendular condition will be formed when many particles join together. The capillary state is formed by further agitation, making the pendular bodies, which serve as the building blocks for more granule development [37, 38].

Primary granule nuclei are formed and stabilized by liquid bridges during the nucleation process. In the advancement of the process, these nuclei can absorb other particles, and the starting phase of the development of the granule begins. The quantity and resilience of nuclei formed can be determined by the liquid binder added. Insufficient amounts of binder give weak and friable nuclei, and excessive binder can result in over-wet conditions and lump formation. The temperature and amount of mixing also affect the equilibrium between the capillary and pendular states. Hence, painstaking optimization is taken in this step to ensure a very robust nucleus form that can be further built into granules of appropriate size and shape.

12.9.2 Transition

There are two ways in whereby nuclei can expand: either by combining two or more nuclei or by adding single particles to the nuclei through pendular bridges. The agitation of the bed will reorganize the coupled nuclei. A significant amount of tiny granules with a reasonably wide size distribution are present during this stage. This point serves as an appropriate endpoint for granules used in the production of tablets and capsules, provided that the size distribution is not too broad. This is because relatively small granules will result in a consistent tablet die or capsule fill, and bigger granules can cause uneven fill and bridge across the die, which may cause issues in small-diameter dies.

In the transition phase of granulation, Baghdadi and colleagues [39] examined the kinetics of granule growth by nuclei aggregating and pendular bridge building. They clarified the principles behind granule expansion and reorganization in response to agitation and particle rearrangement using sophisticated imaging methods and computer modelling. Their study made clear how crucial it is to regulate process variables in order to optimize tablet and capsule manufacturing and get the appropriate distribution of granule sizes [39].

12.9.3 Ball growth

Large, spherical granules are produced by continual granule growth, and the granulating system's mean particle size will eventually increase. Granule coalescence will occur with more agitation, leading to an overmassed, unusable system; however, this will rely on the characteristics of the material being granulated as well as the volume of liquid added. Ball growth is a necessary component of some spheronizing equipment and will occur to some extent in planetary mixers, even though the granules formed may be too large for medicinal applications. Ball growth has three mechanisms, which are as follows:

12.9.3.1 Coalescence

When two or more granules combine, a large granule is formed. This can be employed when there is a sufficiently strong liquid bridging between particles so that the resulting forces resist shearing in the bed. Coalescence makes crucial contributions to increasing the mean particle size, particularly in situations of high liquid content of the mixture and intensive mixing.

12.9.3.2 Breakage

Physically stressed break granules into pieces or are agitated. These fragments can attach to each other to form intact granules, forming secondary layering. Breakage can be controllable in order to stay within a narrower particle size distribution, but simultaneously, excessive breakage produces product inconsistency.

12.9.3.3 Abrasion transfer

Granule attrition occurs due to the pumping up and down motion of the granule bed, eroding that material. Abraded fines are redistributed throughout the system and held on the surface of other granules, resulting in gradual size growth. This mechanism is prevalent in systems that are highly shear and where mixing times are longer, and this also plays a part in the smoothness of the surface of the final granules.

12.9.3.4 Layering

The granule size will increase when another batch of powder mix is applied to a bed of granules because the powder will stick to the granules and form a layer on the surface. This process is only important when making layered granules with spheronizing equipment [4, 40]. Layering is suitable to produce uniformity in layers of APIs or coatings, and this aspect can be utilized to provide modified-release granules.

12.10 Examples

- Bephenium granules: It is used as an anthelmintic.
- Na – amino salicylate and isoniazid granule: It is used in the treatment of tuberculosis.
- Me – cellulose granules: It is used for the preparation of other materials.
- Chloroquine phosphate granules: It is employed as an antimalarial.
- Paracetamol granules: It is prepared to enhance flow properties and compressibility for tablet manufacturing.
- Magnesium carbonate granules: It is used as an antacid and laxative.
- Lactose granules: It is commonly used as a diluent and excipient in pharmaceutical formulations.

12.11 Advantages

- Granules may form a solution easily, but powders cannot.
- Granules can easily fall, i.e., they are free-falling, but powders cannot.
- Granules are physically and chemically more stable, i.e., they can resist the environmental hazards.
- Sometimes we can make a tablet from granules.
- Antibiotics are mostly marketed as granules.

12.12 Characteristics of good granules

- Should be of uniform size.
- It should have a better flow property.
- Should not be disintegrated or distorted on handling.
- Should easily be converted into other dosage forms, e.g., tablets.
- Should form a solution easily.
- Granules must have suitable mechanical properties.

References

[1]   Revision USPCCo, editor. The United States Pharmacopeia: United states pharmacopeial convention: Incorporated; Rockville, Maryland, USA, 1984.
[2]   Aulton ME, Taylor K. Aulton's pharmaceutics: The design and manufacture of medicines: Elsevier Health Sciences; London, UK, 2013.
[3]   Campbell GA, Vallejo E. Primary packaging considerations in developing medicines for children: Oral liquid and powder for constitution. Journal of Pharmaceutical Sciences. 2015; 104(1): 52–62.
[4]   Taylor KM, Aulton ME. Aulton's pharmaceutics e-book: Aulton's pharmaceutics e-book: Elsevier Health Sciences; London, United Kingdom, 2021.
[5]   Adepu S, Ramakrishna S. Controlled drug delivery systems: Current status and future directions. Molecules. 2021; 26(19): 5905.
[6]   Koumbogle K, Gosselin R, Gitzhofer F, Abatzoglou N. Moisture behavior of pharmaceutical powder during the tableting process. Pharmaceutics. 2023; 15(6): 1652.
[7]   Allen L, Ansel HC. Ansel's pharmaceutical dosage forms and drug delivery systems: Lippincott Williams & Wilkins; Philadelphia, PA, USA., 2013.
[8]   Garg T, Rath G, Goyal AK. Comprehensive review on additives of topical dosage forms for drug delivery. Drug Delivery. 2015; 22(8): 969–87.
[9]   Allen LV. The art, science, and technology of pharmaceutical compounding: Washington, DC: American Pharmaceutical Association; 1998.
[10] Eiland LS, Ginsburg DB. PharmPrep: ASHP's NAPLEX review: ASHP; Bethesda, Maryland, USA, 2011.
[11] Hanif F, Majeedullah. Comminution. In: Essentials of industrial pharmacy: Springer; Cham, Switzerland, 2022. pp. 27–44.
[12] Shrewsbury RP. Applied pharmaceutics in contemporary compounding: Morton Publishing Company; Englewood, Colorado, USA, 2015.
[13] Miller RW. Roller compaction technology. In: Handbook of pharmaceutical granulation technology: CRC Press; Boca Raton, Florida, USA, 2005. pp. 187–218.
[14] Abu Fara D, Al-Hmoud L, Rashid I, Chowdhry BZ, Badwan A. Understanding the performance of a novel direct compression excipient comprising roller compacted chitin. Marine Drugs. 2020; 18(2): 115.
[15] Yeung CC, Hersey JA. Powder homogenization using a hammer mill. Journal of Pharmaceutical Sciences. 1979; 68(6): 721–24.
[16] Schönert K. The characteristics of comminution with high pressure roller mills. KONA Powder and Particle Journal. 1991; 9: 149–58.
[17] Gusev A, Kurlov A. Production of nanocrystalline powders by high-energy ball milling: Model and experiment. Nanotechnology. 2008; 19(26): 265302.
[18] Dhangar PD, Sonawane TN, Shaikh AZ, Jain RS. Demonstration of Ball mill and their applications in Pharmacy. Asian Journal of Pharmacy and Technology. 2020; 10(4): 285–88.

[19] Ramos AS, Taguchi SP, Ramos EC, Arantes VL, Ribeiro S. High-energy ball milling of powder B–C mixtures. Materials Science and Engineering: A. 2006; 422(1–2): 184–88.

[20] Kwan CC, Chen YQ, Ding YL, Papadopoulos DG, Bentham AC, Ghadiri M. Development of a novel approach towards predicting the milling behaviour of pharmaceutical powders. European Journal of Pharmaceutical Sciences. 2004; 23(4): 327–36.

[21] Nair PR, Ramanujam M. Circular fluid energy mill. Advanced Powder Technology. 1992; 3(4): 285–98.

[22] Nagato T, Kinoshita N, Endo T, Hasegawa K, Terashita K. Study of design and optimization of tablet coating process by taguchi method. Journal of the Society of Powder Technology, Japan. 2010; 47(6): 394–401.

[23] Ali MY, Tariq I, Ali S, Amin MU, Engelhardt K, Pinnapireddy SR, et al. Targeted ErbB3 cancer therapy: A synergistic approach to effectively combat cancer. International Journal of Pharmaceutics. 2020; 575: 118961.

[24] Shah RB, Tawakkul MA, Khan MA. Comparative evaluation of flow for pharmaceutical powders and granules. Aaps Pharmscitech. 2008; 9: 250–58.

[25] Hadjittofis E, Das S, Zhang G, Heng J. Interfacial phenomena. In: Developing solid oral dosage forms: Elsevier; London, UK., 2017. pp. 225–52.

[26] Mehta A, Barker G. The dynamics of sand. Reports on Progress in Physics. 1994; 57(4): 383.

[27] Al-Hashemi HMB, Al-Amoudi OSB. A review on the angle of repose of granular materials. Powder Technology. 2018; 330: 397–417.

[28] Nandi K, Sen D, Patra F, Nandy B, Bera K, Mahanti B. Angle of repose walks on its two legs: Carr index and Hausner ratio. World Journal of Pharmacy and Pharmaceutical Science. 2020; 9(5): 1565–79.

[29] Gorle AP, Chopade SS. Liquisolid technology: Preparation, characterization and applications. Journal of Drug Delivery and Therapeutics. 2020; 10(3-s): 295–307.

[30] Shanmugam S. Granulation techniques and technologies: Recent progresses. BioImpacts: BI. 2015; 5(1): 55.

[31] Solanki HK, Basuri T, Thakkar JH, Patel CA. Recent advances in granulation technology. International Journal of Pharmaceutical Sciences Review and Research. 2010; 5(3): 48–54.

[32] Jannat E, Al Arif A, Hasan MM, Zarziz AB, Rashid HA. Granulation techniques & its updated modules. The Pharma Innovation. 2016; 5(10, Part B): 134.

[33] Parikh DM. Handbook of pharmaceutical granulation technology. Drugs and the Pharmaceutical Sciences. (vol.. 81). Boca Raton; Florida, USA, 2005.

[34] Cai L, Farber L, Zhang D, Li F, Farabaugh J. A new methodology for high drug loading wet granulation formulation development. International Journal of Pharmaceutics. 2013; 441(1): 790–800.

[35] Peck GE, Soh JL, Morris KR. Dry granulation. In: Pharmaceutical dosage forms-tablets: CRC Press; Florida, USA., 2008. pp. 319–52.

[36] Arndt O-R, Kleinebudde P. Influence of binder properties on dry granules and tablets. Powder Technology. 2018; 337: 68–77.

[37] Kristensen HG, Schaefer T. Granulation: A review on pharmaceutical wet-granulation. Drug Development and Industrial Pharmacy. 1987; 13(4–5): 803–72.

[38] Agrawal R, Naveen Y. Pharmaceutical processing–A review on wet granulation technology. International Journal of Pharmaceutical Frontier Research. 2011; 1(1): 65–83.

[39] Baghdadi YN, Shah RK, Albadarin AB, Mangwandi C. Growth kinetics of nuclei formed from different binders and powders in vertical cylindrical mixing devices. Chemical Engineering Research and Design, 2018; 132. https://doi.org/10.1016/j.cherd.2017.12.045.

[40] Iveson SM, Litster JD, Hapgood K, Ennis BJ. Nucleation, growth and breakage phenomena in agitated wet granulation processes: A review. Powder Technology. 2001; 117(1–2): 3–39.

Muhammad Yasir Ali, Ghulam Abbas, Saeed Ahmad,
Muhammad Mehboob ur Rehman, Abid Mahmood, Nisar ur Rahman,
and Udo Bakowsky

13 Capsules

13.1 Introduction

Capsules are solid dosage forms in which the drug substance is enclosed either in a hard or soft, soluble container or shell made of a suitable form of gelatin. Capsules are cylindrical, spherical, tube-like, or oval-shaped [1]. Recently, hydroxypropyl methylcellulose capsule shells have been commercially available, containing less residual moisture than gelatin capsules, and thus may be useful for filling hydrolytic drugs. Capsules are tasteless and can be administered easily.

Capsules are of two types based on the composition of the capsule shell. They are classified either as hard gelatin capsules, comprising two pieces in the form of cylinders which are closed at one end, or soft gelatin capsules, which are flexible and possess a plasticized gelatin film. Capsules can be filled with a range of formulations, including powders, semisolids, and nonaqueous liquids. Other dosage forms, such as beads, mini-capsules, and mini-tablets, can also be filled in capsule shells to be administered orally.

13.2 Advantages

- Easily dissolves because the outer shell is water-soluble.
- An active drug, if sensitive, can be protected from light and moisture.
- An accurately selected drug can be given.
- Capsules can be formulated easily.
- Fewer adjuncts are necessary compared to other dosage forms.
- Two incompatible drugs can be given.

13.3 Disadvantages

- The cost of production is high, as initially, an empty capsule shell is needed, and then the formulation is prepared.
- For capsule shells, we have to be dependent on shell manufacturers.
- Moisture and temperature can cause distortion.
- The total number of capsules manufactured in 8 h is less than that of tablets.

https://doi.org/10.1515/9783111438108-013

13.4 Gelatin

The raw materials used for the manufacture of both types of capsules are similar. Traditionally, gelatin, water, colorants, and process aids are present. Soft gelatin capsules contain various plasticizers, such as sorbitol and glycerin. The following are some raw materials used for the preparation of the capsule shell.

Gelatin is used for the manufacturing of shells because:
– It is nontoxic.
– It is readily soluble in biological fluids at body temperature.
– It is a good film-forming material. The wall thickness of a hard gelatin capsule is about 100 μm.
– Solutions of high concentration, i.e., 40% w/v, are mobile at 50 °C. Other biological polymers, such as agar, are not.

13.4.1 Manufacturing of gelatin

Gelatin is obtained from the hydrolysis of collagen, which is the main protein constituent of connective tissue [2]. Animal skins and bones are the sources of collagen. We have two types of gelatin, i.e., Gelatin A and Gelatin B. Gelatin A is prepared by acid hydrolysis occurring over 7–10 days, and the substance used for this is animal skin. Gelatin B is prepared by basic hydrolysis of bones, and the duration for this process is 10 times longer. In this process, the bone is first decalcified, giving Ossian a spongy material. Ossian is then solaced in lime pits for several weeks. After hydrolysis, the gelatin is extracted from the treated material using hot water. After extraction, the process of concentrating the gelatin solution is carried out [1].

13.4.2 Characterization of gelatin

13.4.2.1 Bloom strength

It is the most important parameter for the physicochemical characterization of gelatin raw material. It is actually a measure of the stiffness and strength of the gelatin material. The bloom strength depends on various factors, including the source and extraction method used. Long extractions produce high-bloom gelatins, and short extractions produce low-bloom gelatins (Tab. 13.1). Gelatin bloom strength typically ranges from 30 to 300 g. There are three bloom groups of gelatin: low (<150), medium (150–220), and high (>220). A medium bloom strength range of gelatin is preferred for the manufacturing of soft gels.

Tab. 13.1: Examples of pharmacopoeial gelatin's bloom strength and viscosity [3].

S. no.	Gelatin	Source	Bloom (g)	Viscosity at 60 °C (mPa s)
1.	160 LB	Bovine/porcine bone	155–185	3.4–4.2
2.	160 LH	Bovine hide	150–170	3.5–4.2
3.	160 LB/LH	Blend of bovine/porcine bone and bovine hide	150–170	3.5–4.2
4.	200 AB	Bovine bone	180–210	2.7–3.2
5.	200 PS	Pig skin	190–210	2.5–3.1
6.	160 PS/LB/ LH	Blend of pig skin, bovine/porcine bone, and bovine hide	145–175	2.7–3.3

AB, acid bone; LB, limed bone; LH, limed hide; PS, pig skin.

13.4.2.2 Viscosity

Viscosity is related to the bloom strength of gelatin; gelatin material having high bloom strength has high viscosity and high melting temperatures due to the high cross-linking of gelatin chains. As bloom strength depends on the source and method of extraction, so the viscosity of gelatin, as well, i.e., mammalian gelatins show higher viscosity values than marine gelatins, with values of 3.90 cP for bovine skin, 6.37–7.28 cP for pig skin, and 1.87–3.63 cP for different fish gelatins [4]. The optimal viscosity range for softgel manufacturing is 2.8–4.5 mPa s at 60 °C.

13.4.2.3 Molecular weight distribution

In a linear polymer like gelatin, the individual polymer chains seldom possess identical molecular weights and degrees of polymerization, resulting in a distribution around an average value. The molecular weight distribution of gelatin ranges from 10 to 400 kDa, with the most prevalent distributions being approximately 100, 200, and 300 kDa for peptide chains, respectively. The variation in molecular weight distribution of gelatin is influenced by the extraction method employed; for instance, type-A and type-B gelatins exhibit different molecular weight distributions despite having the same bloom strength. Typically, gelatins derived from various extraction techniques are combined to produce a specific gelatin tailored for a particular application.

13.4.2.4 Conductivity, pH, and isoelectric point (pI)

Gelatin has low conductivity, usually less than 1 mS/cm. The gelatins have a slightly acidic pH, usually between 4 and 7, although there are differences depending on the source. For example, the pH of gelatin from bovine skin is 4.8–5.5, and from porcine skin, it is 7.0–9.4 [4]. The pI of gelatin is dependent on the method of hydrolysis used for the extraction of gelatin. Type-A gelatins show a pI of around 9. However, when gelatin is treated with strong acids, deamination occurs, reducing its pI. Type-B gelatins have a lower pI because the alkaline pretreatment causes the deamidation of asparagine and glutamine, which increases the number of aspartic and glutamic acids. The increase in net negative charge changes the pI. The typical pI for Type-B and Type-A gelatins are 4.6–5.2 and 7.0–9.0, respectively (Tab. 13.2).

Tab. 13.2: Examples of commercial gelatin's pH and conductivity [3].

S. no.	Gelatin	Source	Extraction method	Bloom (g)	pH	Conductivity (mS/cm)
1.	Rousselot 160 LB 8	Bovine limed bone	Alkaline	145–175	5.3–6.2	<1
2.	Rousselot 200 AH 8	Bovine hide	Acid	175–205	5.0–6.2	<1
3.	Rousselot 200 BH	Bovine hide	Alkaline	175–205	5.6–6.2	<1
4.	Rousselot 200 H 6	Bovine hide	Acid	175–205	5.0–6.0	<1
5.	Rousselot 250 PS 8	Pig skin	Acid	240–250	4.5–6.9	<0.4
6.	Gelita 170 LB Type B NF SRM Free Bone	Bovine limed bone	Alkaline	155–185	5.3–6.0	<1

AB, acid bone; AH, acid hide; BH, bovine hide; H, hide; LB, limed bone; LH, limed hide; PS, pig skin.

13.4.2.5 Color

The color of gelatin depends on the source of gelatin and the extraction method used. For example, sin croaker and bovine gelatins show a white color, while shortfin scad gelatins show a more yellowish color [5]. In regard to the extraction method, gelatin lightness decreases, and the yellowish color increases with an increase in extraction temperature [6]. Generally, it does not affect the functional properties of gelatin, but it can affect the product's aspects.

13.4.2.6 Foaming properties

It is an indication of the surface activity of gelatin, which has a direct link to the physicochemical and functional properties of gelatin. Foaming properties are studied using parameters such as foaming stability and foaming expansion, which depend on the temperature of extraction, among other parameters, and decrease at high temperatures [6]. The difference in foaming ability is due to the amount of hydrophobic amino acids in the gelatin structure.

13.5 Hard gelatin capsules

The hard gelatin capsules, also referred to as the dry-filled capsules, are the type used by pharmaceutical manufacturers in the preparation of the majority of their capsule products. These capsules are manufactured in two sections: the capsule body and a shorter cap. The two parts overlap when joined [7].

13.5.1 General aspects

The hard gelatin capsules are made from a mixture of gelatin, sugar, and water. Hard gelatin capsule shells are manufactured and supplied empty to the pharmaceutical industry, and they are filled separately; the body is filled with the drug substance. The cap and body are closed by bringing the cap and body together. They are colorless, clear, and essentially tasteless [8].

13.5.2 Gelatin

Gelatin is the most common and well-known material for the manufacture of hard gelatin capsule shells. It is a generic term for a mixture of purified protein fractions obtained from irreversible hydrolytic extraction of collagen obtained from the skin, white connective tissue, and bones of animals. Gelatin is available in different forms, e.g., fine powder, coarse powder, flakes, shreds, or sheets. Gelatin is stable in air, but when it comes in contact with water, it is prone to microbial decomposition. The molecular weight of gelatin ranges from 15,000 to 25,000 Da [8].

13.5.3 Opacifying agents and plasticizers

An opacifier can be added to make an opaque shell. It is used when the filling formulation is a suspension or to avoid photodegradation of light-sensitive ingredients. Titanium dioxide is used as an opacifier [1], and it can be used alone to make a white shell or in combination with other pigments to produce colored shells. The main function of plasticizers is to reduce the rigidity of the polymer and make it more pliable. The ratio (w/w) of dry powder to dry gelatin is the measure of the hardness of the gelatin shell. Commonly used plasticizers are glycerin and polyhydric alcohol. Water itself is a good plasticizer and is naturally present in gelatin [8].

13.5.4 Colorants

The colorants used are of two types: water-soluble dyes or insoluble pigments. Pigments and dyes are mixed together in the form of solutions or suspensions to make a range of colors. Colorant selection is the choice of the manufacturer. The most commonly used dyes are erythrosine (E127), indigo carmine (E132), and quinoline yellow (E104). Pigments used are of two types: black, yellow, and iron oxides (E172), and titanium dioxide (E171), which is white and used to make capsules opaque. Colorants make our dosage forms aesthetic and provide light protection. They are also helpful in the identification of dosage forms [9].

13.5.5 Manufacturing of hard gelatin capsule shells

13.5.5.1 Raw material production

The first step in the preparation of hard gelatin capsules is raw material production. A 35–40% concentrated gelatin solution is prepared using demineralized hot water at 60–70 °C in jacketed pressure vessels. This is stirred until the gelatin has dissolved, and then a vacuum is applied to remove any entrapped air bubbles. Then, the required amounts of pigment suspensions and dye solutions are added. The viscosity is adjusted using hot water, as it controls the thickness of the capsule shell [1].

13.5.5.2 Dipping

The manufacturing machines are approximately 10 m long, 2 m wide, and 3 m high. They consist of pins or pegs of the desired size and diameter. These pegs or pins are made up of manganese bronze. Around 500 pegs are affixed to this machine. Now, these pegs or pins are dipped into the dipping solution for about 12 s. The temperature is maintained at 22 °C, and the humidity of the environment is also controlled. The solution adheres to the outer side of the pins, and then these pins are drawn out [1].

13.5.5.3 Spinning/rotation

Now we rotate or spin the pins. By dipping and then drawing, the pin solution adheres, and also beads of solution are formed at the apex. Through rotation, these pin beads are removed from the solution sheath [7].

13.5.5.4 Drying

The film is dried through a process of adequate heating and air blowing. The pins are moved through air-drying kilns so that we may be able to remove the shell after hardening. Moisture content is strictly controlled [7]. After drying, stripping is done by using bronze jaws. Here we separate the shell. Now the shells are fixed in collars, actually called collets, and the extra portion is trimmed out to obtain the proper size. Trimming is done by using stationary knives. Join the body and cap. They should be aligned, still present in the collects, and then joined.

13.5.5.5 Sorting and imprinting

Sorting should be done because we may have some problems with capsules, for example, holes, different sizes, or distortions. Sorting is done by passing the capsules through light, and any faults are noted for the removal of these defective capsules. Imprinting is an optional step for the manufacturer [1]. We have laser machines for this purpose, which can print a million capsules per hour. Here, we can print the company name or just the trademark, depending on our preference [10].

13.5.5.6 Inspection processes

Inspection is done to remove imperfect capsules. This process has recently been automated using a practical electronic sorting mechanism by Eli Lilly and Company. Capsules are mechanically oriented and transported through a series of optical scanners, and visual imperfections are detected and rejected automatically [10].

13.5.6 Properties of an empty capsule

An empty gelatin capsule contains a significant amount of water, which acts as a plasticizer. The water content in the final capsules should be 13–16% w/w, as it affects the mechanical properties of the capsules. This value can vary depending on the conditions in which they are kept. In cases of low moisture, the capsule loses moisture and becomes brittle. If the water content is lower than the specified range, capsule shells will be brittle, and cracks will appear upon stress application. In cases of excess moisture content beyond the specified range, capsules will lose shape upon stress application due to plastic flow. For hypromellose capsules, the standard moisture content ranges from 3% to 6%, and when they lose moisture, they do not become brittle [11].

13.5.7 Preparation of filled hard gelatin capsules

The preparation of filled hard gelatin capsules may be divided into the following steps:
–   Developing and preparing the formulation, and selecting the size of the capsule
–   Filling the capsule shells
–   Going into the capsule, body, and cap
–   Capsule sealing (optional), cleaning, and polishing of the filled capsules

13.5.7.1 Developing and preparing a formulation

In order to fill a capsule of even the smallest size, a minimum of 65 mg of material is generally required. If the dose of the drug to be placed in a single capsule is inadequate to fill the volume of the capsule, a diluent is necessary to add the proper degree of bulk to produce the proper fill. On the contrary, if the amount of the drug is large enough to be filled in a single capsule, then diluents may not be added. Lactose, microcrystalline cellulose, and pregelatinized starch are common diluents used. In addition to providing bulk, diluents also provide cohesion to the powders, which is beneficial in the transfer of a measured portion of the powder into a capsule. Mg-stearate is a commonly used lubricant in capsule and tablet making to prevent adhesion and facilitate the flow of the drug fill into the encapsulating machinery. However, 1% of this substance in the powder imposes some problems of absorption from the gastrointestinal tract (GIT) due to its waterproofing ability. This problem may be solved by sodium lauryl sulfate as a wetting agent, which bathes the GIT and facilitates absorption.

13.5.7.2 Selection of capsule size

The selection of the capsule is best done during the development of the formulation because the amount of any inert materials to be employed is dependent upon the size or capacity of the capsule to be selected. A properly filled capsule should have its body filled with the drug mixture and its cap fully extended down the body so as to enclose the powder in the body. The cap is not used to hold powder but to retain it, and the capsule size should be selected to meet this requirement.

13.5.7.3 Filling the capsule shell

When filling is done on a small scale, the punch method is used for this purpose. Here is the precise number of empty capsules. The powder to be encapsulated is placed on a sheet of clean paper, glass, or a porcelain plate, and, with a spatula, is formed into a cake having a depth of approximately one-fourth to one-third the length of the body of the capsule. Then the empty capsule body is held between the thumb and forefinger and punched vertically into the powder cake repeatedly until filled. Remember, contact of powder with the finger should be avoided, as it will cause difficulty in capsule handling because of adhesion. The machines for the industrial-scale filling of hard capsules come in a great variety of shapes and sizes, ranging from semiautomatic to fully automatic and ranging in output from 30,000 to 150,000 per hour.

13.5.7.4 Size

Empty gelatin capsules are manufactured in various sizes, usually varying in length and diameter. The size selection depends on the amount of material to be encapsulated (Tab. 13.3).

Tab. 13.3: Size of hard gelatin capsule with fill capacity.

S. no.	Size	Contents (mg)	Filled wt (g)
1	000	950	1.076
2	00	650	0.76
3	0	450	0.54
4	1	300	0.4
5	2	250	0.296
6	3	200	0.24
7	4	150	0.168
8	5	100	0.104

13.5.8 Formulation of fill for hard gelatin capsule

The fill of a hard gelatin capsule may be either powder (or granules) containing the required amount of a drug or a liquid preparation in which the required amount of the drug is dissolved or dispersed. Three main classes of capsules are employed in hard gelatin capsule filling (Tab. 13.4). These are powders and liquids/semi-solids. The formulation considerations for these are given below.

Tab. 13.4: Commercially available drugs as hard gelatin capsules.

S. no.	Brand name	Generic name	Manufacturer's name
1	Tylenol	Acetaminophen	Johnson & Johnson
2	Advil	Ibuprofen	Pfizer
3	Prozac	Fluoxetine	Eli Lily and Company

Tab. 13.4 (continued)

S. no.	Brand name	Generic name	Manufacturer's name
4	Lipitor	Atorvastatin	Pfizer
5	Amoxil	Amoxicillin	GlaxoSmithKline
6	Prilosec, Losec	Omeprazole	AstraZeneca
7	Neurontin	Gabapentin	Pfizer
8	Naprogesic	Naproxen	Sanofi Aventis
9	Flagyl	Metronidazole	Pfizer
10	Pentoloc	Pentoparazole	Takeda Pharmaceuticals
11	Voltaren, Cataflam	Diclofenac	Novartis
12	Ultram	Tramadol HCl	Janssen Pharmaceuticals
13	Augmentin	Amoxicillin + clavulanic acid	GlaxoSmithKline
14	Keflex	Cephalexin	Eli Lily and Company
15	Cipro	Ciprofloxacin	Bayer

13.5.8.1 Powder fill

The following properties should be present in the powder formulation for inclusion in a hard gelatin capsule:
- The powder blend should be uniform and well-mixed.
- It must possess excellent flowability to ensure smooth transfer from the powder bed, through the filling system, and into the capsule body.
- The blend should remain free from clumping, with sufficient compressibility to maintain a stable powder bed.
- The particle size distribution of all ingredients, including the active drug, should be consistent to achieve uniform mixing and minimize segregation.
- Irregularly shaped particles, such as needle-like forms, should be avoided, as they hinder proper capsule filling.
- All formulation components must be highly compatible with one another, as well as with the capsule shell.

13.5.8.2 Liquid/semisolid fill

Liquid/semisolid fills for hard gelatin capsules may be divided into the following categories:

Lipophilic liquids/oils containing therapeutic agents: Liquids commonly used in this category are vegetable oils (sunflower, arachis, olive) and fatty acid esters (glyceryl monostearate).

Water-miscible liquids containing therapeutic agents: Liquids commonly used in this category are polyethylene glycols (PEGs), which are solid at room temperature but liquefy upon heating, and liquid polyoxyethylene-polyoxypropylene block co-polymers (pluronics).

13.5.9 Capsule-filling machines

13.5.9.1 Filling of capsules with powder formulation

Bench-scale filling: The bench-scale filling refers to the small-scale encapsulation process, usually involving 50–10,000 capsules. This approach is commonly applied in hospital pharmacies, as well as in the pharmaceutical industry for trial batches or customized prescriptions. Various simple filling devices are available for this purpose, such as the Feton capsule filler. This equipment is composed of plastic plates perforated with holes that can accommodate either 30 or 100 capsules. During operation, empty capsules are placed into the holes, and the capsule bodies are secured within the plates by tightening screws. The caps are separated, while the capsule bodies remain slightly recessed below the plate's surface. The powder blend is then spread across the surface, filling the recessed capsule bodies. Once filled, the cap plate is positioned over the body plate, and the capsules are closed by applying manual pressure.

Industrial-scale filling: In the pharmaceutical industry, capsule-filling machines are manufactured in a wide range of sizes and designs, with production capacities typically varying between 30,000 and 150,000 capsules per hour. These machines may operate as semi-automatic or fully automatic systems. Fully automatic equipment can function either on an intermittent basis, where the process pauses and indexes to the next position before repeating, or in a continuous mode, as seen in rotary systems.

13.5.9.2 Filling of capsules with pellets

Pellets or granules are often used to formulate modified-release preparations. They are filled using industrial-scale machines that are adapted from powder use. These machines have a dosing system based on a chamber whose volume can be adjusted. Pellets are not compressed. The size of particles is considered while calculating the weight of particles to be filled. Smaller-sized particles cannot fill as much of the available space.

13.5.9.3 Filling of capsules with tablets

Tablets are placed in hoppers and allowed to fall down the tubes. A gate is present at the bottom of the device and allows a set number to pass. The tablets move into the cavity form hopper due to the gravitational effect. A mechanical probe is present in most of the machines, and its function is to check that the correct number of tablets has been transferred. Film-coated tablets are filled in capsules normally to avoid dust generation. The tablets are sized for their free flow into the capsule body. Recently, coated mini tablets have been filled in capsules, having a smaller surface area, thus increasing the uniformity of fill material in the capsules.

13.6 Soft gelatin capsules

13.6.1 General aspects

In the preparation of soft gelatin capsules, the making of the shell and the filling of the shell with the formulation occur spontaneously and occur in a closed machine. Soft gelatin capsules are used in pharmaceuticals, health and nutrition products, cosmetic products, and even recreational products, e.g., paintballs.

Soft gelatin capsules have the following pharmaceutical applications:
– As an oral dosage form for human or veterinary use.
– As a suppository for rectal or vaginal administration of a drug. Nowadays, rectal suppositories are becoming more acceptable for geriatric and pediatric use.
– As single-dose applications of ophthalmic, topical, and otic preparations, as well as rectal ointments for human and veterinary use.

13.6.2 Advantages

– The drug cannot be leaked.
– Prevents oxidation.
– Liquid preparations can be filled.
– Rapid absorption from the GIT because in liquid form the drug is homogeneously distributed, but in powder form (hard gelatin capsule), the drug is not so homogeneously distributed.
– Alternative pathways are available, e.g., vaginal and rectal routes.

13.6.3 Basic components of soft gelatin capsule shell

13.6.3.1 Gelatin

Type A gelatin, typically derived from pork skin, contributes hardness to capsule shells, while Type B gelatin, obtained from connective tissues, imparts flexibility. For this reason, Type B gelatin is preferred in the manufacture of soft gelatin shells [12]. In a standard softgel formulation, gelatin constitutes approximately 40–45% of the shell composition [13]. Although Type A gelatin is less frequently used, multiple shell formulations are available depending on the physicochemical properties of the fill material.

Commercially, gelatin is supplied in various physical forms, including fine or coarse powders, flakes, shreds, and sheets. It is generally stable when stored in dry air; however, upon contact with moisture, it becomes susceptible to microbial degradation. The molecular weight of gelatin typically ranges between 15,000 and 25,000 Da [14].

13.6.3.2 Plasticizers

Plasticizers are incorporated into soft gelatin capsule shells to impart elasticity and flexibility. They generally constitute about 20–30% of the wet molten shell mass. Glycerol is the most widely used plasticizer, though sorbitol and propylene glycol are also commonly employed, often in combination with glycerol. The type and concentration of plasticizer play a crucial role in determining the hardness of the capsule shell and can also influence dissolution, disintegration behavior, and the overall physical and chemical stability of the product. When the concentration of plasticizer exceeds 30%, the shell becomes excessively soft and tacky, whereas concentrations below 20% may produce brittle shells prone to breakage [12, 17].

13.6.3.3 Water

Water constitutes approximately 30–40% of the wet molten shell mass and plays a critical role in both gel preparation and encapsulation. It is required not only during the manufacturing process but also in the final product to maintain appropriate shell characteristics. During production, the gelatin mass typically contains 30–40% w/w water. Following capsule formation, a drying step reduces the water content to about 5–8% w/w. Maintaining this level is essential; excessive drying can lead to brittle shells, while insufficient drying may compromise stability. Gelatin solutions are usually prepared with demineralized hot water, and viscosity is carefully adjusted to achieve the desired shell properties [18].

13.6.3.4 Colorants and/or opacifiers

A colorant (soluble dyes, insoluble pigments, and lakes) or an opacifier (titanium dioxide) may be added to the shell to make our product more visually appealing and to prevent the photodegradation of a photosensitive drug. The color of the capsule shell is kept darker than that of its content [8]. The colorants used are of two types: water-soluble dyes or insoluble pigments. Pigments and dyes are mixed together in the form of solutions or suspensions to make a range of colors. The choice of colorant is up to the manufacturer. The most commonly used dyes are erythrosine (E127), indigo carmine (E132), and quinoline yellow (E104). Pigments used are of two types: black, yellow, and iron oxides (E172) and titanium dioxide (E171). Colorants make our dosage form aesthetic and provide light protection. They are also helpful in the identification of the dosage form.

13.6.3.5 Miscellaneous agents

Preservatives are used to prevent bacterial and mold growth in gelatin solutions during storage. Commonly used plasticizers are potassium sorbate, methyl hydroxybenzoate, ethyl hydroxybenzoate, and propyl hydroxybenzoate. Manufacturers operating under Good Manufacturing Practices guidelines no longer use preservatives [8]. Flavoring agents and sweeteners can be added to improve palatability. Enteric release characteristics can be imparted by using acid-resistant polymers. A chelating agent, such as ethylene diamine tetraacetic acid, may be added to prevent chemical degradation of oxidation-sensitive drugs. Chewable soft gelatin capsules can also be manufactured [3].

13.6.4 Types of soft gelatin capsules

As with the hard gelatin capsule, soft gelatin capsules are usually administered orally. Some soft gelatin capsules can be manufactured to produce several different drug delivery systems, such as chewable softgels, where a highly flavored shell is chewed to release the drug liquid fill matrix; suckable softgels, which consist of a gelatin shell containing flavored medicament which is sucked and a liquid matrix (or air in the capsule shell); twist-off softgels, which are designed with a tag to be snipped off or twisted, allowing access to the fill material; and meltable softgels, which are developed to be used as suppositories or passerines [15].

13.6.5 Manufacturing of soft gelatin capsules

Soft gelatin capsules may be prepared by the plate process, using a set of molds to form the capsules, or by a more efficient and productive method, i.e., the rotary die process.

13.6.5.1 Plate processes

In this process, a hot sheet of gelatin is poured onto the bottom of the plate of the mold, and liquid medication is evenly poured on it. Then, a second sheet of hot gelatin is poured carefully on the medicament, and then the upper plate of the mold is dipped and pressed to form the soft gelatin capsule. Finally, these are washed with an inert or harmless solvent [7].

13.6.5.2 Rotary die process

Water constitutes approximately 30–40% of the wet molten shell mass and plays a critical role in both gel preparation and encapsulation. It is required not only during the manufacturing process but also in the final product to maintain appropriate shell characteristics. During production, the gelatin mass typically contains 30–40% w/w water. Following capsule formation, a drying step reduces the water content to about 5–8% w/w. Maintaining this level is essential; excessive drying can lead to brittle shells, while insufficient drying may compromise stability. Gelatin solutions are usually prepared with demineralized hot water, and viscosity is carefully adjusted to achieve the desired shell properties [16].

 After formation, the capsules are subjected to a series of post-processing steps. Initially, they are washed with an inert solvent, typically by passing them along a conveyor through a naphtha wash, which removes the mineral oil used to lubricate the gelatin ribbons. This is followed by infrared drying, which rapidly eliminates most of the water from the shells. Finally, controlled drying in forced-air tunnels reduces the residual moisture content to approximately 6–10% [10].

13.6.6 Formulation of fill for soft gelatin capsule

The primary fill formulation for soft gelatin capsules is liquids (although powders can also be filled). The therapeutic agent may be either dissolved or dispersed within the fill material. The categories for fill material include [1] lipophilic liquids [2], self-emulsifying systems, and [3] water-miscible liquids (Tab. 13.5).

Tab. 13.5: Commercially available drugs as soft gelatin capsules.

S. no.	Brand name	Generic name	Company name
1	Aquasol A	Retinol (vitamin A)	USV Private Limited
2	Depakene	Valproic acid	Abbot
3	VePesid	Etoposide	Bristol-Myers Squibb
4	Zantac	Ranitidine HCl	GlaxoSmithKline
5	Adalat	Nifedipine	Bayer
6	Unison SleepGel	Diphenhydramine HCl	Unisom
7	Vesanoid	Tretinoin	Roche Pharmaceuticals
8	Avodart	Dutasteride	GlaxoSmithKline
9	SuperEPA	Omega-3	GlaxoSmithKline
10	Fortovase	Saquinavir	Roche Pharmaceuticals

Lipophilic liquids: Lipophilic liquids are commonly used for filling soft gelatin capsules and include both vegetable oils (e.g., soybean oil) and fatty acid esters. Co-solvents and/or surface-acting agents are used because a limited number of therapeutic agents are soluble in these materials.

Self-emulsifying systems: These are lipophilic liquids that contain a nonionic emulsifying agent (e.g., the Tween series). After release into the GIT, emulsification of the fill material takes place and is converted into small droplets. Hence, dissolution enhances and results in increased absorption of the therapeutic agent.

Water-miscible liquids: Water-miscible liquids are high-molecular-weight alcohols, e.g., PEG 400, PEG 600, nonionic surface-acting agents (e.g., Tweens), and pluronics (liquid polyoxyethylene-polyoxypropylene block co-polymers).

Additional formulation excipients. Examples of the categories of excipients include:
– Viscosity-modifying agents
– Surface-active agents
– Colors
– Co-solvents (and other solubilizing agents)

13.7 Quality control of capsules

Quality control is an important function of the pharmaceutical industry. The drugs should be therapeutic and safe, and their effectiveness is predictable and usually consistent. Now, medical devices with better sensitivity are being developed, and advanced analytical methods are being designed for their testing. Capsules are assessed through a specific test, i.e., a quality control test, whether they are manufactured on a large or small scale. The quality of the finished product is assessed [17].

These tests fall into two categories:
– Universal tests
– Specific tests

13.7.1 Universal tests

Universal tests are common and are called generalized tests for all oral dosage forms. These tests are essential to ensure that the finished products that are commercialized, meet the minimum requirements of quality. These tests include:
– Description test
– Identification test
– Assay
– Impurities test

13.7.1.1 Description test

It is not a standard test; rather, it is considered a general test. It is usually concerned with the physical appearance of the finished product to ensure that it complies with the standards of the monographs. Twenty capsules are unpacked and inspected, usually.

Appearance: The products should have a smooth surface and be nonabrasive. The first evidence of instability is evident as the change in body shape of the capsules, along with the firmness or softening, cracking of the capsule surface due to external stress, bending, swelling, or discoloration of the capsule shell. All capsules should be smooth in appearance.

Shape and size: The hard gelatin capsules consist of a range of different sizes. At an industrial scale, the size ranges from 000, designated as the largest (140 mL), up to 5, which is designated as the smallest (0.13 mL), and they are still commercially available. Soft gelatin capsules are available in different shapes, such as ovoid shape (0.06–7 mL), spherical shape (0.04–5 mL), cylindrical shape (0.15–25 mL), and pear shape (0.3–5 mL), etc.

Diameter: A capsule diameter sorter is used to check the diameter, which allows the incoming unit of the capsule to pass through it. The capsule diameter should be within the range of ±0.020 inches (theoretical diameter).

Color: The capsules are fed through a pneumatic conveyor with the help of a diameter sorter automatically. If the capsules' color does not match the reference color, the standard solution of the product is discarded, while other capsules pass the test [18].

13.7.1.2 Identification test

This test is an aid to confirm that the article contains the labeled drug substance by giving a positive identity of the drug substance in the drug product. The analytical technique used must be capable of distinguishing the active pharmaceutical ingredient from all excipients or potential degradation substances that are likely to be present. General tests performed to identify the active pharmaceutical ingredient in the capsules are discussed below [8].

Thin-layer chromatographic identification test: Prepare the test solution as mentioned in each monograph. Insert 10 mL of the prepared solution, 10 mL of the standard solution USP, which is the Reference Standard in the case of a specific drug substance, in a line about 2 cm from the edges of the chromatographic plate, covered with a 0.24 mm layer of silica gel chromatographic mixture. The same solvent is used for identification; at the same time, as mentioned in the individual monographs. Allow the stains to dry, and the chromatogram is held in a solvent system consisting of a mixture of chloroform, water, and methanol. Fill the solvent front up to three-quarters of the length of the plate. Find the spot on a plate by placing it under a UV lamp [8].

Spectrophotometric identification test: These experiments are used for the identification of many chemical compounds. The diagnostic procedure uses IR or UV radiation. The spectrum of infrared radiation absorption provides the most compelling evidence of a substance's identity in a single test. On the contrary, ultraviolet absorption lines have low specificity.

In the infrared absorption method, the test sample line and the corresponding reference standard USP are recorded in the range of 2.6 μm to 15 μm (380 cm^{-1} to 650 cm^{-1}), as mentioned in the monograph. The maxima of the IR spectrum of the test line are at the same wavelength as the corresponding reference standard, USP.

For the ultraviolet absorption spectrum, prepare a test solution of the subject under test in the selected medium as specified in the monograph for the solution. Similarly, prepare a standard solution of the same reference standard, USP. The spectra are compared within a spectral range of 200–400 nm unless otherwise specified. Calculate the input and absorption values using these methods in a single monograph. The maxima and minima of the test solution and standard solution should be at the same wavelength as the acquisition and absorption values in the set limits [8].

Nuclear magnetic resonance: Nuclear magnetic resonance is the most widely used tool for powerful structural elucidation and is used for accurate qualitative and quantitative measurements [19]. The internal reference standard is co-dissolved in the analyte solution. A typical NMR solution is prepared by dissolving the exact weights of both the reference standard and the analyte. Larger weights are used to minimize error. NMR peaks and the concentrations of both the analyte and the reference standard are compared for quantitative analysis. Integrate appropriate peaks that give a quantitative result; this result should be in the range specified in the individual monograph [8].

13.7.1.3 Assay test

The assay is a stability-indicating and specific test to determine the potency of the active ingredient in capsules. In this test, the sample content is tested as specified in the individual monographs, and the amount of the active ingredient is determined in each capsule. According to BP, 20 capsules are tested for the determination of the active ingredient (Tab. 13.6).

Tab. 13.6: Active ingredients' limits.

Weight of API in capsule (g)	Subtract (lower limit) (g)	Addition (upper limit) (g)
14–16	11	11
±5 g	5	5
0.13 or less	0.3	0.6–0.7
More than 0.13	0.3–0.4	0.5
Less than 0.4	1.3	1.6
0.4 or more	0.2–0.4	0.3–0.5/0.7–1.1

Acceptance criteria: The product complies with the test if no more than one individual capsule content is outside the limits of 85–115%, and none is beyond the limits of 75–125% of the average content [8].

13.7.1.4 Bloom strength of gelatin

Gelatin is mixed with water to make a 6.57% solution in conventional bloom bottles. The mixture is stirred for 3 h at room temperature. These bottles are then kept in a 65 °C bath for 20 min. The bloom pots or jars are allowed to cool for 15 min in a temporary room and are conditioned in a water bath for 16 h. The probe is placed just above the sample while performing the gelatin bloom strength test. The probe penetration in gelatin is at the targeted depth of 3 mm with a speed of 0.4 mm/s, and then recedes [8].

13.7.1.5 Impurities

Process contaminants, synthetic products, and other natural or organic impurities may be present in the drug substances and excipients used for the manufacturing of capsules. This contamination is not accepted in drugs and excipients used in the manufacture of that drug. These impurities should be separated. The presence of any label-free impurities in an official item should not be more than 0.1%; otherwise, it differs from the standard. The total amount of all other impurities, in addition to impurities detected by the monographs, should not be more than 2% [8].

13.8 Specific tests for capsules

These tests are further categorized into two major types:
- Physical tests
- Chemical tests

13.8.1 Physical tests

The physical test includes:
- Disintegration test
- Weight variation test

13.8.1.1 Disintegration test

According to the United States Pharmacopeia (USP), complete disintegration is defined as the stage at which any remaining residue of the dosage unit – apart from fragments of insoluble coating or capsule shell – appearing on the test apparatus screen or adhering to the lower surface of the disk (if employed), exists as a soft mass without a discernible firm core. This test is designed to confirm that capsules break down within the specified time when placed in a liquid medium under standardized experimental conditions [20].

The disintegration testing apparatus is composed of a basket-rack assembly containing six open-ended transparent tubes and a 1,000 mL beaker with a low base, having a height between 139 and 160 mm for immersion of the medium. The diameter of the beaker ranges from 96 to 115 mm, and the surrounding fluid is thermostatically maintained at 35–39 °C. During the test, the basket-rack moves up and down at a speed of 29–32 strokes per minute. The liquid volume within the vessel is adjusted so that the highest position of the basket remains submerged. Disks are employed only when mentioned in the individual monograph. Each tube can hold a cylindrical disk with a thickness of 9.45–9.65 mm and a diameter of 20.65–20.85 mm. These disks are manufactured from transparent plastic material with a specific gravity of 1.18–1.20. Every disk contains five perforations measuring 1.8–2.1 mm in diameter, one located at the center and four arranged around it in a circular pattern. When specified, a disk is placed in each tube, and the experiment is performed accordingly [21].

Method: According to the British Pharmacopoeia (BP), a capsule is placed in each tube of the basket-rack assembly, which is then immersed in a beaker containing 60 mL of water maintained at 36.5 °C. Since hard gelatin capsules may float on the water surface, the use of an additional disk is required to keep them immersed. The apparatus is operated for 30 min, after which the basket-rack assembly is withdrawn from the medium (Tab. 13.7). This procedure is applied to the disintegration testing of both hard and soft gelatin capsules [22].

Tab. 13.7: Disintegration testing conditions and interpretation (BP).

S. no.	Types of capsules	Medium	Temperature	Limit (BP)
1	Hard gelatin	Water/0.1 M HCl/ artificial gastric juice R	37 ± 2 °C	30 min or as per individual monograph
2	Soft gelatin	Water/0.1 M HCl/ artificial gastric juice R	37 ± 2 °C	30 min or as per individual monograph
3	Enteric-coated (gastro-resistant)	0.1 M HCl → phosphate buffer, pH 6.8	37 ± 2 °C	2 h in HCl (no disintegration); 60 min in buffer → disintegrate

Acceptance criteria: The acceptance criteria state that capsules meet the disintegration test requirements if no residue remains on the screen of the apparatus. However, if small fragments of the shell are observed, they must exist only as a soft mass without a firm core. When a disk is used, it may retain some shell fragments. If one or two capsules fail to disintegrate, the test should be repeated with 12 additional capsules. To comply with the requirements, at least 16 out of the total 18 capsules tested must show complete disintegration [23].

13.8.1.2 The weight variation test

For hard gelatin capsules: The 20 capsules are individually weighed to determine the mean weight. Each capsule should fall within 90% to 110% of this average weight. If any capsule does not comply, another set of 20 capsules is weighed individually. The contents of each capsule are then carefully removed using a small brush, and the empty shells are weighed separately [22]:

Net content weight = The weight of the shell – gross weight

No more than two capsules should show a deviation greater than 5% from the mean content, provided that this difference does not exceed 25% of the total weight. If more than two but not more than six capsules deviate from the mean by 10–25%, then the content of an additional 40 capsules must be tested. The average content of all 60 capsules is then calculated, and the deviations are reassessed against this new mean. Out of the 60 tested capsules, no more than six may show a deviation greater than 10% of the mean content, and in no case should the deviation exceed 25% [24].

For soft gelatin capsule: For soft gelatin capsules, the total weight of the contents is determined in a similar manner as for hard gelatin capsules. Each capsule is individually weighed, then carefully opened, and the contents are removed using suitable solvents. The solvents are allowed to evaporate completely at room temperature, leaving behind the empty capsule shells. Each shell is then weighed individually, and the net content is calculated by subtracting the shell weight from the total capsule weight.

According to the official criteria, no more than two capsules may differ from the mean content by more than 5%, and in no case should this deviation exceed 25% of the total weight. If more than two but not more than six capsules show a deviation of 10–25% from the mean, the contents of an additional 40 capsules must be analyzed. A new mean is then calculated for all 60 capsules, and deviations are reassessed relative to this revised mean. Among the 60 capsules tested, no more than six may deviate by more than 10% from the average content, and none should show a variation greater than 25% [25].

13.8.2 Chemical tests

13.8.2.1 Dissolution test

The dissolution test for capsules is carried out using the Type II (paddle) apparatus. The temperature is maintained at 37 ± 2 °C. The apparatus is operated for 4.5 h, and if multiple time intervals are specified, samples are withdrawn at the designated times with a tolerance of ±2 min. Commonly used dissolution media include water, acidic buffer (pH 1.3), and phosphate buffer (pH 6.8). The pH of the medium must be within ±0.05 units of the value specified in the individual monograph. Both the USP and BP dissolution apparatuses are acceptable for performing this test.

During testing, the capsule contents are released into the dissolution basket, which is then immersed in the medium. The basket is rotated at the prescribed speed mentioned in the individual monograph (Tab. 13.8). The dissolution medium is placed in a 1,000 mL glass vessel, and the temperature is carefully controlled at 37 ± 0.5 °C throughout the procedure [26, 27].

13.8.2.2 Acceptance criteria for immediate-release hard gelatin and hard gelatin capsule

Q is the amount of active ingredient dissolved as specified in the individual monograph and is expressed as the percentage of labeled content of a particular dosage unit, e.g., 5% and 15%.

Acceptance criteria for prolonged-release capsules have limits that include each of the Q value, which is the quantity dissolved at the specified dosing interval [8].

Tab. 13.8: Acceptance criteria for the dissolution of immediate-release capsules.

S. no.	Stage	Number tested	Acceptance criteria
1	S1	6	Each unit is not less than $Q + 5\%$
2	S2	6	Average of 12 units (S1 + S2) is equal to or greater than Q; no unit is less than Q 15%
3	S3	12	The average of 24 units (S1 + S2 + S3) is equal to or greater than Q, and no more than 2 units are less than Q 15%, and no unit is less than Q 25%.

13.8.2.3 Dissolution test of modified-release capsules

There are two stages of the dissolution test of modified-release capsules [8].

Acid stage

– Place 750 mL of 0.1 M hydrochloric acid in the vessel and assemble the apparatus.
– Allow the medium to equilibrate to a temperature of $37 \pm 0.5\ °C$.
– Place one dosage unit in the apparatus, cover the vessel, and operate the apparatus at the specified rate.
– After 2 h of operation in 0.1 M hydrochloric acid, withdraw an aliquot of the fluid and proceed immediately as directed under the buffer stage.
– Perform an analysis of the aliquot using a suitable assay method.

13.8.2.4 Acceptance criteria for the acid stage of dissolution

The test specifications are met if the units comply with the table below. Testing is continued through three levels unless the results of both the acid and final buffer stages correspond to the previous level [8].

Buffer stage

– Complete the operations of adding the buffer and adjusting the pH within 5 min.
– With the apparatus operating at the specified rate, add to the fluid in the vessel 250 mL of 0.20 M solution of tri-sodium phosphate dodecahydrate R that has been equilibrated to $37 \pm 0.5\ °C$.
– Adjust, if necessary, with 2 M hydrochloric acid or 2 M sodium hydroxide to a pH of 6.8 ± 0.05.

- Continue to operate the apparatus for 45 min, or for the specified time.
- At the end of the time period, withdraw an aliquot of the fluid and perform the analysis using a suitable assay method.

Testing is continued through three levels. The Q value in the table is 75% of the dissolved amount, unless prescribed in the monograph.

13.8.2.5 Uniformity of mass

Weigh the intact capsule, open the capsule without the loss of any shell part, and remove the contents completely. In the case of a soft gel capsule, first wash the shell with solvents and allow it to stand until the color of the solvents is no longer seen. The shell is weighed, and the mass of all contents is the main difference between subsequent weighings. Repeat the process for the remaining 19 capsules [8].

Not more than two capsule masses must deviate from the mean mass by more than the percentage deviation given in the table below. None of the capsules must deviate by more than twice the percentage deviation [8].

13.8.2.6 Test of moisture permeation

The rate or extent of moisture permeation is determined by utilizing units that are packed together with a color-indicating pellet desiccant. The packed unit is then subjected to a known relative humidity for a specified duration. The desiccant is monitored through color differentiation. A change in color signifies the absorption of moisture. To calculate moisture content, measure the weight of the pellet before and after the test, and compute the moisture by subtracting the pre-test weight from the post-test weight. An excess of moisture can result in the formation of sticky capsules.

13.8.2.7 Capsule stability testing

Upon storage, gelatin capsules rapidly attain equilibrium with atmospheric conditions. For the stability of capsules, in particular, soft gelatin capsules, relative humidity, and temperature specifications are to be adjusted, which is made of mineral oil containing gelatin as shell and with glycerin in dry form, dry gelatin is in a ratio of 0.6 to 1, and the dry gelatin-to-water ratio is 1:1. They are dried at 20–40% of relative humidity at 20–44 °C [28].

Physical stability is associated with the formation and the capsule shell losing its ability to hold, in the case of soft gelatin capsules. The upper controlled-release capsules have satisfactory chemical stability at temperatures above cold to over 60 °C. The moisture content of the capsules increases with the rise in humidity [8].

13.9 Storage of capsules and packing

Place the capsules in a tightly closed container made of glass or plastic. Store them in a cool place. These containers have an advantage over cardboard-type boxes because they are lightweight and protect the capsules from dust and moisture. Cotton pieces are placed under and over the capsules (packed in vials) to avoid the shaking of the capsules. For the packaging of hygroscopic-type capsules, place a desiccant such as silica gel, and an anhydrous form of calcium chloride can be placed in order to prevent further absorption of moisture [29].

Different packaging of capsules includes:
- Plastic cap-screwed bottle
- Clam blister shell (having one piece of plastic which folds over, locks itself without providing any heat or any other processes)
- Blister-type packing (heat-sealed-type blister on cardboard)
- Plastic pail or bucket (an economical form of packaging for bulk quantities of units)
- Plastic-locked zip pouch

References

[1]   Aulton ME, Taylor K. Aulton's pharmaceutics: The design and manufacture of medicines: Elsevier Health Sciences; Edinburgh, England, 2013.

[2]   Jones RT. Gelatin: Manufacture and physico-chemical. Pharmaceutical Capsules (2nd edition). Pharmaceutical Press; London, England, 2004; 23.

[3]   Naharros-Molinero A, Caballo-González MÁ, De la Mata FJ, García-Gallego S. Shell formulation in soft gelatin capsules: Design and characterization. Advanced Healthcare Materials. 2024; 13(1): 2302250.

[4]   Said NS, Sarbon NM. Physical and mechanical characteristics of gelatin-based films as a potential food packaging material: A review. Membranes. 2022; 12(5): 442.

[5]   Cheow C, Norizah M, Kyaw Z, Howell N. Preparation and characterisation of gelatins from the skins of sin croaker (*Johnius dussumieri*) and shortfin scad (*Decapterus macrosoma*). Food Chemistry. 2007; 101(1): 386–91.

[6]   Nagarajan M, Benjakul S, Prodpran T, Songtipya P, Kishimura H. Characteristics and functional properties of gelatin from splendid squid (*Loligo formosana*) skin as affected by extraction temperatures. Food Hydrocolloids. 2012; 29(2): 389–97.

[7]   Allen L, Ansel HC. Ansel's pharmaceutical dosage forms and drug delivery systems: Lippincott Williams & Wilkins; Philadelphia, USA, 2013.

[8]   Mohammed AL. Capsules: Types, manufacturing, formulation, quality control tests and, packaging and storage – a comprehensive review. World Journal of Pharmaceutical and Life Sciences. 2020; 6: 93–104.

[9]   Jones B. Colours for pharmaceutical products. Pharmaceutical Technology International. 1993; 4: 14–20.

[10] Lachman L, Lieberman HA, Kanig JL. The theory and practice of industrial pharmacy: Lea & Febiger Philadelphia; USA, 1976.

[11] Jones DS. FASTtrack Pharmaceutics dosage form and design: Pharmaceutical press; London, England, 2016.

[12] Guillén G, Giménez B, López Caballero M, Montero García P. Functional and bioactive properties of collagen and gelatin from alternative sources: A review. Food Hydrocolloids. 2011; 25(8): 1813–1827.

[13] Naharros-Molinero A, Caballo-González MÁ, De la Mata FJ, García-Gallego S. Shell formulation in soft gelatin capsules: Design and characterization. Advanced Healthcare Materials. 2024; 13(1): 2302250.

[14] Gómez-Guillén M, Giménez B, López-Caballero M, Montero M. Functional and bioactive properties of collagen and gelatin from alternative sources: A review. Food Hydrocolloids. 2011; 25(8): 1813–27.

[15] Hoag S. Capsules dosage form: Formulation and manufacturing considerations. In: Developing solid oral dosage forms: Elsevier; Edinburgh, England, 2017. pp. 723–47.

[16] Franc A, Vetchý D, Fülöpová N. Commercially available enteric empty hard capsules, production technology and application. Pharmaceuticals. 2022; 15(11): 1398.

[17] Taylor MK, Ginsburg J, Hickey AJ, Gheyas F. Composite method to quantify powder flow as a screening method in early tablet or capsule formulation development. American Association of Pharmaceuticals Scientists PharmSciTech. 2000; 1: 20–30.

[18] Pervaiza F, Zahra SA, Qaiser F, Fatima SK, Khane RM Characterization and Evaluation of Capsules and Study of QC tests for Capsules.

[19] Chiwele I, Jones BE, Podczeck F. The shell dissolution of various empty hard capsules. Chemical and Pharmaceutical Bulletin. 2000; 48(7): 951–56.

[20] Brown WE. Compendial requirements of dissolution testing – european pharmacopoeia, japanese pharmacopoeia, united states pharmacopeia. In: Pharmaceutical dissolution testing: CRC Press; Florida, USA, 2005. pp. 87–98.

[21] Huynh-Ba K, Moreton RC. Development of United States Pharmacopeia-National Formulary (USP–NF) monographs and general chapters. In: Specification of drug substances and products: Elsevier; Edinburgh, England, 2025. pp. 185–204.

[22] Santillo M, Lagarce F. Quality requirements and analysis. In: Practical pharmaceutics: An international guideline for the preparation, care and use of medicinal products: Springer; London, England, 2023. pp. 785–807.

[23] Smeets O, Santillo M, Van Rooij H. Quality requirements and analysis. In: Practical pharmaceutics: An international guideline for the preparation, care and use of medicinal products: Springer; London, England, 2015. pp. 707–29.

[24] Xin C, Xun Y, Zhou Y, Wang F, Li F, Wu Q, et al. Qualitative and quantitative assessment of related substances for ketoconazole cream based on national drug sampling inspection in China. BMC Chemistry. 2025; 19(264): 1–17. Available at SSRN 5230801.

[25] Osei-Asare C, Owusu FWA, Apenteng JA, Entsie P, Adi-Dako O, Kumadoh D, et al. Evaluation of the microbial quality of commercial liquid herbal preparations on the Ghanaian market. INNOSC Theranostics and Pharmacological Sciences. 2023; 6(2): 1–7.

[26] Upton R, Agudelo I, Cabrera Y, Caceres A, Calderón A, Calzada F, et al. A US Pharmacopeia (USP) overview of Pan American botanicals used in dietary supplements and herbal medicines. Frontiers in Pharmacology. 2024; 15: 1426210.

[27] Niazi SK. Advice to the US FDA to allow US pharmacopeia to create Biological Product Specifications (BPS) to remove side-by-side analytical comparisons of biosimilars with reference products. Pharmaceutics. 2024; 16(8): 1013.

[28] Schalley CA, Castellano RK, Brody MS, Rudkevich DM, Siuzdak G, Rebek J. Investigating molecular recognition by mass spectrometry: Characterization of calixarene-based self-assembling capsule hosts with charged guests. Journal of the American Chemical Society. 1999; 121(19): 4568–79.

[29] Banker GS, Siepmann J, Rhodes C. Modern pharmaceutics: CRC Press; Florida, USA, 2002.

Daulat Haleem Khan, Muhammad Yasir Ali, Udo Bakowsky,
Mulazim Hussain Asim, Ghazala Ambreen, Asia Naz Awan,
Saeed Ahmad

14 Tablets

14.1 Introduction

The solid medicaments can be administered orally in the form of powders, granules, cachets, tablets, or capsules for attaining local therapeutic effects in the mouth, throat, and digestive tract or for systemic effects in the body. These formulations have the amount of drug that can be administered as a single unit; these are referred to as solid unit dose forms in general. In addition to conventional solid dosage forms, tablets offer several benefits over other dosage forms. The most common dosage type is a tablet; over 70% of all medications are given out in the form of tablets [1].

The USP and NF describe the tablets as "solid dosage forms of medicinal substances usually prepared with the aid of suitable pharmaceutical adjuncts."

Indian Pharmacopoeia defines pharmaceutical tablets as solid, flat, or biconvex dishes, unit dosage form, formulated by compressing a single drug or combination of drugs either by using diluents or not. Tablets may vary in shape, size, weight, hardness, thickness, hardness, disintegration characteristics, and other aspects depending upon the amount of medicinal substance and the intended mode of administration [2].

"Official tablets may also be defined as circular discs with either flat or convex faces."

The majority of the tablets are used in the oral administration of drugs and may also be used for other routes, e.g., sublingual, buccal, or vaginal but the type of adjuncts, also called excipients, may be different in such cases [3].

Active constituents present in the tablets may be water-soluble or -insoluble. For example, antacids are usually water-insoluble, and some other types of drugs like antihypertensive drug (atenolol) are water-soluble [4].

14.1.1 Properties

- The tablets should be exquisite, unique, and free from faults like chips, fractures, discoloration, and contamination.
- The tablets should be durable enough to survive shocks throughout manufacture, packing, shipping, and distribution.
- Physical stability is essential for maintaining physical qualities throughout time.

https://doi.org/10.1515/9783111438108-014

- The medication agent(s) should release predictably and consistently in the body.
- The therapeutic agent(s) must be chemically stable throughout time to prevent it from deteriorating over time [5].

14.1.2 Advantages and limitations of tablets as dosage form

Tablets are the most popular dosage forms among all the available ones, with 70% abundance over other drug delivery systems.

Advantages

- Among all oral dosage forms, tablets being the unit dosage form provide the highest level of dose accuracy and the least amount of content fluctuation.
- These dosage forms provide different ranges of drug release patterns along with variation in duration of action. Tablets can be designed for rapid or controlled release of drugs. This reduces the need for frequent dosing which leads to better patient compliance.
- Tablets are convenient to use and have elegant dosage forms. They have greater physicochemical and microbial stability over all other oral dosage forms.
- Tablets can be subjected to a coating process for taste masking and color identification.
- They have the easiest and cheapest packaging and handling.
- For identification, tablets can be easily embossed or debossed with the brand's name.
- Tablets are lighter in weight and compact dosage forms with suitability for large-scale production.
- Some tablets can be grooved making it easy to break into many portions. This allows the patient to take the dosage as prescribed by the physician.
- Tablets can be designed to release therapeutic agents at specific sites in the gastrointestinal (GI) system that reduce side effects, increase absorption, and provide a local impact (e.g., ulcerative colitis). This may not be readily accomplished by the other oral dosage forms.
- Tablets can contain many therapeutic agents, even if they are not physically or chemically compatible. The composition and design of tablets can efficiently regulate the release of these medicinal agents.

Disadvantages

- Difficult swallowing for young and unconscious individuals.
- Some drugs are amorphous and have low density, making them difficult to compress into dense compacts.
- Drugs with poor wetting, delayed dissolution, and high absorption in the GI tract may be challenging to synthesize and produce into tablets while maintaining sufficient bioavailability levels.
- Encapsulation or coating may be necessary for bitter or odorous medications as well as those sensitive to air. In such circumstances, capsules may be the most cost-effective option.
- Certain substances, such as aspirin, can irritate the GI mucosa.
- Slow breakdown and dissolution may cause difficulties with bioavailability.

14.2 Tablet ingredients

Tablets contain several inert components in addition to active pharmaceutical ingredients (APIs); these inert substances are referred to as excipients or additives. Depending on their role in the final tablets, they can be categorized. Those that contribute to giving the formulation acceptable processing and compression qualities are in the first group, which includes lubricants, glidants, binders, and diluents. The second set of excipients aids in giving the completed tablet more appealing physical attributes that ultimately enhance patient compliance and includes disintegrants, surfactants, colors, and, for chewable tablets, tastes, and sweeteners; for controlled-release tablets, this group includes hydrophilic polymers or hydrophobic compounds including waxes or other biodegradable polymers. To increase stability and shelf-life, additional materials or antioxidants may occasionally be added. Even though these additional elements have been referred to be "inert," it has become clear that the dosage forms that comprise these excipients and their characteristics are closely related. Preformulation studies show how they affect the techniques used to formulate the dosage forms as well as stability and bioavailability.

There are certain criteria that all the excipients should follow, including

- They should be nontoxic and physiologically inert.
- They should be accepted by all the country's regulatory agencies.
- They must be free from any microbial contamination.
- They should be stable both physically and chemically with themselves and with active pharmaceutical ingredients and other tablet components.
- They should not alter the bioavailability of drugs in the tablets.
- They must have low cost and should be available in all countries in a variety of grades where manufacturing of the product is conducted.
- They should not be contraindicated in certain groups of the population.

It is not necessary to use all the additives mentioned above in the formulation of all tablets; some of them may require a few, and others require all the additives.

14.2.1 Diluents

Diluents are also called fillers that are added to tablets to form the bulk of the tablets, i.e., they increase the mass of the tablets when the drug dosage itself is not sufficient to make up this bulk. These are specially added to individual drugs when the active ingredient is in very small amounts to make it convenient and render the manufacturing process more reliable and practical for forming tablets, e.g., 10 mg methyltestosterone tablet. But some drugs do not require diluents because the doses of these drugs are high, e.g., aspirin and certain antibiotics. Diluents are added in a tablet formulation to improve cohesion, permit direct compression, and promote the flow of the powder to the hopper [6].

The selection of diluents plays a critical role in tableting. Diluents must be compatible with the drug and nonreactive with the ingredients; otherwise, they reduce the bioavailability of the drug. For example, the use of calcium phosphate with tetracycline product reduces its bioavailability to half because divalent and trivalent cations make insoluble complexes with most acidic or amphoteric antibiotics. This greatly affects the absorption, which is why these drugs should not be coadministered with milk [7].

It is advised to employ water-soluble diluents when drugs have limited water solubility to prevent potential bioavailability issues. When manufacturing tablets are used in low dosages for therapeutic usage, including cardiac glycosides, highly adsorbent materials, like bentonite and kaolin, are avoided.

Examples

- Lactose anhydrous (powdered and is sprayed over the active constituents)
- Starch (seldom used, may be obtained from potato, maize, tomato, etc.)
- Mannitol
- Sorbitol
- Sucrose
- Sodium chloride
- Kaolin and other purified clays
- Calcium carbonate
- Calcium sulfide
- Dicalcium phosphate
- Mannitol, sorbitol, and fructose are mainly used in chewable tablets

In tablet formulations, the most widely used diluent is *lactose* because it does not react with the drug or used in anhydrous form or hydrous form. Compared to hydrous lactose, anhydrous lactose has several advantages since it does not undergo the Maillard reaction, which, with some drugs, can cause discoloration and browning. However, the major drawback of the anhydrous form is that it absorbs moisture from humid environments, so tablets formulated using this lactose should be properly packaged to avoid moisture exposure. The hydrous form of lactose is employed in the wet granulation process. *Spray-dried lactose* is employed for direct compression after mixing with the active ingredient. This form of lactose loses some of its direct compression characteristics if it is dried so much that its moisture level reaches below 3%. Combining amine bases or alkaline salts with lactose and an alkaline lubricant can cause tablets to discolor (darken) over time due to the presence of formaldehyde; that is why, when employing spray-dried lactose, neutral or acidic lactose must be taken into account [8].

14.2.1.1 Starch

Starch, a polysaccharide made of amylose and amylopectin, serves as a diluent, binder, and disintegrant in tablet formulations. A pregelatinized grade with modified starch granules yields a free-flowing powder. Sta-Rx 1500 is a directly compressible starch, requires the addition of lubricants when 5–10% drugs are combined with it. Emdex and Celutab are two hydrolyzed starches that can be used in place of mannitol in chewable tablets because of their sweetness.

Alternative fillers include dextrose, sucrose, sorbitol, and mannitol, which have a pleasant flavor and are commonly used in lozenges and chewable tablets. *Dextrose* has hydrous and anhydrous forms and can replace spray-dried lactose to reduce tablet darkness. *Mannitol* is the most expensive sugar, which has a cooling effect when sucked or swallowed, due to its negative heat of solution. It is used in combination with sorbitol to decrease the overall cost of diluents. Sucrose, a sugar, is also employed in tablet formulation, but most of the manufacturers avoid it because of diabetic patients. They are available in different tradenames such as DiPac, Nu Tab, and Sugar Tab; these all absorb moisture when exposed to humid environments.

14.2.1.2 Microcrystalline cellulose (Avicel)

Avicel is made by hydrolyzing cellulose and then spray-drying it. The particles generated are aggregations of smaller cellulose fibers. Aggregates with varying particle sizes and flowabilities can be created by adjusting preparation conditions.

It is frequently mixed with other compounds because it is a costly substance when used as a diluent in large concentrations. Microcrystalline cellulose also acts as a disintegrant, gives good flow properties to the material, and makes it best suited for direct compression. They are available in two grades: PH 101 (powders) and PH 102 (granules). To reduce the tablet mottling and hardening of tablets, 5–15% microcrystalline cellulose is used in the granulation process.

14.2.2 Binders and adhesives

A binder, sometimes known as an adhesive, is applied to a drug-filler mixture to create granules and tablets with the necessary mechanical strength. Binders are used to hold powder together as granules and to assist in ultimately holding the compressed tablets together. The amount of binder used significantly impacts the features of compressed tablets. Using too much or too powerful a binder might result in hard tablets that are difficult to dissolve and can wear down punches and dies quickly.

Examples

- Gum acacia and gum tragacanth
- Starch paste (10% starch solution in water, 2% acacia, most widely used)
- Simple syrup (50–85% solution)
- Gelatin solutions (10–20%)
- Alcohol-glucose solutions (50% alcohol + 50% glucose + 25% water)
- Cellulose derivatives, e.g., methyl cellulose 400 CPS, Na-CMC (carboxy methylcellulose)
- Polyethylene glycol 4000 or 6000
- Polyvinyl pyrrolidine (PVP)
- Honey (but starch is preferred to it because of its cheap and easy availability)

Water, alcohol, and acetone or a mixture of these are not binders in their own right but may act like binders due to their solvent action upon ingredients such as sucrose, lactose, and starch. The process converts powdered materials to granules and retains residual moisture, allowing them to adhere when they undergo compression. Binders are employed in the formulation in different ways. They can be mixed as a dry powder with other excipients before the wet granulation. They can be used as a granulation solution during wet granulation. They may be employed as a dry powder in the other components before the compaction process. Starch, sucrose, and gelatin are commonly used solution binders that are generally considered to be effective. Dry binders often used include microcrystalline cellulose and cross-linked polyvinylpyrrolidone. Polymers like polyvinylpyrrolidone and cellulose derivatives, especially hydroxypropyl methylcellulose, are increasingly employed as binders due to their increased adhesive characteristics.

14.2.2.1 Starch paste

Corn starch paste is a popular binder. Concentration levels might range from 10% to 20%. To prepare, disperse cornstarch in cold purified water to generate a 5–10% w/w suspension. Warm in a water bath with continuous stirring until a transparent paste develops. Some starch is not completely hydrolyzed during paste production. Starch paste may be used as both a binder and a way to introduce additives into granules.

14.2.2.2 Gelatin

Gelatin is a 10–20% solution that should be produced fresh and used warm to avoid solidification. The gelatin is mixed with cold, filtered water and left to stand until it is hydrated. After heating in a water bath to dissolve the gelatin, the solution is weighed to get the required concentration and volume [2].

14.2.3 Disintegrants

These are the substances that are added to tablet formulation when required, to induce the tablets to disintegrate upon contact with water in the GI tract after oral administration. They are added before the addition of binders. Tablets, when added to water, swell due to the presence of disintegrants because they cause the breakage of tablets into very small particles. These smaller fragments have a greater surface area, which ultimately leads to increased dissolution of the drug. Tablets that are hydrophobic or made with a strong compression force have a low rate of water absorption and breakdown, which is unacceptable. In these situations, properly selected disintegrants are used to break the tablet within the specified time range given in the pharmacopoeias.

Examples

- Starch (may be used for all other purposes alone, e.g., binder, lubricant, and diluents)
- Primo gel (sodium-starch glycolate)
- Cellulose
- PVP (use only super PVP, i.e., PVP having the greatest no. of cross-linkages between its molecules)
- Vee gum (Mg-Al silicate)

Disintegrants work through many methods to achieve their effects.

14.2.3.1 Disintegrants that improve water uptake

Disintegrants facilitate liquid passage into tablet pores, causing fragmentation. Surface-active agents are a common chemical that promotes liquid penetration. These chemicals increase the hydrophilicity of medication particle surfaces, allowing liquids to penetrate tablet pores and moisten the solids. Other chemicals may enhance liquid penetration by attracting water to the tablet's pores through capillary forces. Adding a surface-active chemical can improve water penetration into tablet aggregates during disintegration. A de-aggregation and wetting procedure can create a fine dispersion of wetted drug particles, increasing the surface area accessible for breakdown and consequently enhancing the rate of dissolution.

14.2.3.2 Disintegrants that rupture the tablet

Tablets might rupture due to the expansion of the disintegrant particles during water sorption. Non-swelling disintegrants may break tablets by several processes according to research. One method involves particle repulsion upon contact with water, while another involves restoring deformed particles to their natural shape after contact with water, such as those damaged during compaction.

Many effective swelling disintegrants have been created that swell considerably during water intake, successfully breaking tablets into fragments. The most frequent and effective disintegrants that work by a swelling process are modified cellulose or starch. Typically, the formulation contains high-swelling disintegrants at relatively small concentrations, ranging from ~1% to 5% by weight [9].

The rate of disintegration depends upon the physical and chemical properties of the materials of tablets such as binder or lubricants, tablet hardness, and surface area. Carbon dioxide evolution can also cause compressed tablets to disintegrate. When tablets containing a combination of sodium bicarbonate and an acidulant, such as tartaric or citric acid, are mixed with water, they effervesce. To ensure a speedy and thorough breakdown of water, add enough acid to produce a neutral or slightly acidic response.

14.2.4 Lubricants

Lubricants, glidants, and anti-adherents have overlapping functions. Lubricants are added to granules before compression to reduce friction between the walls of the die cavity and the walls of the tablets. In the absence of lubricants, tablet material is stuck to the punches and causes capping or fragmentation of tablets, which leads to defective tablets. That is why lubricants are added to almost all tablet formulations. Improper selection as well as excessive quantity of lubricants make the tablet poorly

disintegrate and thus delay the dissolution process of the drug product. Some examples are given below:
- Calcium stearate
- Magnesium stearate
- Starch
- Sodium chloride
- Talc
- Cocca butter
- Natural fats
- Hydrogenated vegetable oils
- Petroleum
- Paraffin wax

Lubrication involves two mechanisms: fluid lubrication and boundary lubrication. Fluid lubrication involves a layer of fluid separating moving surfaces to minimize friction. Fluid lubricants are rarely employed in tablet formulations. Liquid paraffin is sometimes utilized in effervescent tablet formulations. Boundary lubrication is a surface phenomenon in which sliding surfaces are separated by a thin layer of lubricant. Compared to fluid lubrication, boundary lubrication has a greater friction coefficient and wear on solid materials. Boundary lubricants include all chemicals that interact with sliding surfaces such as adsorbed gases.

Most of the lubricants contain hydrophobic characteristics that retard the disintegration and dissolution of tablets. To mitigate detrimental consequences, hydrophilic alternatives to hydrophobic lubricants have been proposed. For example, polyethylene glycol and surface-active agents. Under-lubricated blends show compression sticking problems, while over-lubricated blend adversely affects tablet hardness and tablet strength [10].

14.2.5 Glidants

Glidants are intended to promote the flow properties of powders or granules by reducing the interparticle friction. It has been suggested that glidant particles are present in the spaces between particles, which results in a reduction of friction. Glidants are typically hydrophobic and are added just before the compression process. The most common glidant is colloidal silicon dioxide, which is employed at low concentrations of 1% or less. Talc (asbestos-free) and magnesium stearate are also employed and may act as both a lubricant and a glidant. The concentration of glidant must not be too high or too low, which adversely impacts the tablet disintegration and dissolution [11].

14.2.6 Antiadherents

Antiadherents reduce the friction between powder and punch faces, preventing particles from adhering. The improper moisture content causes the powder particles to stick to punches known as sticking or picking. Adherence can cause a thin coating of powder to accumulate on the punches, resulting in an uneven tablet surface with false patterns or symbols. Some lubricants, including magnesium stearate, also have antiadherent characteristics. Other substances with low friction-reducing properties can also function as antiadherents such as talc and starch.

14.2.7 Granulating agents

These are the substances that are added to the powder during the granulating process to convert fine powders into granules. The quantity of granulating agents should be controlled because an insufficient quantity of granulating agents may lead to poor adhesion and soft tablets. Excessive quantity may lead to the formation of tablets with greater hardness.

14.2.8 Adsorbents

Adsorbents are used when there is a need to add a liquid or semisolid ingredient to the formulation. They can sorb liquid components onto dry powder in the tablet formulation. The oil or liquid components can be incorporated into the powder in concentrations ranging from 0.5% to 0.7%, e.g., kaolin and bentonite. These materials are made up of hydrated aluminum silicate. The particle size of bentonite is smaller than kaolin because of the colloidal nature of the former.

14.2.9 Flavoring agents

These agents are added to the formulation to mask the bitter or unpleasant taste of APIs in the tablets. This can be done by coating the tablet with suitable and compatible flavors. Generally, they are added in the form of sprays of alcohol or an ethereal solution of volatile oil (flavorants) on the granules before compression of the finished tablets. After spraying flavorants, granules are tumbled or sealed in containers to allow the penetration of flavorants. Flavoring agents are thermolabile, so they can be employed in the formulation before the heating operation, e.g., fruit flavors [12].

14.2.10 Sweeteners

Sweetening agents are used mainly in chewable tablets to limit the use of sugars in these tablets. Added to the tablets used orally. Mannitol is 75% sweet as sucrose. Some artificial sweeteners are also used, including saccharin, which is 500 times sweeter than sucrose. However, due to its carcinogenic nature, its use is limited. Aspartame is also used as an artificial sweetener, but it is moisture-sensitive due to which causes it to lack stability, e.g., mannitol, sucrose, lactose, and saccharin (but now banned).

14.2.11 Coloring agents

Colored tablets are designed to enhance the look or distinguish the product. Colorants are used to impart elegance to tablets as well as the identify of tablets of the same or different pharmaceutical companies. Some drug formulations color the tablet, resulting in an unappealing, scattered look. To address this issue, a suitable coloring ingredient is incorporated into the tablet formulation. To ensure an even distribution of color throughout a tablet, the wet granulation process involves adding a water-soluble color to the granulation liquid. In tablet preparation, two forms of colors are used, which are FD&C and D&C dyes, which are applied as the lake forms or as a solution in granulating liquids. Lakes have been defined by the FDA as the "Aluminum salts of FD&C water-soluble dyes extended on a substratum of alumina." Lakes are insoluble and colored by dispersion, e.g., Amaranth dye (No. 2 in FD and C), lakes (are water-insoluble but made soluble by dissolving in water-soluble substances), iron oxide (red), and titanium dioxide (Tab. 14.1).

Tab. 14.1: Some common tablet excipients.

S. no.	Excipients	Examples
1.	Diluents	Calcium phosphate, carboxymethylcellulose calcium, cellulose, dextrin, lactose, microcrystalline cellulose, PR-gelatinized starch, sorbitol, and starch
2.	Binder and adhesives	Acacia, alginic acid, carboxymethylcellulose, cellulose, dextrin, gelatin, liquid glucose, and magnesium aluminum silicate, maltodextrin, methylcellulose, povidone, sodium alginate, starch, and zein
3.	Disintegrant	Alginic acid, carboxymethylcellulose, cellulose, colloidal silicon dioxide, croscarmellose sodium, crospovidone, potassium polacrilin, and povidone
4.	Lubricant	Calcium stearate, glyceryl palmitostearate, magnesium oxide, poloxamer, polyvinyl alcohol, sodium benzoate, and sodium lauryl sulfate, sodium stearyl sulfate, stearic acid, talc, and zinc stearate
5.	Glidant	Magnesium trisilicate, cellulose, starch, talc, and tribasic calcium phosphate

Tab. 14.1 (continued)

S. no.	Excipients	Examples
6.	Anti-adherent	Corn starch, metallic stearate, and talc
7.	Adsorbent	Kaolin and bentonite
8.	Flavoring agents	Ethyl maltol, ethyl vanillin, menthol, and vanillin
9.	Sweetener	Mannitol, sucrose, lactose, and saccharin
10.	Coloring agent	FD&C or D&C dyes or lake pigments, iron oxide (red), and titanium dioxide

14.3 Types of tablets

Tablets are solid dosage forms of medicinal substances usually prepared with suitable pharmaceutical adjuvants. Various types of tablets can be described as in [13].

14.3.1 Oral tablets for ingestion

– Standard compressed tablets
– Multiple compressed tablets
    – Compression-coated tablets
    – Sugar-coated
    – Film-coated tablets
    – Gelatin-coated tablets
    – Enteric-coated tablets
– Targeted tablets
    – Floating tablet
    – Colon targeting tablet
– Chewable tablets
– Dispersible tablets
– **Tablets used in the oral cavity**
– Lozenges and troches
– Sublingual tables
– Buccal tablet
– Dental cones
– Mouth dissolved/rapidly dissolving tablets
– **Tablets administered by other routes**
– Vaginal tablet
– Rectal tablet

- Implants
- **Tablets used to prepare solution**
- Effervescent tablets
- Hypodermic tablet
- Dispensing/soluble tablet
- Tablet triturate

14.3.1.1 Standard compressed tablets

These tablets are prepared by a simple compression cycle. They are designed in such a way that they provide rapid disintegration in the GI tract that results in rapid drug release. Their manufacturing involves the compression of granules or powders in various shapes and sizes. Compressed tablets typically include a variety of pharmaceutical excipients in addition to the medicinal agent, such as the following:

- Diluents (e.g., lactose) provide bulk to tablets.
- Binders (e.g., sucrose) that promote adhesion of particles.
- Disintegrants (e.g., starch) promote the rapid breakdown of tablets after oral administration.
- Lubricants (e.g., starch) promote the smooth flow of granules into the tablet die, preventing sticking problems.
- Miscellaneous adjuncts such as colorants and flavorants.

Compressed tablets are mostly employed for oral administration. May be used as buccal administration, vaginal administration, and sublingual administration [14].

14.3.1.2 Multiple compressed tablets

These tablets are prepared by subjection to more than a single compression. These tablets have two or more layers. The inner layer is called the core of the tablet, while the outer layer is called the shell of the tablet. The first layer is generated by lightly compressing the drug-containing powder mix/granules. The other drug-containing powder/granule mix is squeezed over the gently compressed first layer to generate the next layer. A distinct therapeutic ingredient may be included in each layer, which may be divided into separate layers due to chemical or physical incompatibilities, and staged drug release (drug release is not at the same rate because every layer has its own concentration of drug) [15].

14.3.1.3 Compressed coating tablets

This tablet is easy to use repeatedly. The outer layer administers the first dose, whereas the inner core delivers the medicine later on. This method is effective for releasing two APIs: one in the coat for rapid release and another in the core for sustained release. This idea allows for both loading and maintenance doses of a single medication.

14.3.1.4 Sugar-coated tablets

It is a compressed tablet coated with sugar. The coating should be water-soluble and quickly dissolves after swallowing. Usually, simple syrup is used for coating. The coat may be colored or colorless. The sugar layer acts as a barrier against unpleasant tastes or odors and shields the medicine within from the external environment. It also offers a sophisticated, glossy look. Patient compliance rises as a result of the medicine's pleasant flavor, often utilized in the creation of mineral and vitamin combinations. These tablets have some disadvantages like (a) the time of manufacture is greater as compared of uncoated tablets; (b) the size of the tablet may increase up to 50%; and (c) the weight of the tablet is increased.

14.3.1.5 Film-coated tablets

This sort of coated tablet does not require the medicine to be coated. To increase tablet strength, film coating is utilized instead of sugar coating. These compressed tablets have very thin coats. This approach uses polymers such as hydroxypropyl cellulose (HPC), hydroxypropyl methylcellulose (HPMC), and ethylcellulose. It is also faster than the sugar-coating method. While it is more durable, less bulky, and easier to apply than sugar coating, it is not as visually appealing or beautiful. The coating is intended to rupture and reveal the core tablet at the desired position in the GI system. These tablets have some advantages like less time-consuming during manufacture, bad taste is masked, moisture of the environment is blocked, the bulk of tablets as compared to sugar-coated tablets is less, and the size may be increased up to 5%.

14.3.1.6 Gelatin-coated tablets

The gel cap (innovator product) is a capsule-shaped compressed tablet that reduces the size of coated products by one-third compared to capsules filled with powder. Gelatin-coated tablets are easier to swallow and more resistant to tampering than unopened capsules.

14.3.1.7 Enteric-coated tablets

Enteric-coated tablets are resistant to acidic environments, preventing medication release in the stomach. It easily releases drugs in alkaline intestinal environments. Drugs have to pass through the stomach before being released, resulting in delayed-action tablets. Those drugs that irritate the stomach and are destroyed by the acidic environment of the stomach can be given in this type by using an enteric coating [16]. Polyvinyl acetate and phthalates are used for coating these tablets. Examples include ecotrin tablets and caplets (GlaxoSmithKline Beecham) [17].

14.3.1.8 Targeted tablets

There are two types of tablets in this category:

Floating tablets: These are intended to increase the duration the dosage form stays in the GI system. This prolongs GI residence time and maximizes medication absorption in a specific location of the GI tract. These tablets have low density. It can expand in a gastric environment. To achieve a better reaction, the medicine should be kept floating in the stomach during diarrhea. Drug distribution is controlled. It reduces mucosal irritation by delivering drugs gradually. Treats GI diseases such as gastro-esophageal reflux. Improved patient compliance through easier administration [18].

Colon targeting tablet: It delivers a therapeutic dose of the drug to a specific target region, such as the colon, to achieve the appropriate concentration in the body. It is ideal for medicines with instability, low solubility, short half-life, high volume of distribution, poor absorption, limited specificity, and low therapeutic index. The pH in the colon ranges from 6.4 to 7, and the existence of microbial flora influences medication release.

Drug release techniques in this field include coating with pH-sensitive polymers like Eudragit S100 and L100, biodegradable polymers that are sensitive to colonic bacteria, and bioadhesive polymers like polycarbophil redox-sensitive polymers. It ensures correct medication transport to the lower GI tract, preventing release in the upper GI tract [19].

14.3.1.9 Chewable tablets

Chewable tablets must be broken and eaten between the teeth before consumption. These tablets are provided to children with trouble swallowing and adults who detest swallowing. These tablets are designed to dissolve in the mouth at a moderate rate, with or without chewing. Chewable tablets are commonly used to target specific areas of the body rather than the entire body. They are made up of a gum core that may or may not be coated. The core consists of insoluble gum bases, fillers, waxes, antioxidants, sweeteners, and flavorings. The gum base proportion ranges from 30% to 60%. Mannitol as a flavoring is also incorporated into these types of tablets. Antiflatulents may be given in this form [20].

14.3.1.10 Dispersible tablets

European Pharmacopoeia defines dispersible tablets as uncoated or film-coated tablets that can be homogeneously dispersed in water before ingestion. A dispersible tablet is typically dispersed in 5–15 mL of water (e.g., a tablespoonful or a glass) before being presented to the patient. Dispersible tablets should dissolve within 3 min in water at 15–25 °C. The dispersion from a dispersible tablet should pass through a sieve screen with a nominal mesh aperture of 710 μm. The introduction of an acid/base couple can improve the dispersion qualities of dispersible tablets by causing the base to release carbon dioxide when the couple's components are dissolved in water [21].

14.3.2 Tablets used in the oral cavity

14.3.2.1 Lozenges and trouches

Lozenges are flavored medicinal dosage forms designed to be sucked and held in the mouth or throat. There are two types of lozenges: hard candy (boiled) and compressed tablet (TROUCHES). Lozenges can be used to provide local drugs to the mouth or throat as well as for systemic drug absorption. A pastille is a soft lozenge made of gelatin, glycerin, or a combination of acacia, sugar, and water. Compressed lozenges do not include disintegrants. Other additions (binders and fillers) should have a pleasing flavor or feel during breakdown. Gelatin is commonly used as a binder in compressed lozenges, with sorbitol, mannitol, and glucose as fillers [2].

14.3.2.2 Sublingual tablets

The drug is absorbed immediately via the mouth's mucosal lining when administered beneath the tongue, resulting in instant systemic action. Tablets are typically tiny, flat, and oval and are minimally compacted to maintain their softness. To ensure effective medication absorption, the pill should disintegrate promptly. It dissolves in small amounts of saliva. Sublingual, or "under the tongue," is a method of giving drugs through the mouth. This allows for quick absorption through blood vessels beneath the tongue rather than the digestive tract. Tablets for sublingual administration (e.g., nitroglycerine) have a rapid-release character.

14.3.2.3 Buccal tablets

These medications are meant to be dissolved in a buccal pouch. Tablets are not meant to dissolve. The pill is dissolved by placing it near the parotid duct entrance. Buccal pills are commonly used for hormone replacement treatment. Long-acting buccal tablets contain viscous natural or synthetic gums to generate a hydrated surface layer, allowing medication to slowly diffuse and be absorbed through the buccal mucosa. Mucoadhesive polymers, including PANA and carbopol 934, are employed. Tablets of buccal administration (e.g., progesterone tablets) have a slow drug release character [22].

14.3.2.4 Dental cones

These tablets are intended to fit loosely into the empty socket after tooth extraction. The objective of this tablet is to inhibit bacterial growth in the socket with a slow-releasing antibacterial component or to minimize bleeding using an astringent or coagulant tablet. This product dissolves or erodes gently in 20–40 min with a tiny amount of serum or fluid present. Typically utilized carriers include sodium chloride, sodium bicarbonate, or amino acids.

14.3.2.5 Mouth-dissolved/rapidly dissolving tablets

Mouth-dissolving tablets are solid dose forms that dissolve quickly under the tongue. Mouth-dissolving tablets have a pleasant mouthfeel and do not require water to consume. MDT dissolves quickly in saliva, taking about 15 s to 3 min. True fast-dissolving MDT dissolves quickly in saliva, often within seconds. Fast-disintegrating tablets include chemicals that speed up tablet disintegration in the oral cavity, taking approximately 1 min to fully dissolve. This product is a top option due to its high hardness, consistent dosing, and ease of administration [23].

14.3.3 Tablets administered by other routes

14.3.3.1 Vaginal tablet

These ovoid-shaped tablets are placed into the vagina with a special plastic tube inserter. Following insertion, the tablet retains and slowly dissolves, releasing the therapeutic substance and providing the local pharmacological action (e.g., in the treatment of bacterial or fungal infection). These tablets are usually antiseptic, astringent, or steroidal. Vaginal pills can enhance the systemic absorption of medicinal drugs.

Similarly, buccal/sublingual tablets are critical that the tablet dissolves in vivo rather than disintegrates because disintegration reduces tablet retention inside the vagina.

14.3.3.2 Rectal tablets

This is an ancient and appropriate method of therapy. Rectal fluid volume, buffer capacity, pH, and surface tension can vary greatly, even within a single person, leading to variable absorption through this route. Rectal tablets do not require refrigeration. These tablets have improved product stability, even at room temperature.

14.3.3.3 Implants

These tablets are implanted in bodily cavities and have a long-lasting impact, ranging from days to months to years. The tablets are tiny and cylindrical. These implants are surgically implanted under the skin and gradually absorbed over several months or years. A specialized injector with a hollow needle and plunger is utilized to administer the rod-shaped tablet. For other shapes, surgery is required. The formulations are sterile and include no excipients. It is used for the treatment of chronic diseases, e.g., arthritis. These tablets primarily provide growth hormones to food-producing animals. The ear is the ideal place for administering drugs.

14.3.4 Tablets used to prepare solution

14.3.4.1 Effervescent tablets

Effervescent tablets have sodium bicarbonate and an organic acid, often tartaric or citric, in addition to the medication. In the presence of water, these additives react, releasing carbon dioxide, which acts as a disintegrator and causes effervescence. Effervescent tablets are soluble except for minor amounts of lubricant. Easy to drink after dissolving in water and masks the unpleasant taste of the drug.

14.3.4.2 Hypodermic tablet

Hypodermic tablets are soft, soluble, and contain one or more drugs in a tablet. It is used to prepare injection solutions. Since stable parenteral solutions are now available for the majority of pharmacological compounds, there is no reason to utilize hypodermic tablets for injection. Avoid using these solutions as they are not sterile.

These pills are still widely produced for oral delivery. The official compendia have never recognized hypodermic tablets [16].

14.3.4.3 Dispensing/soluble tablet

Dispensing tablets offer a practical way to introduce potent drugs into powders and liquids, eliminating the need to weigh small quantities. These tablets are designed for quick compounding and should not be used as a dosage form. These tablets are highly toxic if administered orally by mistake. A material incorporated in dispensing tablets includes mild silver proteinate, bichloride of mercury, and quaternary ammonium compounds [2].

14.3.4.4 Tablet triturate

Tablet triturates are compact, cylindrical tablets that are either molded or compressed. These products often contained potent drugs blended with lactose and a binder such as powdered acacia. Tablet triturates are often soft and friable. All ingredients are water-soluble. Any water-insoluble material is avoided. They are present in the sublingual form. It can be helpful in the preparation of other dosage forms, e.g., solid or liquid dosage forms.

14.4 Tablet manufacturing

The main rationale for granulating the powder before tableting is:
- To increase the bulk density of the powder mixture and thus ensure that the required volume of powder can be filled into the die.
- To improve the flowability of powder.
- To improve mixing homogeneity and reduce segregation by mixing small particles.
- To improve the compatibility of the powder.

14.4.1 Wet granulation method

This is the most widely used method for tablet manufacturing; it consists of the following procedures [24].

14.4.1.1 Mixing

If the size of the drug is large, then first reduce the size to form powder. Add excipients and pass through a sieve 40–60. To ensure batch uniformity, the drug powders must be mixed thoroughly before moistening. This is generally done in a motor-driven powder blender or mixer. A small quantity of disintegrants is also added here. Mixing is done for 15–20 min.

14.4.1.2 Preparation and addition of blenders

If we must use starch as a binder, then make it a paste in water (10%). Starch is first added to cold water, and then it is heated to form a paste. Now mix this paste with the powders of Step 1.

14.4.1.3 Wet mixing

Mix the binder in this step, prepared in the previous step. The amount of binder should be sufficient to make the powder wet. If an excess of binder is added, then we have a very hard type of tablet. But if the binder is present in very small amounts, then we have tablets that have less hardening. These tablets are easily breakable and so difficult to transport. Moreover, excessive binder addition may cause a longer disintegration time of tablets.

14.4.1.4 Screening of wet mass/wet screening

Granules will not readily flow into a punching machine cavity with which they are comparable in size due to crowding at the entrance of that cavity. On the other hand, very fine granules often exhibit poor flow characteristics. Both conditions lead to excessive weight variation. In addition, tablets that are too thin relative to their diameter are prone to breakage when handled. For this reason, the next step is wet screening. Here products of the previous step are passed through mesh 6–8.

14.4.1.5 Drying

Drying wet granules may be done in different ways. Usually, the drying temperature of granules is 60 °C but may be lower if the thermolabile substances are to be dried. If we use tray dryers, air exchange is essential to prevent the saturation of the oven atmosphere with the solvent vapor. Agglomeration of granules and migration of solute are minimized, and oven drying of granules is promoted. May be 24 h is required for the drying of large batches of granules. Among the newer methods of drying in use today is fluidization conducted in fluid driers. In this method, granules are dried by the use of warm air. As the air is blown upward, the granules also move upward and then are taken back to their normal position. This process is continued up to the maintenance of the temperature of the air that is blown and that which comes out, because with the removal of the solvent temperature of the air is decreased, and if the temperature is constant, then it means that there is no air solvent present in the granules. If the effectiveness of the binder depends upon the solvent presence, then some solvent is allowed to remain in it. Approximately 100 kg of substance can be dried in 20–30 min by this method.

14.4.1.6 Screening of dried granules/dried screening

Due to the presence of drying, there may be some large lump formation. So, screen these lumps through a mesh 12–20. Sizing of the granules is necessary so that a small die cavity for the production of small tablets may be filled by the flowing granules.

14.4.1.7 Lubrication

This is done so that the granule is covered with lubricant. Lubricant may be dusted over the spread-out granulation through a fine mesh screen. Among the most used lubricants are talc, magnesium stearate, stearic acid, and calcium stearate. But out of this, magnesium stearate is most used.

14.4.1.8 Compression

Finally, granules are compressed in a tablet compressor for the final touch.

Advantages of wet granulation

–   Incompatibility of two ingredients may be successfully handled in some cases by granulating the troublesome ingredients, separating, and then mixing the granules.
–   Mixing of all ingredients is proper.

Disadvantages of wet granulation

– A long process to be performed.
– At every stage, we need some equipment.
– In the case of lubrication, we have to use mechanical vibration, due to which the size of the granule is distributed.
– Material is lost.

14.4.2 Dry granulation method

When tablet ingredients are sensitive to moisture or are unable to withstand the elevated temperature during drying, and when the tablet ingredients have sufficient inherent binding or cohesive properties, slugging or dry granulation may be used to form granules. It eliminates several steps but still includes weighing, mixing, slugging, dry screening, lubrication, and compression. The active ingredients of the part of lubricants are blended. As powdered materials have a considerable amount of air, under pressure, this is expelled, and a fairly dense piece is formed. The more time allowed for this air to escape, the better the tablet or slug.

When the slugging is used, large tablets are made as slugs, about 7/8 to 1 inch is the proper size of tablets in this method. This slug must be hard enough to be broken up without producing an excessive amount of powder. The slugs are broken up by hand or by a mill (e.g., chilsonator (Fitzpatrick), Roller compactor (vector), and compactor mill (Allis Chalmers)) and passed through a screen of desired mesh for sizing. Lubricant is added in the usual manner, and tablets are prepared by compression. Formally, the alkali metal and ammonium halides, hexamine, and KCl were the commonly used medicinal substances that could be dry granulated. More recently, the list has been extended by commercial production of suitably granular forms of dried yeast, dry cascara extract, aspirin, some ferrous salts, and a few other substances.

14.4.3 Direct compression method

Some granular chemicals like KCl and methamine possess free-flowing as well as cohesive properties that enable them to be compressed directly in a tablet machine without the need for either wet or dry granulation. In direct compression of tablets, the tableting excipients used must be materials with properties of fluidity and compressibility.

In addition to the use of excipients of special properties, forced or induced feeders that have been developed permit the preparation of certain additional tablets by direct compression because the de-aerating action of the feeder on light, bulky powder makes them dense and enables them to fill the cavities of the die under moderate pressure [15].

14.5 Unit operations in tablet manufacturing

A tablet press or compression machine is the instrument used in the manufacturing of tablets. The tableting powder or granules are compressed with extreme precision under enormous pressure to shape them into a single tablet. The equipment operates on the principle of compression and performs compressing operations on granules or powder by the simultaneous action of punches and die [25].

The following are various types of tablet manufacturing equipment.

14.5.1 Single-punch tablet machine

Single-punch tablet machines are the most basic models on the market. There are several models to choose from. Even though most of them are automated, there are a few that may be controlled by hand. With a single punch, compression is achieved. The die cavity fills once the feed or hopper shoe with the granulation is placed over it. Retraction of the feed shoe allows it to remove any extra granulation from the die cavity. To compress the granules inside the die cavity, the upper punch lowers. To expel the tablet, the lower punch lifts, and the upper punch raises. The compressed tablet is pushed off the die platform by the feed shoe as it moves back to fill the die cavity.

The capacity of the die cavity determines the weight of the tablet; the bottom punch may be adjusted to change the amount of granules, which changes the weight of the tablet. Sturdier models are needed for tablets bigger than 1/2 inch in diameter. This also holds true for pills that need to be extremely hard such as compressed lozenges. The heavier models work well for slugging since they can withstand significantly higher pressures [26].

14.5.2 Rotary tablet machines

The benefits of rotary machines for higher productivity are numerous. The tablet granules pass via a feed frame, a huge steel plate that spins beneath a head that is continually rotating and holding many sets of punches and dies. This technique encourages a consistent die fill, which results in an exact tablet weight. Compression occurs as the upper and lower punches travel over a set of rollers. The material in the die cavity is slowly squeezed from the top and bottom by this movement, which allows the trapped air to escape. The tablet is raised and ejected by the lower punch. While the machine is operating, tablet weight and hardness may be adjusted without the use of tools. Within the feed hopper, the internal flow of the granulation is one of the factors that cause the variance in tablet weight and hardness during compression.

The majority of rotary machine types have an excess pressure release mechanism that absorbs the shocks and undue strain from each compression and protects the machine. It is easy to remove the punches and dies for cleaning and inspection and to insert different sets to create a wide range of sizes and forms. Most of the rotary tablet presses use force during main compression or variation in tablet height at precompression to determine the weight of the tablet. Using an upper and lower punch, they compress a volume of grains that are trapped in a die between two rollers. The force required to compress tablets can be altered by varying the space between the rollers. The compression force will not change after the roller spacing is determined. Several factors influence the exact number of granules that fall in each die. Granule size, size distribution, and punch length are a few examples of factors that may have an impact. Furthermore, too fast rotational speed gives the granules too little time to drop into the die. Tablets with varying weights and densities are produced via variations in the number of grains in each die. The efficacy of the dose form may be impacted by uneven tablet characteristics brought on by variations in the maximum compression force [26].

14.5.3 High-speed rotary tablet machines

Models of the rotary tablet machine that can compress tablets at high production rates have gradually developed over time. This has been achieved by enhancing feeding mechanisms, adding dual compression points to certain models, and increasing the number of stations – that is, sets of punches and dies – in each turn of the machine head. Double rotary machines are ones with two compression points, and single rotary machines are those with only one compression point. The tablet chute produces half of the tablets at an angle of 180°.

The actual speed still depends on the physical properties of the tablet granulation and the rate that is consistent with compressed tablets having adequate physical characteristics, even though these models are mechanically capable of operating at production rates of more than 5,000 tablets per minute.

Ensuring sufficient die filling is the primary challenge in high-speed machine operation. The die cavity beneath the feed frame has too little filling time when filling up quickly to maintain the necessary die packing and uniform flow. To replenish the dies in the extremely short filling time allowed on the high-speed machine, a variety of techniques for force-feeding the granules into the dies have been developed. Material can be partially compacted prior to final compaction in presses equipped with pre-compression rollers. This facilitates the material's partial deaeration and particle alignment before to ultimate compression. This prevents laminating and capping from trapped air and aids in the direct compacting of materials [27].

14.5.4 Multilayer rotary tablet machines

Additionally, variants of rotary tablet machines that can create multiple-layer tablets have been developed; these machines can create one, two, or three layers of tablets. Tablets with stratified content have several benefits. By dividing the layers that hold incompatible medications with a layer of inert material, the pharmaceuticals can be combined into a single tablet. It has made it possible to create medication with a time delay and provides a multitude of options for creating color combinations that give the product a unique identity.

14.6 Coating of tablets

Coating is the process of applying a dry outer layer to a dosage form to provide benefits such as product identification and medication release. When a coating solution is applied to a batch of tablets in a coating pan, the tablets' surfaces become coated with a sticky polymeric film. The tablets are then allowed to dry, resulting in a non-sticky dry surface. The coating process requires characteristics such as the spray pattern, drop size, and nozzle spacing (in addition to several other non-spray-related parameters), which must all be carefully regulated to achieve equal dispersion of the coating material [28].

14.6.1 Objectives of coating

An additional step that may be employed in tablet manufacturing is tablet coating, which increases the cost of tablets, but there are some purposes for coating tablets:
- An enteric coating protects the drug from breaking down in the stomach.
- Non-steroidal anti-inflammatory drugs may irritate certain sites inside the GI system such as the stomach; coating prevents this irritation.
- To provide controlled release throughout the GI system.
- To target medication release to a specific region in the GI system, for example, drug delivery to the colon for the treatment of inflammatory diseases.
- To disguise the taste of medications.
- Improve the tablet's appearance.

14.6.2 Coating process

It is most desired that the coating should be consistent and not fracture under stress. As a result, a variety of processes were developed for applying the coating to the tablet surface. Typically, coating solutions are sprayed over uncoated tablets while they are being agitated in a pan, fluid bed, etc. As the solution is applied, it forms a thin coating that adheres to each tablet. The liquid component of the coating solution is subsequently evaporated by blowing air across the tumbling pans. The coating can be created in a single application or in layers by spraying many times. Rotating coating pans are widely utilized in the pharmaceutical sector [29].

14.6.3 Sugar coating

Sugar coating has been the conventional method for coating medicinal items, particularly tablets.

The process involves applying a sugar solution with color numerous times to create uniform, beautiful, and glossy tablets. The duration of sugar coating is high, from a few hours to a few days. Tablets with deep convex surfaces and thin, rounded edges are ideal for sugar coating. Sugar-coated tablets should be resistant to breaking, chipping, and abrasion. Sugar coatings are often lengthy and aggressive. The process involves applying sucrose-based coatings on tablet cores using appropriate equipment. While traditional panning equipment and hand application of syrup have been widely employed, newer specialized equipment and automated approaches are gaining traction. Sugar coatings are made from substances that quickly dissolve in water.

Sugar coatings can be applied to a variety of pharmaceutical materials, but are often used to coat tablets. Sugar-coated tablets are often designed for quick release. During the sugar-coating process, the sealing phase (described below) includes applying a polymer-based coating to the surface of uncoated tablets. Sugar-coated products can have delayed or extended release by using specialized polymers that are insoluble in water [30]. The basic sugar-coating process involves the following steps.

14.6.3.1 Sealing

When the seal coat is applied, it prevents moisture from entering the tablet core. It is used in pan-ladling to avoid over-wetting a specific area of the tablet bed. Without a seal covering coat, tablets might absorb too much moisture, causing softening or disintegration of the tablets and affecting the chemical and physical stability of the product. Spray methods may avoid the need for a seal phase, ensuring the end product's physical and chemical stability.

Seal coating agents (sealants) include shellac, zein, CAP, and PEG 4000 (dissolved in alcohol or acetone). Shellac is more effective due to its polymerization, although it increases tablet disintegration and dissolution times. Zein, an alcohol-soluble protein derivative from corn, does not increase dissolution time with age.

14.6.3.2 Sub-coating

To round out the tablet edges, sugar coatings are frequently put in rather large quantities in the tablet core, increasing the weight by as much as 50–100% on average. A large portion of this material build-up is accomplished during the sub-coating stage by including bulking agents such as calcium carbonate into the sucrose solutions. Moreover, polysaccharide gums, like gum acacia, can be added as a binder to lessen brittleness and antiadherents, such as talc as a dusting powder, to stop tablets from adhering to one another. The hot air is blown into the coating pan for proper drying of the coat. The alternative addition of sugar solution and dusting powder (talc), along with drying, is repeated until the desired thickness is obtained.

14.6.3.3 Smoothing (color coating)

The subcoating step is infamous for leaving a rather rough surface finish. Sub-coated tablets are often smoothed out by adding a sucrose coating solution that is often colored with titanium dioxide to get the necessary amount of whiteness. This makes it easier to apply the coloring layer, which needs a smooth surface. A coloring agent is added to the sucrose syrup. Water-soluble dyes have been employed traditionally, but pigments have gradually taken their place to speed up the coating process and reduce issues with color migration. No color should be added until the tablet surface becomes smooth because premature application leads to mottling.

14.6.3.4 Finishing

Without dusting powder, three to four coats of syrup are applied quickly, and each coat is dried with a circulation of cold air. This creates a smooth, durable coat.

14.6.3.5 Polishing

To obtain the desired luster and glossy finish, the polishing of tablets is done with the application of waxes. It is done in a standard coating pan or canvas-lined polishing pan. Suitable waxes include beeswax, carnauba wax, or candelilla wax applied as finely ground powders or as suspensions/solutions in an appropriate organic solvent.

14.6.4 Film coating

Sugar-coating is time-consuming and has been superseded by film-coating technology. Film coating is a single-step process. It involves spraying a solution of polymer, pigments, and plasticizer onto a spinning tablet bed, creating a thin, homogeneous layer on the tablet surface [24]. The thickness of the film coat ranges between 20 and 100 μm. The polymer used for drug release depends on the intended location (stomach/intestine) and release rate [2]. Although film coatings can be categorized in a variety of ways, it is customary to do so based on how the applied coating is intended to affect drug release characteristics. Thus, film coatings can be categorized as follows: non-functional and functional.

Non-functional film coatings, also called immediate release film coating, change the look of tablets and protect them from external influences, whereas functional film coatings, called modify or delay drug release, enhance stability (e.g., gastro-resistant). Film coating techniques include organic solvent-based, aqueous, and solvent-free coating (Tab. 14.2). Organic solvent-based coatings, while not optimal, are utilized on hydrophobic or lipophilic polymers. Aqueous film coating is widely utilized for its safety and environmental benefits. The solvent-free coating offers a cost-effective and time-saving alternative to solvent-based techniques [31].

Tab. 14.2: Materials used in film coating.

S. no.	Material	Type	Uses	Examples
1.	Film Former	Enteric	To control the release of the drug	Cellulose acetate phthalate (CAP), acrylate polymers, hydroxypropyl methylcellulose phthalate (HPMCP), polyvinyl acetate phthalate (PVAP)
		Non-enteric	Have suitable mechanical properties and coatings that are relatively easy to apply.	Ethyl cellulose, hydroxy propyl methyl cellulose (HPMC), povidone (PVP), and acrylate polymers
2.	Solvents	–	To dissolve or disperse the polymers. It pertains to the chemical modification of the basic polymer that alters the physical properties of the polymer.	Water, IPA, and methylene chloride Chloroform, acetone, ethanol, and methanol Glycerol, propylene glycol, PEG 200–6000 grades
3.	Plasticizer	Internal plasticizing External plasticizing	1–50% by weight of film former It is incorporated with the primary polymeric film former, changing the flexibility, tensile strength, or adhesion properties of the resulting film.	Castor oil, glycerin, diethyl phthalate (DEP), dibutyl phthalate (DBP), and tributyl citrate (TBC), and surfactant (TWEEN, SPAN)

Tab. 14.2 (continued)

S. no.	Material	Type	Uses	Examples
4.	Colorants	Inorganic materials	For light shade: a concentration of less than 0.01% may be used. For dark shade: a concentration of more than 2.0% may be required.	Iron oxides Anthocyanins, caramel, carotenoids, turmeric, and carminic acid
5.	Opaquant extenders	–	Formulations to provide more pastel colors and increase film coverage	Titanium dioxide, silicate (talc and aluminum silicates), and carbonates (magnesium carbonates)

14.6.4.1 Basic process requirements for film coating

The basic criteria for a film-coating process are consistent regardless of the equipment employed. These include:
-   Proper atomization of spray liquid or application to tablet cores.
-   Proper mixing and agitation of the tablet bed.
-   Spray coating relies on each core passing through the spray zone. This differs from sugar coating, which involves spreading syrup from tablet to tablet while tumbling in the coating pan before drying.
-   Enough energy in the form of hot drying air to evaporate solvent. This is especially significant when applying aqueous-based coatings, which need higher energy input.
-   Effective exhaust systems for removing dust and solvents from air.

14.6.4.2 Ideal characteristics of film coating polymers

Solubility
-   The solubility of polymers is crucial for two reasons: it establishes how the coated product will behave in the digestive system, including how quickly the drug will be released and whether there will be a delay in the start of drug release.
-   It establishes the coating's solubility in a selected solvent system, a factor that can significantly impact the final coating's functional characteristics.

For immediate-release products, film coatings should include polymers with strong solubility in aqueous solutions. This will help the active component dissolve quickly from the completed dosage form after consumption. Typically, these coatings are applied as solutions in a suitable solvent system, such as water or a strong presence being exhibited. However, film coatings that are intended to alter the rate or onset of drug release from the dosage form often have little to no solubility in aqueous fluids;

these coatings are typically applied as aqueous polymer dispersions or as polymer solutions in organic solvents [32].

Viscosity

The ease of applying a film coating is limited by its viscosity. Coating liquids with high viscosity (more than 500 mPa s) might be difficult to transfer from storage to spray guns and atomize into small droplets. Polymers used as solutions in a certain solvent should have low viscosity at the desired concentration. This will make spraying the coating solution easier and more reliable, especially for large-scale film-coating equipment.

Permeability

Film coatings' functional qualities rely heavily on appropriate permeability, which is mostly determined by the polymer used. Coating permeability is important for
– Masking the bad taste of active ingredients in dosage forms.
– Enhancing dose stability by reducing exposure to ambient vapors and gases, especially water vapor and oxygen.
– Adjusting the rate of active component release from the dose form.

The characteristics of different polymers and film-coating formulas differ significantly.

14.6.5 Compression coating

Tableting is a process of compacting granular materials around a prepared core using a specialized equipment. Compression coating is a dry operation. Coating a tablet core with organic solvents or water can give flavor masking or delayed or enteric qualities, making it advantageous in certain situations.

Compression coating involves modifying the usual tableting procedure. Tablet cores are first prepared before being physically transported to a bigger die that is partially loaded with coating powder. The tablet core is positioned centrally in a partly filled die, followed by more coating powder and a second compaction event. Compression coating is a sophisticated mechanical technique that needs precise formulation and processing of the coated layer. Large or irregularly shaped granules might cause the core to tilt in the second die, leading to uneven or partial coating and a visible core on the tablet surface [33].

14.7 Tablet processing problems

Issues with tablet processing may arise from issues with the compression apparatus, the formulation, or both. As a result, we may group the issues into the following categories.

14.7.1 The defects related to the tableting process

14.7.1.1 Capping

It is a typical problem that may happen while making tablets. Formulation-related problems include having too many fines in granulation, having moisture levels that are too low or too dry to allow proper binding, not drying granules enough, using the wrong or too little binder, using the wrong or too little lubricant, and compressing granules at a temperature that is too low. To fix the problem, you can use a 100–200 mesh filter to get rid of the fines, wet the granules enough, add hygroscopic materials like PEG 4000, sorbitol, or methylcellulose, dry the granules properly, increase the amount of binder, or add dry binders like powdered sugar, hydrophilic silica, gum acacia, powdered sorbitol, PVP, or pre-gelatinized starch. Other options are to improve lubrication, switch to a new lubricant, and compress at room temperature. Some machine-related problems include improperly completed dies, punches that are too deeply concave or beveled, the bottom punch not being set up correctly during ejection, the sweep-off blade not being adjusted correctly, and the turret rotating too quickly. Polishing dies, using the right materials, making sure punches are level, aligning the bottom punch and sweep-off blade accurately, and slowing down the turret to give the dwell time more time, are all ways to fix the problem.

14.7.1.2 Lamination

Lamination happens when there are oily or waxy granules, too much hydrophobic lubricant, or magnesium stearate. Causes connected to the machine include the edges of the tablet relaxing quickly after being ejected and the tablet decompressing quickly. To fix the problem, you may change the way you mix, add materials that absorb or adsorb, change or decrease the lubricants, use tapered dies with a 3°–5° outward taper, use pre-compression, and lower the final compression pressure and turret speed.

14.7.1.3 Cracking

Cracking may happen when granules are too big or too dry, when tablets expand, or when granulation is done in the cold. Some problems with machines include that air is trapped, which makes tablets expand, and deep concavities that create breaks when the tablets are ejected. Some ways to fix this are to compress granules with more fines, get the right amount of binder and moisture, improve granulation, use dry binders, and compress at room temperature. Using tapered dies and unique take-off mechanisms on machines may also help keep things from breaking.

14.7.2 Problems with excipients

14.7.2.1 Chipping

When granules are excessively dry or too tightly bonded, they may chip, which causes them to adhere to the punch faces and chip at the bottom. Worn die grooves, barrel-shaped dies, inward-folded punch edges, or too much concavity are some of the problems that might happen with machines. Fixes include drying things out properly, adding lubricants, wetting granules to make them more flexible, adding hygroscopic materials, polishing dies, bending them into cylinders, sharpening punch edges, and making flat punches less concave.

14.7.2.2 Sticking

It happens when granules are not dried correctly, when there is not enough or the right kind of lubricant, when there are too many binders, when substances are hygroscopic or oily, or when granules are weak. The problem is caused by machine variables such as deep concavity, poor compression pressure, or too much speed. Some ways to fix this are to dry the granules completely, change the amount of moisture, improve the lubrication, change the kind or amount of binder, manage the humidity during compression, add absorbents, and optimize granulation. To fix machines, you may make the concavity less, the pressure higher, and the compression speed slower.

14.7.2.3 Picking

Picking happens when the granules are too moist, there is not enough lubrication, the materials melt too easily when heated, the drug concentration is too high and the melting temperatures are too low, the granules become too hot, or there is too much binder. Some things about the machine that might cause problems include rough punch faces, deep lines, and not enough pressure. Some ways to fix the problem are to dry the press and granules properly, use colloidal silica and more lubrication, use materials with a high melting point, put the press and granules in the fridge, compress them at room temperature, cool them down before compressing them, use less binder, and switch to other binders. Machine options include cleaning punch faces, adjusting font design, employing chromium plating, lowering sharpness, and using the highest compression pressure possible.

14.7.3 Problems caused by more than one thing

Mottling happens when colorful medications are combined with white or colorless excipients, when dyes move about during drying, when dyes are not mixed properly, or when colored binder solutions are not mixed properly. Using the right colorants, altering the solvent systems, changing the binders, decreasing the drying temperatures, reducing the particle size, mixing completely, and adding dry color additives during powder blending together with fine adhesives like acacia or tragacanth before granulation are all ways to fix the problem.

14.7.3.1 Problems with the machine

Double impression occurs when the upper or lower punches may move freely during tablet ejection, particularly when the punch faces are etched. You may fix the problem by using keyed tools that stop rotation or by utilizing newer presses that have mechanisms that stop rotating.

14.7.3.2 Problems with the coating process

Cratering: It looks like a volcanic craters that show the surface of the tablet. This usually happens when the coating solution is applied too quickly and the drying process is not done well. To fix the problem, you may use better drying conditions and lower spray rates by making the solution thicker.

Picking: Picking during coating happens when film sections are pushed apart because tablets stay together. This generally happens because the coating solution was applied too thickly or the film did not cure properly. Some solutions are to raise the temperature of the drying air, make the drying process more efficient, or change the viscosity of the solution. Pitting happens when there are pits on the surface of the tablet core, but no film breaks. This is caused by the air being too hot for drying. You may prevent this by not preheating and carefully managing the temperature of the air that comes in so that the core stays below the melting point of the additions.

Blooming: It makes the coating dull when it is stored at high temperatures because of low molecular weight or too much plasticizer. Some solutions are to minimize the amount of plasticizer used and use alternatives with a greater molecular weight.

Blushing: It is haziness or white patches that happen as the coating temperature goes up, and the sorbitol lowers the gelation temperature of cellulose derivatives. To fix the problem, you may reduce the temperature of the drying air and stay away from sorbitol in products that include hydroxypropyl cellulose, methyl cellulose, or similar polymers.

Bridging: It happens when the coating covers over logos or writing because the solution was not applied correctly, it was too thick, it was not atomized properly, or it shrank during drying. It might also happen as the film pulls away at the corners. One way to fix the problem is to change the kind or amount of plasticizer.

Orange peel: Orange peel or roughness creates a matte, uneven surface that looks like an orange peel. This happens because the coating dries quickly and is quite thick. To fix the problem, you need to use light drying conditions and add solvents to the solution to make it less thick.

14.8 Evaluation of tablets

14.8.1 Shape and diameter

Tablets' shape and diameter depend upon the tools used. The less concave the punches, the flatter the resulting tablet. Conversely, the more concave the punches, the more convex the resulting tablets. In the punching machine, we have two punches, that is, the upper punch and lower punch, along with a middle part called the die.

The punches having raised impressions will produce a recessed impression on the tablet, and the punches having recessed etching will produce tablets having monograms. Monograms may be on both sides or only on one side. This engraving is mainly done for the identification of different tablets [34].

14.8.2 Thickness of tablets

The thickness of a tablet is determined by the amount of fill permitted to enter the die and the amount of pressure applied. Moreover, if the granules have a uniform size, then the thickness of the tablet will not vary unless the pressure is constant. If the size of granules is not constant, then we have varying thicknesses of tablets, and the resulting tablet is difficult to pack.

Take 10 tablets randomly from the batch and measure the size with the help of the vernier caliper. If the thickness varies by ±5, then variation is allowed. Tablet gauges may also be used for this purpose [35].

14.8.3 Hardness

The resistance of the tablet to chipping, abrasion, or breakage depends upon the tablet's hardness. In the past, thumb rules were applied for the hardness of the tablet. In the 1930s, Monsanto constructed a device for the measurement of hardness, called the

Monsanto Hardness Tester. This instrument measures the force required to break the tablet when force is applied diametrically. The force is measured in kilograms. A hardness of 4 kg is the minimum for a satisfactory tablet. Another instrument is the Pfizer hardness tester. The force required to break the tablet is recorded on a dial and may be expressed as kilograms or pounds of force. Manufacturers, e.g., Key Erweka and others, make similar apparatuses to the Heberlein or Schleanger apparatus (most widely used nowadays). If the tablet is too hard, it may not disintegrate in the required period or may not withstand the handling during coating or packing and shipping operations [36].

14.8.4 Friability

Friability is a measure of the resistance of the tablets to shipping and abrasion. The purpose of having a friability test is to make sure that the tablets formed can withstand mechanical stresses during their manufacturing, distribution, and handling by the end user (Fig. 14.1). Several tablets are weighed and placed in the tumbling apparatus (friabilitor). Roche friabalitor consists of a drum of transparent synthetic polymer with polished internal surfaces and is subject to minimum static build-up. One side of the drum is removable. The tablets are tumbled at each turn of the drum by a curved projection that extends from the middle of the drum to the outer wall. The drum is attached to the horizontal axis of a device that rotates at 25 rpm. After each revolution, tablets fall. After a given revolutionary period, tablets are weighed, and the loss of weight is measured. This loss should be in the range of 0.5–1%. The compressed tablets shall not lose more than 1% of weight after the test. Then tablets are of standard hardness [37].

14.8.5 Weight variation

The weight of the tablet is adjusted by the combined effects of factors, e.g., number of tablets and amount of active drug. Due to these reasons, the diameter of dies should be controlled to produce the required size tablets, which in turn will control the weight of tablets, but no apparatus is more accurate to produce a tablet of the required weight. There are BP and USP-allowed errors in the tablet weights. If the tablet size is higher then it will have less range of allowed error. But if the tablet size is low, the allowed error is high.

Let us say a manufacturer is manufacturing a tablet of 500 mg then the allowed error is ±5%, i.e., weight if is in the 475–525 range then it is in the allowed range (Tab. 14.3). If tablets of weight in the range of 450–550 are manufactured then the machine has some problems and all the process of tablet manufacturing should be repeated by guiding these tablets, so to avoid this problem we have to check the process after every 30 min [38].

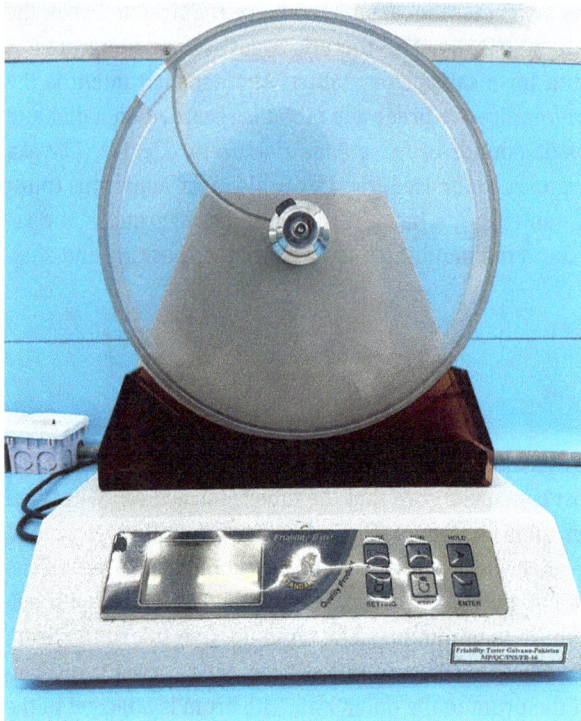

Fig. 14.1: Tablet friability tester (special thanks to Mass Pharmaceuticals, Pakistan).

Tab. 14.3: USP limits for the weight variation test for uncoated tablets.

S. no.	Average tablet weight (mg)	Percentage deviation
1.	130 mg or less	±10%
2.	130 mg to 324 mg	±7.5%
3.	More than 324 mg	±5%

14.8.6 Disintegration test

Complete disintegration is defined as that state in which any residue of the unit, except fragments of insoluble coating or capsule shell, remaining on the screen of the test apparatus or adhering to the lower surface of the discs, if used, is a soft mass having no palpably firm core.

The apparatus consists of a basket-rack assembly, a 1,000-mL, low-form beaker, 138–160 mm in height and having an inside diameter of 97–115 mm for the immersion fluid, a thermostatic arrangement for heating the fluid between 35 and 39 °C and a device for raising and lowering the basket in the immersion fluid at a constant frequency rate between 29 and 32 cycles per minute, through a distance of not less than 53 mm and not more than 57 mm. The volume of the fluid in the vessel is such that at the highest point of the upward stroke, the wire mesh remains at least 15 mm below the surface of the fluid and descends to not less than 25 mm from the bottom of the vessel on the downward stroke. At no time should the top of the basket rack assembly become submerged. The time required for the upward stroke is equal to the time required for the downward stroke, and the change in stroke direction is a smooth transition rather than an abrupt reversal of motion. The basket-rack assembly moves vertically along its axis (Fig. 14.2). There is no appreciable horizontal motion or movement of the axis from the vertical [39].

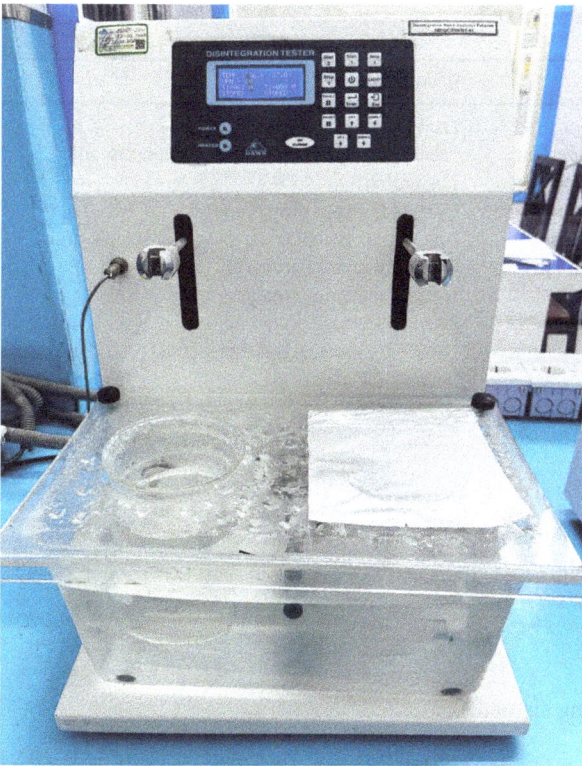

Fig. 14.2: A typical disintegration tester (special thanks to Mass Pharmaceuticals, Pakistan).

Place one dosage unit in each of the six tubes of the basket, and if specified, add a disc. Operate the apparatus using water as the immersion fluid unless another liquid is specified and maintain its temperature at 35–39 °C. At the end of the specified time, lift the basket from the fluid and observe the dosage units: all of the dosage units have disintegrated completely. If one or two dosage units fail to disintegrate, repeat the test on 12 additional dosage units. The requirements of the test are met if not less than 16 of the 18 dosage units tested are disintegrated [40].

14.8.7 Dissolution test

The USP-NF provides several official methods for carrying out dissolution tests of tablets (Tab. 14.4). Tablets are grouped into uncoated, plain-coated, and enteric-coated tablets. The selection of a particular method for a drug is usually specified in the monograph for a particular drug product. Buccal and sublingual tablets are tested by applying the uncoated tablet procedure [41].

Tab. 14.4: Dissolution apparatuses.

S. no.	Apparatus	Name	Drug products
1.	Apparatus 1	Basket apparatus	Tablet, capsules
2.	Apparatus 2	Paddle apparatus	Tablets, capsules, modified drug products, and suspensions
3.	Apparatus 3	Reciprocating cylinder	Extended-release drug products
4.	Apparatus 4	Flow through cell	Drug products containing low-water-soluble drugs
5.	Apparatus 5	Paddle over disk	Transdermal drug products
6.	Apparatus 6	Rotating cylinder	Transdermal drug products
7.	Apparatus 7	Reciprocating disk	Extended-release drug products
8.	Diffusion cell (Franz)	(Non-USP-NF/BP)	Ointments, creams, and transdermal drug products (topical)

14.8.7.1 Basket apparatus

The assembly consists of the following:
– a vessel
– a motor
– a metallic drive shaft
– a cylindrical basket (stirring element)

The *vessel* is cylindrical, made of glass or other inert, transparent material, with a hemispherical bottom and a capacity of 1 L. Its height is 160–210 mm, and its inside diameter is 98–106 mm. Its sides are flanged at the top. A fitted cover may be used to retard evaporation. The vessel is partially immersed in a suitable water bath of any convenient size or heated by a suitable device such as a heating jacket. The water-bath or heating device permits maintaining the temperature inside the vessel at 37 ± 0.5 °C during the test and keeping the dissolution medium in constant, smooth motion. No part of the assembly, including the environment in which the assembly is placed, contributes significant motion or vibration beyond that due to the smoothly rotating stirring element. The shaft is positioned so that its axis is not more than 2 mm at any point from the vertical axis of the vessel and rotates smoothly and without significant wobble that could affect the results. A speed-regulating device is used that allows the shaft rotation speed to be selected and maintained at a specified rate, within ±4% deviation. Shaft and *basket* components of the stirring element are fabricated of stainless steel, type 316 or equivalent. A basket having a gold coating of about 2.5 μm (0.0001 in) thick may be used. The dosage unit is placed in a dry basket at the beginning of each test. The distance between the inside bottom of the vessel and the bottom of the basket is maintained at 25 ± 2 mm during the test [42].

14.8.7.2 Paddle apparatus

The assembly consists of the following:
– vessel
– motor
– metallic drive shaft
– paddle

The paddle apparatus uses the assembly from Apparatus 1, except that a *paddle* formed from a blade and a shaft is used as the stirring element. The shaft is positioned so that its axis is not more than 2 mm from the vertical axis of the vessel, at any point, and rotates smoothly without significant wobble that could affect the results. The vertical center line of the blade passes through the axis of the shaft so that the bottom of the blade is flush with the bottom of the shaft. The distance of 25 ± 2 mm between the bottom of the blade and the inside bottom of the vessel is maintained during the test. A suitable two-part detachable design may be used, provided the assembly remains firmly engaged during the test. The paddle blade and shaft may be coated with a suitable coating to make them inert. The dosage unit is allowed to sink to the bottom of the vessel before the rotation of the blade is started (Fig. 14.3). A small, loose piece of non-reactive material, such as not more than a few turns of wire helix, may be attached to dosage units that would otherwise float [41].

14.8.8 Content uniformity and tablet assay

As all the tablets have both active and inactive ingredients, different tablets have different percentages of drug, e.g., if we have 250 mg of active drug in a 500 mg tablet, then

$$\text{\%age of drug} = 250/500 \times 100 = 50\%$$

Now, 50% active drug should be present in the tablet uniformly; otherwise there may be effects produced on the bioavailability. Take 30 tablets randomly from the batch,

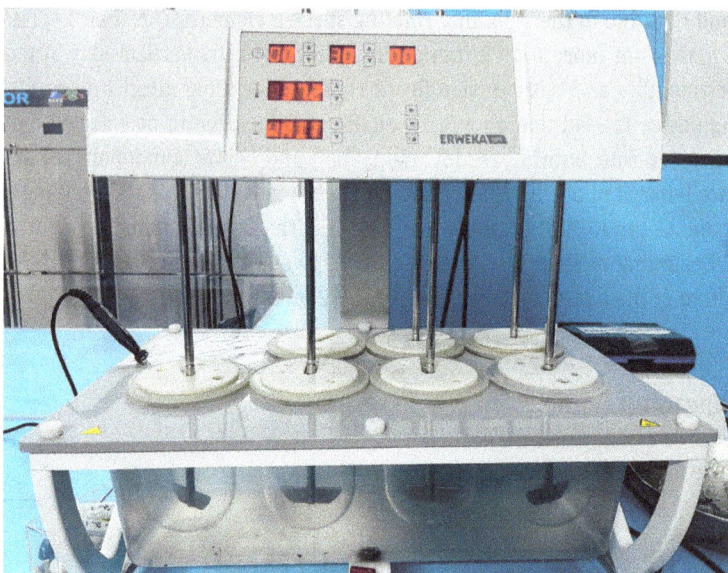

Fig. 14.3: A typical paddle dissolution apparatus (special thanks to Mass Pharmaceuticals, Pakistan).

and out of these 30, 10 tablets are assayed first separately. If the contents of not less than 9 of the tablets are within the limits of 85–115% of the labeled potency, then the contents are uniformly distributed. This is called content uniformity. If tablets have contents out of range of 75–125% then perform the content uniformity test with the other 10 tablets from the remaining 20 tablets. If still variations are present, then the apparatus has a problem, and the tablets are manufactured again. The test is not required for multivitamin and trace-element preparations and in other justified and authorized circumstances [43].

References

[1]   Ubhe TS, Gedam P. A brief overview on tablet and it's types. Journal of Advancement in Pharmacology. 2020; 1(1): 21–31.

[2]   Lachman L, Lieberman HA, Kanig JL. The theory and practice of industrial pharmacy: Lea & Febiger Philadelphia; 1976. pp. 210–212.

[3]   Kim J, De Jesus O. Medication routes of administration. in: StatPearls, StatPearls Publishing, Treasure Island (FL), 2023. http://www.ncbi.nlm.nih.gov/books/NBK568677/ (accessed April 17, 2023). 2021.

[4]   Eriksen SP, Irwin GM, Swintosky JV. Antacid properties of calcium, magnesium, and aluminum salts of water-insoluble aliphatic acids. Journal of Pharmaceutical Sciences. 1963; 52(6): 552–56.

[5]   Harbir K. Processing technologies for pharmaceutical tablets: a review. International Research Journal of Pharmacy. 2012; 3(7): 20–23.

[6]   Pandey V, Reddy KV, Amarnath R. Studies on diluents for formulation of tablets. International Journal of Chemical Science. 2009; 7(4): 2273–77.

[7] Guerra W, Silva-Caldeira PP, Terenzi H, Pereira-Maia EC. Impact of metal coordination on the antibiotic and non-antibiotic activities of tetracycline-based drugs. Coordination Chemistry Reviews. 2016; 327: 188–99.

[8] Bharate SS, Bharate SB, Bajaj AN. Interactions and incompatibilities of pharmaceutical excipients with active pharmaceutical ingredients: A comprehensive review. Journal of Excipients and Food Chemicals. 2016; 1(3): 3–26.

[9] Al-Achi A, Gupta MR, Stagner WC. Integrated pharmaceutics: Applied preformulation, product design, and regulatory science: John Wiley & Sons; 2022.

[10] Faldu B, Zalavadiya B. Lubricants: Fundamentals of tablet manufacturing. International Journal of Research in Pharmacy and Chemistry. 2012; 2(4): 921–25.

[11] Parekh BV, Saddik JS, Patel DB, Dave RH. Evaluating the effect of glidants on tablet sticking propensity of ketoprofen using powder rheology. International Journal of Pharmaceutics. 2023; 635: 122710.

[12] Al-Achi A. Tablets: A brief overview. Journal of Pharmacy Practice and Pharmaceutical Sciences. 2019(1): 50, 49–52.

[13] Tiwari S, Mahapatra S. Unit dosages form tablet: An overview. International Research Journal of Humanities, Engineering Pharmaceutical Sciences (IJHEPS). 2015; 1: 8–36.

[14] Sahoo P Tablets. 2007.

[15] Dogra S, Shah I, Upadhyay U The most popular pharmaceutical dosage form: Tablet.

[16] Patil PR, Bobade VD, Sawant PL, Marathe RP. Emerging trends in compression coated tablet dosage forms: A review. International Journal of Pharmaceutical Sciences and Research. 2016; 7(3): 930.

[17] Allen L, Ansel HC. Ansel's pharmaceutical dosage forms and drug delivery systems: Lippincott Williams & Wilkins; 2013.

[18] Kaur B, Sharma S, Sharma G, Saini R, Singh S, Nagpal M, et al. A review of floating drug delivery system. Asian Journal of Biomedical and Pharmaceutical Sciences. 2013; 3(24): 1–6.

[19] Singh A, Dabral A. A review on colon targeted drug delivery system. International Journal of Pharmaceutical Sciences and Research. 2019; 10(1): 47–56.

[20] Renu JD, Jalwal P, Singh B. Chewable tablets: A comprehensive review. The Pharma Innovation Journal. 2015; 4(5): 100–05.

[21] Nandhini J, Rajalakshmi A. Dispersible tablets: A review. Journal of Pharmaceutical Advanced Research. 2018; 1(3): 148–55.

[22] Lieberman HA, Rieger MM, Banker GS. Pharmaceutical dosage forms: Disperse systems. (No Title). 1988.

[23] Masih D, Gupta R. Mouth dissolving tablets – a review. Pharmaceutical and Biosciences Journal. 2013; 1(1): 18–24.

[24] Ahmed SAN, Patil SR, Khan MS, Khan MS. Tablet coating techniques: Concept and recent trends. International Journal of Pharmaceutical Sciences Review and Research. 2021; 66(1): 43–53.

[25] Begum SG, Bai AS, Kalpana G, Mounika P, Chandini JA. Review on tablet manufacturing machines and tablet manufacturing defects. Indian Research Journal of Pharmacy and Science. 2018; 5(2): 1479–90.

[26] Fox SC. Remington education pharmaceutics: Pharmaceutical Press; 2014.

[27] Konkel P, Mielck JB. Associations of parameters characterizing the time course of the tabletting process on a reciprocating and on a rotary tabletting machine for high-speed production. European Journal of Pharmaceutics and Biopharmaceutics. 1998; 45(2): 137–48.

[28] Arora R, Rathore KS, Bharakatiya M. An overview on tablet coating. Asian Journal of Pharmaceutical Research and Development. 2019; 7(4): 89–92.

[29] Toschkoff G, Just S, Knop K, Kleinebudde P, Funke A, Djuric D, et al. Modeling of an active tablet coating process. Journal of Pharmaceutical Sciences. 2015; 104(12): 4082–92.

[30] Himaja V, Sai K, Karthikeyan R, Srinivasa B. A comprehensive review on tablet coating. Austin Pharmacology & Pharmaceutics. 2016; 1(1): 1–8.

[31] Dumpa M, Kamadi M, Vadaga A. Comprehensive review on tablet coating problems and remedies. Journal of Pharma Insights and Research. 2024; 2(1): 042–9.

[32] Porter SC. Coating of tablets and multiparticulates. Aulton's Pharmaceutics: The Design and Manufacture of Medicines. 2007; 3: 500–14.

[33] Windheuser J, Cooper J. The pharmaceutics of coating tablets by compression. Journal of the American Pharmaceutical Association. 1956; 45(8): 542–45.

[34] Sultan T, Xu X, Rozin EH, Sorjonen J, Ketolainen J, Wikström H, et al. Effect of shape on the physical properties of pharmaceutical tablets. International Journal of Pharmaceutics. 2022; 624: 121993.

[35] Diarra H, Mazel V, Busignies V, Tchoreloff P. Investigating the effect of tablet thickness and punch curvature on density distribution using finite elements method. International Journal of Pharmaceutics. 2015; 493(1–2): 121–128.

[36] Seitz JA, Flessland GM. Evaluation of the physical properties of compressed tablets I: Tablet hardness and friability. Journal of Pharmaceutical Sciences. 1965; 54(9): 1353–57.

[37] Osei-Yeboah F, Sun CC. Validation and applications of an expedited tablet friability method. International Journal of Pharmaceutics. 2015; 484(1–2): 146–55.

[38] Ahmed S, Islam S, Ullah B, Biswas SK, Azad AS, Hossain S. A review article on pharmaceutical analysis of pharmaceutical industry according to pharmacopoeias. Oriental Journal of Chemistry. 2020; 36(1): 1–10.

[39] Amaral Silva D, Webster G, Bou-Chacra N, Löbenberg R. The significance of disintegration testing in pharmaceutical development. Dissolution Technologies. 2018; 25: 30–38.

[40] Gupta MS, Kumar TP. Characterization of orodispersible films: An overview of methods and introduction to a new disintegration test apparatus using LDR-LED sensors. Journal of Pharmaceutical Sciences. 2020; 109(10): 2925–42.

[41] Salve PM, Sonawane SV, Patil MB, Surawase RK. Dissolution and dissolution test apparatus: A review. Asian Journal of Research in Pharmaceutical Sciences and Biotechnology. 2021; 11: 229–236.

[42] Uddin R, Saffoon N, Sutradhar KB. Dissolution and dissolution apparatus: A review. International Journal of Current Biomedical and Pharmaceutical Research. 2011; 1(4): 201–07.

[43] Blanco M, Alcalá M. Content uniformity and tablet hardness testing of intact pharmaceutical tablets by near infrared spectroscopy: A contribution to process analytical technologies. Analytica Chimica Acta. 2006; 557(1–2): 353–59.

Part 4: **Semisolid dosage forms**

Muhammad Yasir Ali, Khurram Waqas, Muhammad Asjad Ur Rahman,
Muhammad Arsal Rafiq, Zeenat Arshad, Abd-Ur-Rehman Khan,
Usman Saleem

15 Ointments, creams, and miscellaneous

15.1 Ointments

Ointments are semisolid preparations intended for use on the skin or mucous membranes [1]. Some of these ointments are designated for ophthalmic applications, known as ophthalmic ointments. They can be classified as either medicated or non-medicated. Non-medicated ointments are utilized for their emollient and lubricating properties or as a base for medicated ointments.

15.1.1 Ointment bases

The term ointment bases refers to the non-medicated carriers that incorporate active pharmaceutical ingredients (APIs) to produce ointments. A well-chosen base is essential for ensuring proper drug release, effective adherence to the site of application, minimal irritation, and, in some cases, additional therapeutic actions like emolliency or a protective effect. Selecting the correct ointment base is essential in the realm of pharmaceutical formulation, as it significantly affects the patient experience and various elements of drug delivery. This fundamental aspect has a considerable impact on the solubility of the drug, which in turn influences how effectively the active ingredients are distributed and dissolved within the base. Furthermore, the base is crucial for managing the rate of drug release, which directly impacts the therapeutic effectiveness of the medication and its absorption through the skin. The comfort of the patient is of utmost importance, given that the properties of the base such as its greasiness, fragrance, and ease of washing affect overall acceptability and compliance with treatment. In addition, careful selection of the base is necessary to ensure the long-term integrity and efficacy of the pharmaceutical product as well as the stability of the entire formulation.

15.1.1.1 Hydrocarbon bases (oleaginous bases)

These bases are a mixture of hydrocarbons and are obtained from petroleum. They are not absorbed from the skin. They are immiscible with water, not water washable, difficult to remove from skin, and sticky and greasy in nature. They are used as emollients and bases for hydrophobic drugs. They are inert and used as occlusive dressing;

https://doi.org/10.1515/9783111438108-015

that is, they are applied to stop secretions. However, not suitable for water-soluble drugs and infected skin wounds.

Petrolatum: It is a mixture of semisolid hydrocarbons obtained from petroleum. It melts at a temperature between 38 and 60 °C. It is also called yellow petroleum, petroleum jelly, or soft paraffin. It may be amber color or white-colored. It can be used for an emollient effect and can be retained for a long time. It may be used alone as vaseline (Tab. 15.1).

Tab. 15.1: An example of a traditional petrolatum-based ointment.

S. no.	Ingredients	Concentrations (w/w)
1.	White petrolatum USP	50–80%
2.	Lanolin	1–5%
3.	Natural and /or synthetic waxes	2–10%
4.	Oil-soluble emulsifier	1–3%
5.	API	As required
6.	Antioxidants	0.1–0.5%
7.	Fragrance/essential oils	0.1–1%

White petrolatum: White petrolatum, also called white soft paraffin and petroleum jelly, is a white to yellowish semisolid mixture of hydrocarbons dewaxed from paraffinic residual oil, predominantly with carbon numbers higher than C25. The major constituents inside the white petrolatum are n-paraffin, iso-paraffin, and naphthene. White petrolatum can be used as an API, such as a skin protectant, as well as an inactive pharmaceutical ingredient (excipient) in a wide range of topical semisolid pharmaceutical and cosmetic formulations (e.g., ointments, creams, and lotions). Although white petrolatum has been widely used for over a century, it is still challenging to ensure formulation product quality due to variable rheological properties under different shear stress and thermal conditions. In addition, quality control of white petrolatum (excipient) is challenging since the source of the crude petroleum varies, and therefore the distillation as well as purification processes to prepare white petrolatum also vary from manufacturer to manufacturer as well as from batch to batch.

The spatial and crystal microstructure of white petrolatum is still not clear. Several techniques, such as a polarized light microscope, X-ray diffraction, scattering, and thermal analysis, have been utilized to analyze the microstructure (microcrystalline properties) of white petrolatum.

Lanolin: Lanolin is obtained from sheep's wool and is extensively used in pharmaceuticals and cosmetics. It is generally considered to consist of a mixture of naturally formed esters derived from higher alcohols and higher fatty acids. It can act as an emollient, occlusive dressing and can absorb a reasonable amount of water to incorporate water-soluble APIs.

Waxes: Waxes are the layer of the fatty component on the surface of plant leaves, insect bodies, and animal skins, while technologists use the term to refer to any products that contain fatty materials obtained from plants, insects, marine, or mineral origin that are of commercial value. Waxes can also be described as hydrophobic organic substances of medium chain length. Regardless of their definition, there is no dispute that waxes have a wide range of applications.

Hard paraffin: It is a mixture of solid hydrocarbons obtained from petroleum or shale oil. It is a colorless, odorless, wax-like substance. It solidifies between 50 and 57 °C and is used to stiffen ointment bases.

Mineral oil: It is a mixture of liquid hydrocarbons obtained from petroleum. It is transparent, colorless, and odorless. It cannot be used alone but is used in combination with petroleum or a drug. It is called liquid petroleum or liquid paraffin.

15.1.1.2 Absorption bases

Absorption bases, unlike hydrocarbon types, are hydrophilic. They are anhydrous (or hydrous water-in-oil (w/o) emulsion) and will absorb water; therefore, they can be used for dry skin conditions. They are not water-removable. On the other hand, they can sometimes cause irritation and are greasy in nature; therefore, acceptability by patients may decrease. There are two types of absorption bases: (a) non-emulsified bases; for example, anhydrous lanolin and hydrophilic petroleum. They usually form w/o emulsion: (b) emulsified bases; for example, hydrous lanolin and cold cream.

They are less occlusive, more emollient, and easily applicable than the hydrocarbon bases.

Anhydrous lanolin: It is also called wool fat. It is obtained from sheep. It contains not more than 0.25% water. When mixed with water can absorb water as much as twice its weight in water. This incorporation of water results in the formation of a w/o emulsion:

Water + drug → drug solution + lanolin → w/o-medicated ointment

Hydrophilic petroleum: It is composed of cholesterol, stearyl alcohol, white wax, and white petroleum. This base is non-greasy and is water miscible in nature. It can absorb about three times its weight of water.

Emulsified bases: These are those bases present in w/o emulsion form; for example, hydrous lanolin and cold cream.

Hydrous lanolin: Lanolin is prepared from wool fat (70%) and water (30%). The BP specifies a wool fat content (75%) and water (25%). It is w/o emulsion that contains

30% water, but additional water can be incorporated by mixing. So it can be used as an emollient effect alone [2].

Cold cream: It is semisolid, white w/o emulsion and contains cetylester wax, white wax, mineral oil, sodium borate, and purified water. This sodium borate combines with fatty acids to form soap that acts as an emulsifier agent. These bases are also helpful in the incorporation of aqueous solutions of certain drugs; for example, sodium sulfacetamide, into hydrocarbon bases. First, make an emulsion of this solution in absorption bases, then mix these bases into hydrocarbon bases [3].

15.1.1.3 Water-soluble bases

These are anhydrous, soluble in water, water-removable, greaseless, non-rancidifiable, and non-occlusive. Completely water-soluble bases have been developed from the macrogol [polyethylene glycol]. The formula of polyethylene glycol is $CH_2OH \cdot (CH_2OH_2)_n \cdot CH_2OH$. We have different numbers of their PEG [polyethylene glycols]; that is, PEG 200, 300, 400, 600, etc. PEGs ranging from 200 to 1,000 are viscous liquids, but PEGs ranging from 1,500 to so on are solid in nature and PEGs between their two groups are greasy semi-solids. When we need to add water or powdered water-soluble drugs to these bases, a portion of such bases is used to solubilize drugs, which is then mixed with other ingredients. However, if higher concentrations of drug solution are to be incorporated, then a portion of the liquid PEGs can be replaced by stearyl alcohol.

15.1.1.4 Water-removable bases

These are o/w emulsions that are capable of being used on the skin with water. So they are also called water-washable bases. It is observed that when the medicinal agents are incorporated in these bases, they are better absorbed; for example, hydrophilic ointments.

Hydrophilic ointments: These are o/w emulsions. They contain sodium lauryl sulfate and stearyl alcohol as emulsifying agents. The oleaginous phase is white petroleum, aqueous water, and propylene glycol.

15.1.1.5 Choice of bases

If rapid release of the drug is required, then we use water-washable and water-soluble bases. If a drug is rapidly hydrolyzed, then we stabilize this by using hydrocarbon bases. If the skin is dry, then an occlusive ointment base is required for wetting it. If the skin is wet, then we should dry it by using a suitable base (usually water water-washable base is used).

15.1.2 Preparation of ointments

Ointments are prepared by levigation and fusion methods, as discussed below.

15.1.2.1 Levigation method

Levigation of solid: This procedure can be carried out on a small scale by mixing the ointment ingredients with a mortar and pestle, a glass slab, and a spatula. When made this way, a long, broad-blade spatula is used to use a smaller spatula to avoid ointment buildup. A hard rubber spatula is the better choice if the spatula's components are reactive. Trituration is the first step of reducing the drug or powder size to an extremely fine consistency. Then, put the ointment bases or levigating agents (like mineral oil, PEGs, or glycerine) on one side of the glass slab and the powder on the other. Mix some of the powder into the base until the mixture is consistent. Next, mix the ointment base with a tiny bit of the other powder. Continue doing this until all of the powders and ointment bases have been well combined. When a very small amount of the drug (particularly strong drugs) or powder is available, it should be combined with a small amount of the ointment base, making sure that all of the powder is used. The remaining base should also be added gradually, using a geometric method [4].

Levigation of solutions: It is not levigation but just mixing the drug solution and the ointment base. Remember that water-soluble or hydrophilic bases will be suitable for the absorption and incorporation of the drug solution. When we have to mix in hydrophobic bases, a portion of these bases is replaced by the hydrophilic bases. Then the drug solution or drug [if it is ready in aqueous form] is mixed with the hydrophilic base and then finely mixed with the hydrophobic base.

15.1.2.2 Fusion method

All or some of the components are melted and then combined and congealed. The components that are heat-labile are not mixed before melting are directly in the melted base, and when their bases are cooled down to a low temperature. This method of ointment preparation can be adopted in two different ways: (a) melt the material that has a high melting point, then add other materials; (b) melt the material that has a low melting point and then add others and gradually increase the temperature. The drug/drug solution may be added in either of the phases, depending upon their solubilities.

By this method, both the oleaginous as well as the emulsified bases can be used in the preparation of ointments. In the case of an emulsified base, separately heat the oily and water phase to 70–75 °C. But the temperature of the water should be at least 3–4 °C higher than the oily phase. Add water to the oily phase by constant stirring. If the drug is heat labile, it is added after mixing water, but if it is heat stable, then it is added before adding water directly into the oily phase, and then water is mixed.

15.1.3 Preservation of ointment

If ointments are contaminated, then due to the presence of these contaminated microbes, the product may induce infection in the patient, particularly if it is contaminated by *Pseudomonas aeruginosa* and *Staphylococcus aureus*.

Some topical products are manufactured by sterile techniques through microbial filtration in laminar–flow hoods. However, the majority of the topical products are not intended to be sterile and do not undergo such aseptic procedures, but must still meet acceptable limits for microbial contents.

15.1.4 Packaging

Ointments are usually packed either in jars or tubes. The jars may be of glass, uncolored or colored green, amber, blue, or opaque, and porcelain white. Plastic jars may also be used. In a tube, open end and cap end, two ends, sealed, then rotated. The product added is still in liquid form. Tubes for ophthalmics used have 3.5 g of product.

Ointment jars may be filled with a flexible spatula. An ointment prepared by the fusion method can be poured into a jar in liquid form.

Tube-filled ointments have some advantages over jars because they are conveniently handled by the patient. Also, they are less contaminated during use or even packaging than jars.

15.1.5 Characteristics of an ideal ointment

An evaluation of an ointment formulation's physical and chemical properties is necessary to ensure its integrity. The components should remain chemically stable over time. The ointment's texture needs to be consistent and visually appealing. The skin should feel smooth and silky after application. To avoid discomfort or an uneven dispersion of the active ingredient, there should be no grit or particulate matter. The melting point of a topical ointment base is a crucial property. The base should ideally melt or soften at body temperature or nearly so. Effective use requires this attribute. Because the ointment liquefies slightly when it comes into contact with the skin, it spreads and penetrates easily. Additionally, the ointment base's excipients need to be nonirritating. The ointment bases should not provoke adverse skin reactions, redness, or itching. For the patient to be comfortable and accept the medication, this is crucial. The API must be finally divided for incorporation. It must be reduced to extremely fine particles. The entire ointment base must then be thoroughly mixed with these particles. A constant dosage of the drug is present in every application of the ointment, thanks to uniform dispersion.

15.2 Creams

Creams are semisolid or viscous liquids that are either w/o emulsions and contain one or more soluble substances [5]. Shaving cream, vanishing cream, and hand creams are o/w emulsions, and cold creams are w/o emulsions in nature. Active ingredients are dispersed in either phase or added when the emulsion has been formed and allowed to cool. Unlike oral and parenteral dosage forms, topical dermatological formulations often require many excipients. Each excipient should be justified by function and need. If a novel excipient is used, the FDA will probably require additional safety data, which will increase the nonclinical study burden (time and cost). In the study, excipients such as solvents, preservatives, antioxidants, surfactants, and other agents are used to overcome solubility, stability, or skin penetration challenges.

15.2.1 Vanishing creams

Vanishing creams are o/w emulsions and usually contain large amounts of water (70–85%) and stearic acid (15–30%). When they are applied stearic acid layer along with the API is left behind after the evaporation of water. These are non-greasy in nature, have good spreadability, and impart emollient and smoothing effects on skin.

15.2.1.2 Example

15.2.1.2.1 Prepare vanishing cream with anionic surfactant and/or nonionic surfactant

The oil phase (stearic acid, cetyl alcohol, and stearyl alcohol) is mixed by melting in a china dish on constant stirring. Components of the aqueous phase (glycerine, methyl paraben, propyl paraben, potassium hydroxide, and tween 60) should be mixed in a separate container and warmed to about the same temperature as the oil phase. The preservatives propyl paraben and methyl paraben were added just before mixing, not during the process of heating. The aqueous phase is added to the oil phase drop by drop with constant stirring. Then perfume, if added, is incorporated when the formulation begins to solidify (Tab. 15.2).

Tab. 15.2: Ingredients for stearic acid-based vanishing cream.

S. no.	Ingredients	Anionic surfactant	Nonionic surfactant
1.	Stearic acid	30%	14%
2.	Cetyl alcohol	1.0%	1%
3.	Stearyl alcohol	1.0%	–
4.	Glycerine	10%	–
5.	Methyl paraben	0.07%	0.07%
6.	Propyl paraben	0.03%	0.03%
7.	Potassium hydroxide	0.09%	–
8.	Tween 60	–	1.5%
9.	Purified water (q.s.)	100%	100%

15.2.2.1 Example

15.2.2.1.1 Prepare 100 g vanishing cream with mineral oil and lanolin

The oil phase (stearic acid, lanolin, cetyl alcohol, and mineral oil) is mixed by melting in a china dish on constant stirring. Components of the aqueous phase (potassium hydroxide, propylene glycol, and water) should be mixed in a separate container and warmed to about the same temperature as the oil phase. The aqueous phase is added to the oil phase drop by drop with constant stirring (Tab. 15.3).

Tab. 15.3: Ingredients for stearic acid and mineral oil-based vanishing cream.

S. no.	Ingredients	Concentration (g)
1.	Stearic acid	9
2.	Lanolin	2
3.	Cetyl alcohol	1
4.	Mineral oil	5
5.	Potassium hydroxide	0
6.	Propylene glycol	5
7.	Purified water	77.5

15.2.2 Cold cream

The invention of cold cream is credited to Galen, a physician in the second century from Greece. This cold cream is thick and softens when it touches the skin. It is perfect for dry skin on the elbow, feet, and knees and also perfect for natural ways of removing makeup and avoiding eczema in dry parts of your body. The combination of fats and water in this product helps moisturize – the cream gets its name because it is cold to the touch – with people using it to soften their skin, soothe sunburns, and protect faces from wintry weather, too. The emulsion is of a "water in oil" type, unlike the "oil in water" type emulsion of vanishing cream, so-called because it seems to disappear when applied to the skin. Lubricating creams, night creams, or massage creams are a type of cold cream with the addition of lanolin (wool fat) and its derivatives. The name "cold cream" derives from the cooling feeling that the cream leaves on the skin. Cold creams were usually made as w/o emulsions. After the creams are applied to the skin, much of the water evaporates, leaving the remaining oil to act as a solvent that cleanses the skin of cosmetics and other grime. There may also be some surfactant activity.

Make pieces of both waxes and then melt them because in large sizes, it is difficult to melt. Then, sodium borate reacts with all three oily substances to make soap. Melt both waxes with mineral oil by using steam. Then add sodium borate to hot water (approximately 70%). This mixture is added to the other mentioned above mixture. Mix continuously with constant stirring until a solid mass is obtained. The abovementioned formula was first used by Galen, hence called Galen Cerate [6].

15.2.2.1 Example

15.2.2.1.1 Prepare 100 g cold cream with beeswax, mineral oil, and lanolin

To start, mix beeswax, mineral oil, cetyl alcohol, lanolin, and span 60 in a china dish to make an oil phase. Add borax to purified water to make an aqueous phase and then heat both the phases separately in a water bath, nearly at 55–60 °C. Dropwise add the boric acid solution into the oil phase while stirring the mixture continually with a glass rod. After that, we continually stirred the entire mixture using a glass rod until it took on a semisolid form (Tab. 15.4).

15.2.3 Role of ingredients

15.2.3.1 Stearic acid

It is used as a cream base to formulate an o/w stable emulsion after reacting with potassium hydroxide. Potassium stearate, obtained as a result of the reaction between these ingredients, acts as an emollient and gives a white, creamy texture to the cream.

Tab. 15.4: Ingredients used in cold cream.

S. no.	Ingredients	Concentrations		
		Formulation 1	Formulation 2	Formulation 3
1.	Beeswax (white wax)	12 g	14 g	10 g
2.	Mineral oil	50 g	50 g	45 g
3.	Borax	0.5 g	0.7 g	0.8 g
4.	Cetyl alcohol	12.5 g	–	–
5.	Lanolin	–	–	2 g
6.	Span 80	–	–	1 g
7.	Purified water (q.s.)	100 g	100 g	100 g

Potassium stearate, on the other hand, gives excellent spreadability to the vanishing creams. After application, water evaporates and the remaining ingredients make a matte and smooth thin layer over the skin. Furthermore, it can act as a drug delivery system for localized effects.

15.2.3.2 Beeswax

Beeswax is frequently included in the development of skin moisturizers and body creams since it supports moisturizing and softening, as well as encourages cell repair, making it a fantastic ingredient to employ when creating products specifically for dry and rough skin types. Beeswax serves as an emollient (skin softening) and humectant (attracts water and aids in absorbing it in the skin) in this form of body care and increases the viscosity of cold cream. Besides this, beeswax also shows healing properties. It is a common ingredient of most cosmetic products like skin moisturizers, body creams, lip gloss, lip balm, eye shadow, and lipstick. Beeswax also has medicinal benefits in its use in body products. Beeswax has anti-inflammatory, antiviral, and antibacterial properties. These properties make beeswax very helpful for the healing of smaller wounds and injuries and can also be used for slight skin irritations. Beeswax also acts as a stiffening agent, increasing the viscosity of the cold cream.

15.2.3.3 Mineral oil

Liquid paraffin is a mineral oil that is essential for skin care since it keeps the skin moisturized. It serves as a barrier to the skin, preventing moisture loss. Liquid paraffin is also used in cosmetics. The ingredients can be found in a variety of beauty products such as detergent creams, cold creams, moistened creams, oils, and cosmetics. It can be used to treat dry skin. To aid in the skin's ability to retain moisture, liquid paraffin is frequently used in cream formulations for skin care. The skin has a built-in barrier that protects against moisture loss and keeps the skin healthy. Skin disease symptoms may be lessened with liquid paraffin.

15.2.3.4 Borax

The main component of borax is boric acid salt. This serves as an emulsifying agent by reacting with the beeswax's free acids. A chemical called an emulsifying agent makes water and other substances miscible. Borax keeps the preparation's pH stable and inhibits bacterial development. It is known as sodium tetra borate in scientific jargon, and it is a mixture that is combined from the depths of the earth and also found as deposits in lake bottoms and mountain runoffs. Wax and borax are both common ingredients in creams, gels, and lotions for cosmetic usage. Borax's alkaline nature makes it the ideal ingredient in cleaner products, where it is famously used to help wash off the oil or grease from the hands. However, some of these properties might irritate the skin and lead to rashes. Boric acid and sodium borate both inhibit or stop bacterial development, preventing the spoiling of cosmetics and personal care items.

15.2.3.5 Methyl paraben

Methyl paraben is a form of paraben. Parabens are chemicals that are frequently employed as preservatives to extend the shelf life of items. Methylparaben functions as a preservative to stop the growth of dangerous microorganisms. To stop the growth of mold and other dangerous bacteria, they are added to food or cosmetics. Many products that contain methylparaben also contain one or two other types of parabens in their ingredients. Foods, medications, and a wide range of cosmetic products all contain parabens. Parabens may be present in a variety of cosmetics including makeup, moisturizer, hair care, and shaving products.

15.3 Paste

Paste, like ointment, is intended for external application to the skin. Paste is the ointment having a greater solid mass, as a result of which they are thicker and stiffer than the ointment. Because of their large percentage of solids, they are more absorptive and less greasy than ointment. As they are more absorptive and have stiffness, they have a very low tendency to flow and thus can effectively be employed to absorb serous secretion and secretion from lesions. Due to their stiffness and impenetrability, pastes are not generally applied to hairy skin [6]. During their preparation, they become so hard that it is difficult to mix with a spatula, so we have to use heavy-duty machinery.

15.3.1 Composition of pastes

Depending on the particular formulation and intended application, a paste dosage form contains a variety of ingredients. Nevertheless, the following elements are frequently seen in pharmaceutical paste formulations:

Active pharmaceutical ingredient

The pharmaceutical element(s) that have therapeutic action are known as the API. This may involve a single medication or a mix of medications.

Vehicle

The vehicle gives the paste the required stability and consistency. Water, glycerin, propylene glycol, and several kinds of oils are examples of common vehicles.

Thickening agents

Thickening agents give the paste its desirable consistency by making it more viscous. Examples include artificial polymers (like carbomer) and natural gums (like xanthan gum and acacia gum).

Humectants

Humectants keep the paste from drying out by assisting in its moisture retention. Common humectants include glycerin and propylene glycol.

Preservatives

Preservatives to stop microbiological development and increase the paste's shelf life, preservatives are added. Parabens and benzalkonium chloride are two examples.

Flavoring agents

Flavoring agents enhance the paste's flavor and increase its palatability. Flavoring agents are added to oral paste formulations. Fruit, vanilla, and mint are common tastes.

Sweetners

Sweeteners hide the flavor of bitter or disagreeable-tasting active substances, and sweeteners may be added. Sucrose, sorbitol, and saccharin are a few examples.

Colorants

Colorants are added to the paste to change its color or to make distinct formulations stand out from one another. Usually, they are food-grade pigments or colors.

Emulsifying agents are used to keep the water and oil components of a paste blended if they are present. Polysorbates and lecithin are common emulsifiers. Stabilizers help the paste stay both chemically and physically stable. They aid in keeping the formulation from separating or deteriorating over time.

15.3.1.1 Example

15.3.1.1.1 Prepare 50 g Lassar's paste

Given that the paste contains a significant proportion of powder (45–70%) and is a stiff formulation, it is advised not to use a levigating agent like mineral oil. The quantity of mineral oil needed to levigate the powdered medications would lead to a softer end product. Instead, a portion of these ingredients should be mixed with soft paraffin directly. Triturate zinc oxide, starch, and salicylic acid separately on a glass slab. Melt the white soft paraffin in a beaker. Mix zinc oxide, starch, and salicylic acid mixture and then add white paraffin slowly, dropwise, accompanied by trituration. Continuous triturating all the ingredients till they become homogenous. Zinc oxide pasties are used as protectants and astringents. It is a smooth, white, odorless paste. It has covering and protective properties, gives consistency to topical products, and is said to have cooling and slightly astringent properties. It is also widely used as a complete sunblock due to its UV-reflecting properties. Salicylic acid is bacteriostatic and fungicidal. It also possesses keratolytic properties and may exert a solubilizing effect on the stratum corneum, with the dissolution of the intracellular cement (Tab. 15.5).

Tab. 15.5: Composition of Lassar's paste.

S no.	Ingredients	Concentration
1.	Zinc oxide	24 g
2.	Starch	24 g
3.	Salicylic acid	2.0 g
4.	White soft paraffin	50 g

15.4 Plasters

Plasters are semisolid or solid adhesive masses spread upon a suitable packing material and are intended for external application to a part of the body to provide prolonged contact at that site [7]. They are made up of two layers including backing layers and adhesive layers. Plasters are adhesive at body temperature and may be used to provide protection or mechanical support (non-medicated) or to provide a localized or systemic effect (medicated plasters). Backing layers are a material or layer to which masses (medicated or non-medicated) are employed. It may be of paper, cotton, or silk. Linen or plastic. This can be cut to get the size or area of the site of the application. The adhesive layer is the adhesive material consisting of rubber or a synthetic resinous material.

15.4.1 Salicylic acid plaster

Salicylic acid has astringent properties. Therefore, to treat corns and warts, salicylic acid plaster is applied to thickened skin. Salicylic acid's keratolytic action makes it easier to remove the thicker skin (keratolysis is the term used to describe the lysis of the keratinized layer of skin). Plasters usually contain 10–40% salicylic acid [8]. The skin should be moistened with warm water for 15–30 min before these plasters are applied. This hydration improves salicylic acid absorption. It is crucial to remember that this treatment should not last longer than 14 days. The plasters have holes so that cutting is simple. These holes, however, are used to check for secretions if they have already been cut; if they are, the old plaster needs to be replaced.

15.5 Lotions

Lotions are liquid preparations (suspensions or dispersions) intended for external application to the skin. Usually, lotions contain finely divided powdered substances that are insoluble in the dispersion medium and are suspended through the use of suspending agents and dispersing agents [9]. Other lotions have as the dispersed phase liquid substances that are immiscible with the vehicle and are usually dispersed using emulsifying agents or other suitable stabilizers. Most commonly, the vehicles of lotions are aqueous base systems. They are applied directly to the skin without rubbing with the help of some absorbent material such as cotton wool, or cotton wool soaked in the lotion and then applied to the infected skin. Lotions are intended to be applied to the skin for protection (maybe for cooling or soothing) or therapeutic value, depending upon the constituents used. Lotions that have a lubricating effect are intended to be used in areas where the skin rubs against itself such as between the fingers, thighs, and under the arms [10]. Because of its fluidity, it is uniformly applied over the skin and then dried soon after the application, leaving a thin coat of its medicinal components on the skin's surface. The addition of alcohol to lotion hastens its drying and produces a cooling effect, whereas the addition of glycerin to a lotion keeps the skin moist for a sufficiently long time and does not allow the preparation to dry. Preservatives are added to the lotions because if not added, bacteria and molds grow in the lotions. The composition of lotions includes emollients (water, cetyl esters, lanolin oil), emulsifiers (stearyl alcohol and cetyl alcohol), fragrances, humectants (glycerin), occlusive (dimethicone and lanolin oil), and solvents (water, alcohol, and benzyl alcohol).

15.5.1 Examples

15.5.1.1 Prepare 100 mL of calamine lotion

Ligate the calamine and zinc oxide with glycerin to reduce the size of the particles. Now add a small amount of bentonite to hot water and wait for mixing by stirring. Then add a further small amount of bentonite by sprinkling and mixing. In this way, all bentonite is mixed. Pour the levigated material into the bentonite solution, then add sodium citrate (in solution form), and finally add phenol. Finally, make up the volume with water (Tab. 15.6). It is used in the treatment of itching, pain, sunburn, insect bites, and minor skin irritants [11].

Tab. 15.6: Formulation of calamine solution.

S. no.	Ingredients	Concentration
1.	Calamine	15 g
2.	Zinc oxide	5 g
3.	Bentonite	3 g
4.	Glycerine	5 g
5.	Sodium citrate	0.5 g
6.	Phenol	0.5 mL
7.	Purified water (q.s.)	100 mL

15.5.1.2 Hydrocortisone lotion (suspension)

Adrenocortical steroid and topical anti-inflammatory agent. Hydrocortisone (HC) lotion in 1.0% and 2.5% concentrations is the most effective remedy in the relief of pruritus, and the reduction of the inflammation of atopic dermatitis, neurodermatitis, dermatitis venenata (industrial, household, cosmetic, and plant), and anogenital dermatitis [12]. HC has been applied topically to treat a variety of conditions due to its anti-pruritic and anti-inflammatory properties. Several different membranes have been used to study the permeation of drugs, including synthetic membranes, animal skin, and human skin donated from either cadavers or plastic surgery. Cellulose acetate membrane with 0.45-mm pore size was one of several synthetic membranes that have been used to determine the in vitro drug release profile of HC from creams, ointments, and lotions using the diffusion Franz cell system. Ethyl alcohol and propylene glycol increase the skin penetration of the drug. The presence of ethyl alcohol and propylene glycol increased the solubility of HC in the gel formulation, and their combined effects caused the permeability coefficient of HC from gel to be greater than in suspensions (creams and lotions). The gel formulation is a single phase with a high percentage of water. HC inside the gel moves toward the membrane according to its concentration gradient, and the concentration is maintained on the surface of the membrane, keeping the drug-release rate high. About the diffusion of HC through the synthetic membrane, the gel formulation had a statistically significantly higher diffusion rate compared to other formulations. The permeability coefficient of HC from the gel is close to the permeability coefficients observed for HC in the cream and the lotion [13].

15.5.1.3 Selenium sulfide lotion

Selenium sulfide is a powerful anti-dandruff active molecule that is extraordinarily effective at helping eliminate cases of severe dandruff [14]. Hyperkeratosis presents as a thickening of the skin. Selenium sulfide is an agent that is typically used for the management of seborrheic dermatitis and tinea versicolor [15]. Topical selenium sulfide is an effective therapy for pityriasis versicolor. However, it is not accepted readily by all subjects because it can be messy, time-consuming, and malodorous. Sodium sulfacetamide lotion 10% or selenium sulfide lotion 2.5% was applied once daily as a thin coating to the entire neck, torso, and upper extremities. The treatment also included all other areas affected with pityriasis versicolor. Many treatment options for pityriasis versicolor exist. Selenium sulfide shampoo is considered to be the conventional first-line therapy at this time. Most clinical trials in the treatment of pityriasis versicolor compare various experimental agents with selenium sulfide lotion 2.5%. However, selenium sulfide can be irritating to the skin, often resulting in local discomfort and pruritus [16].

15.6 Liniments

Liniments are alcoholic or oleaginous solutions or emulsions of various medicinal substances intended for external application to the skin, generally with rubbing [17]. Alcoholic or hydroalcoholic liniments are generally for producing cooling, but liniments of oleaginous phases are counter-irritant and rubefacient (the agents that are mild irritants and which cause redness of the skin) or are used when penetrating action is required. These oleaginous liniments are generally applied to the skin when massage is desired. These are also used as astringents. Comparatively, at the time of massage, oily liniments are more useful and exhibit milder action [18]. Oleaginous liniments are less irritant to the skin than alcohol. For oleaginous liniments, the solvent may be a fixed oil such as almond oil, peanut oil, cottonseed oil, or volatile substances such as wintergreen oil or turpentine, or it may be a combination of fixed or volatile oils. These bases or vehicles for liniments are selected depending upon the liniment use (counter-irritant, rubefacient, or massage), and it also depends upon the solubility of components in that base. The composition of liniments can vary widely depending on the specific product and its intended use, but here are some common components; for example, APIs include counterirritants, solvents include alcohol and water, emollients include glycerin and mineral oil, thickeners include carbomer and xanthan gum, preservatives include parabens and phenoxyethanol, fragrances include essential oils, and stabilizers include propylene glycol.

15.6.1 Herbal liniments

Liniments are for external use for aches and pains. Herbal liniments are normally used as warming massage mediums to relieve soreness in muscles and ligaments. Heat-inducing herbs such as cayenne are normally used in the preparation of liniments together with alcohol for extraction or a mixture of alcohol and/or oil. Liniments should not be used on cuts or broken skin. The stability of liniments is similar to that of herbal oils (if oil was used as a vehicle in the preparation process) [19]. Liniments are for external use for aches and pains. Herbal liniments are normally used as warming massage mediums to relieve soreness in muscles and ligaments. Liniments should not be used on cuts or broken skin [20]. Turpentine and wintergreen oils are irritants, and methyl salicylate, if used in liniment, is a local analgesic and is an irritant too. One thing about liniments of emulsion should remember that these are prepared by general methods of preparation (dry gum, wet gum, or bottle methods), but their mode of application and vehicle types are different from those of ordinary emulsions. The labeling includes used for external purposes only, shake well before use (if it is of a dispersed system type or contains insoluble ingredients). After use, tighten the container and keep it out of reach of children.

15.7 Topical gels

Gels are semisolid systems consisting of a dispersion of small or large molecules in a liquid vehicle rendered jelly-like through the action of an added substance such as carboxymethylcellulose [21]. Gels are a relatively newer class of dosage forms created by entrapment of large amounts of aqueous or hydro-alcoholic liquid in a network of colloidal solid particles, which may consist of inorganic substances such as aluminum salts or organic polymers of natural or synthetic origin. Depending upon the nature of colloidal substances and the liquid in the formulation, the gel will range in appearance from entirely clear to opaque. Most topical gels are prepared with organic polymers such as carbomers, which impart an aesthetically pleasing, clear, sparkling appearance to the product and are easily washed off the skin with water. Gels are two-component semisolid systems rich in liquids. In a typical polar gel, a natural or synthetic polymer builds a three-dimensional matrix throughout a hydrophilic liquid. Typical polymers used include the natural gums tragacanth, carrageenan, pectin, agar, and alginic acid; semi-synthetic materials such as methylcellulose, hydroxyethyl cellulose, hydroxypropyl methylcellulose, and carboxymethylcellulose; and the synthetic polymer, carbomer 934, may be used. Such gels release medicaments well; the pores allow relatively free diffusion of molecules, which are not too large. Gels are compatible with many substances and may contain penetration enhancers for anti-inflammatory and anti-nauseant medications. Examples include benzamycin topical gel (erythromycin and benzoyl peroxide), cleocin topical gel (clindamycin), desquam-X 10 gel (benzoyl peroxide), solaquin forte gel (hydroquinone), a bleach for hyperpigmentation, compound W gel (salicylic acid), and a keratolytic gel.

Depending on what their intended purpose is, topical gels can contain a variety of substances. Typical ingredients in topical gels include the following: The primary chemicals responsible for the therapeutic effect are known as active pharmaceutical ingredients or APIs. As examples, consider: analgesics (such as lidocaine and diclofenac), anti-inflammatories, such as diclofenac and ibuprofen, gelling agents, such as hydroxyethyl cellulose and carbomer, solvents (such as propylene glycol, water, and alcohol), preservatives such as propylparaben and methylparaben, emollients (such as mineral oil and glycerin), and enhancers of penetration (propylene glycol, dimethyl sulfoxide (DMS), etc.). A few examples of marketed gels are presented in Tab. 15.7.

Tab. 15.7: Examples of gels.

S. no.	Ingredients	Proprietary	Gelling agent
1.	Becaplermin	Regranex gel	Na CMC
2.	Benzoyl peroxide	Desquam-X gel	Carbomer 940
3.	Clindamycin	Cleocin T gel	Carbomer
4.	Tretinoin cellulose	Retin-A	Hydroxypropyl cellulose

15.8 Topical tinctures

Tinctures are normally alcohol and water extracts of plant materials. The plant part that is used can vary from plant to plant (e.g., some will be bark, some root, some seed, some leaf, some fruit, or just flower). Some tinctures use a combination of plant parts. Tinctures are made by soaking the relevant plant part in alcohol, which extracts the active nutrients to form a concentrated liquid. This makes them readily available for your body to absorb. There is the added advantage of the alcohol in a tincture being a preservative, allowing the extract to be kept for several years. Most commercially produced tinctures have a minimum alcohol content of 25% v/w [22]. These tinctures are used for external purposes. The composition of topical tinctures includes active compounds like alkaloids, flavonoids, terpenoids, and phenolic compounds like water, preservatives include sodium benzoate and potassium sorbate. Emollients include almond oil, thickeners include xanthan gum, antioxidants include Vitamin E, and fragrances include peppermint oil.

15.8.1 Examples

15.8.1.1 Prepare 100 mL of iodine tincture

The sodium iodide reacts with the iodine to form sodium tri-iodide (Tab. 15.8). The free iodine is not soluble in water, but this complex (tri-iodide) is soluble in water. This complex is also formed when we use potassium iodide in place of sodium iodide. The main emphasis is to dissolve iodine because if not dissolved and present in free form, then it may form a complex with ethyl alcohol, resulting in the loss of its antibacterial activity. The anti-infective agent is applied to the skin in general household first-aid procedures. Tincture of iodine is often found in emergency survival kits, used both to disinfect wounds and to sanitize surface water for drinking. It can be used as a skin disinfectant and as a water disinfectant. It is used to disinfect minor cuts, burns, and scratches [23].

Tab. 15.8: Iodine tincture formulation.

S. no.	Ingredients	Concentration (%)
1.	Iodine	2
2.	Sodium iodide	2.4
3.	EtOH	50
4.	Purified water (q.s.)	100

15.8.1.2 Prepare 100 mL of compound benzoin tincture

Compound benzoin tincture is a mixture made of our naturally occurring resins. It can also contain myrrh and angelica and occasionally balsam of Peru. Balsam of Peru, benzoin, styrax, and tolu are all balsamic oleoresins. Benzoin can cross-react with balsam of Peru, styrax, eugenol, vanilla, alpha-pinene, benzyl alcohol, and benzyl cinnamate, as they are all similar in chemical structure [24, 25]. All these make up 20% of the tincture. As storax is a highly viscous, sticky liquid, it is weighed first, then the drug mixture is macerated in the wide-mouthed container. Generally, storax is weighed in the container in which maceration is carried out. This maceration is done with alcohol (Tab. 15.9).

Tab. 15.9: The official U.S.P. formula for compound benzoin tincture.

S. no.	Ingredients	Concentration (%)
1.	Benzoin	10
2.	Aloe vera leaf	2
3.	Storax (styrax)	8
4.	Balsam of Peru	4
5.	Alcohol, 95% (q.s.)	100

Compound benzoin tincture serves as a protectant; it strengthens the skin in the management of bedsores, ulcers, cracked nipples, and fissures of the lips and anus. Additionally, it is utilized as an inhalant for bronchitis and respiratory issues. For inhalation, mix water with a few drops of compound benzoin tincture. It functions as a carrier for podophyllum, which is employed in treating venereal warts (genital warts). Podophyllum can also act as a teratogenic agent; if absorbed systemically, it may lead to paresthesia, loss of sensations, peripheral effects, lethargy, confusion, and even coma as a central nervous system effect.

15.8.1.3 Thimerosal tincture

A range of different ingredients, like organic, water-soluble, mercurial, and antibacterial agent (thimerosal) at 0.1% concentration with water, acetone, and alcohol, making up about 50% of the vehicle, and monoethanolamide and ethylenediamine acting as stabilizers and chelating agents, are added. The inclusion of these chelating agents is necessary because certain metabolic impurities, like copper, may be present during the preparation or could be introduced later from different sources in the tincture. It is used as a first aid household tincture antiseptic for preoperative purposes. It is orange-red-colored tincture (stains skin red) and resembles iodine tincture, which imparts a reddish-brown stain on skin [26]. Thimerosal tincture is used as a household and preoperative tincture and acts as an antiseptic.

15.8.1.4 Properties of tinctures

Sometimes water may be added to make a cosolvent system with ethanol to increase the solubility of different ingredients and the wetting properties of the tincture. The pH of the tincture ranges from 4 to 6, while the density ranges from 0.85 to 0.9 g/cm^3. The pH value represents important data concerning the stability of intermediate formulations and the choice of the adjuvant used in the final formulation. The presence of volatile ingredients in the tinctures makes them susceptible to flames; therefore, handling should be done accordingly [27].

15.9 Collodions

Collodion is a word derived from the Greek "call," which means to stick. These are liquid preparations composed of pyroxylin dissolved in a solvent mixture usually composed of alcohol and ether with or without medicinal substances. Pyroxylin is chiefly tetranitrocellulose obtained by a mixture of nitric acid and sulfuric acid on cotton. As pyroxylin is insoluble in water, we use ethyl alcohol and ether as a solvent mixture. Collodions are intended for external use when applied to the skin with a fine camel's hair brush or glass applicator. The solvent is rapidly evaporated, leaving behind a thin layer of pyroxilin, or if it is medicated collodion, then a thin layer of medicament is left on the site of application.

15.9.1 Composition of collodions

Collodion is a cosmetic ingredient that is a solution of nitrocellulose dissolved in ethanol and ether and that is reported to function as a binder and a film former.

15.9.1.1 Pyroxylin (nitrocellulose)

Nitrocellulose, a cellulose-based ingredient, is used in cosmetics, reportedly functioning as a dispersing agent nonsurfactant, and a film former. Adhering closely to the skin or wound surface. This film provides a protective barrier against mechanical irritation, moisture, and microbial contamination. Medicated collodions containing nitrocellulose are employed to deliver active ingredients in localized treatments, such as salicylic acid collodion for the removal of corns and warts, or flexible collodion for protecting small cuts and abrasions [28].

15.9.1.2 Solvents

The most common solvents used are a mixture of alcohol and ether. Ethanol is often used due to its ability to dissolve pyroxylin effectively. Ether acts as a cosolvent with alcohol, aiding in the dissolution of pyroxylin and facilitating the formation of a thin, flexible film upon application.

15.9.1.3 Plasticizers

Plasticizers are added to collodions to impart flexibility and improve film-forming properties. Common plasticizers include camphor and castor oil, which also contribute to the film's adhesion and durability.

15.9.1.4 Medicinal agents

Collodions may also contain medicinal agents such as topical anesthetics, antiseptics, or anti-inflammatory drugs.

15.9.1.5 Collodion membranes

Collodion was employed as early as 1872 to produce organic membranes in sheet form. Bigelow and Gemberling prepared collodion membranes by pouring a few milliliters of collodion on a clean, dry piece of plate glass and tilting it to and fro to spread the solution. The dry membrane was peeled from the glass. Membranes as large as 20 cm. in diameter, varying in thickness from 0.2 to 0.4 mm. were prepared. Bartell and Carpenter prepared membranes on a mercury surface and found that it was necessary to control the whole casting and drying process in a closed box to produce collodion membranes of similar properties. It has been suggested that the collodion membranes be removed from the surface of the mercury before drying is complete [29].

15.9.2 Examples

15.9.2.1 Simple collodions

It is a clear viscous liquid that is prepared by dissolving pyroxylin in a 3:1 mixture of ether and alcohol (Tab. 15.10). The film formed after the application of simple collodions is useful in holding the edges of an incised wound together. However, its presence on the skin is uncomfortable due to its inflexible nature.

Tab. 15.10: Composition of simple collodion.

S. no.	Ingredients	Concentration
1.	Pyroxylin	40 g
2.	Ether	750 mL
3.	Ethanol	250 mL

15.9.2.2 Flexible collodions

The castor oil renders the product flexible, permitting the movement of the body parts as such fingers, and toes, whereas the camphor makes the product waterproof [30]. When this is applied over an incised skin area it makes the area waterproof and protects it from external stress (Tab. 15.11).

Tab. 15.11: Composition of flexible collodion.

S. no.	Ingredients	Concentration (%)
1.	Camphor	2
2.	Castor oil	3
3.	Simple collodions q.s	100

15.9.2.3 Salicylic acid collodions

It is a keratolytic product and is used for the removal of the corn and wart. The collodion is used carefully because it is irritating to normal skin and thus should be applied dropwise (contains 10% salicylic acid in flexible collodion). When first drop is dried, then second one should be applied. For the protection of normal skin, first apply petroleum to the healthy skin around the corn or warts and then apply the product [31].

15.10 Poultice

It is a semisolid or viscous mass dosage form intended for external use for the sake of supplying warmth to inflamed parts of the body. A poultice is an ancient form of topical medication also known as a cataplasm. It is a soft mass of vegetable constituents or clay, usually heated before application [32]. Poultices are made by mashing fresh herbs, wrapping them in a gauze, and applying them to an affected area of the skin after the temperature is suitable for application. Poultices may be used externally to relax muscles or to ease minor skin eruptions, poison ivy, insect bites, superficial wounds, and inflammation. Since they are normally made from fresh herbs, they should be used immediately and cannot be stored.

15.10.1 Poultice materials

A variety of materials have been used to prepare poultices. In most cases, cellulose fibers or mixtures of clay minerals and aggregates with or without the addition of cellulose fibers are used. Frequently used clay minerals are montmorillonite or kaolinite, but attapulgite, sepiolite, and others are also reported. Sometimes paper pulp, cellulose tissues, or textiles are used [33].

15.10.2 Mode of application

First, heat the poultice until it can only just be tolerated on the back of the hand. Spread a thick layer of this on the some dressing and over the affected area with muslin so that the removal of the dressing, when we want, becomes easy. Apply the dressing to the muslin. Apply a thick layer of cotton wool to retain the heat, and then, at the end, cover it also with the same dressing [34].

15.10.3 Types of poultice

15.10.3.1 Herbal poultice

Herbal poultices involve the use of various plants and herbs known for their medicinal properties. These poultices are created by crushing or blending fresh or dried herbs into a paste and applying them directly to the affected area; for example, comfrey, chamomile, and lavender.

15.10.3.2 Clay poultice

Clay poultices utilize natural clays such as bentonite or kaolin, known for their absorbent and detoxifying properties. These poultices are created by mixing the clay with water or other liquids to form a paste, which is then applied to the skin; for example, bentonite clay and kaolin clay.

15.10.3.3 Bread poultice

Bread poultices are made from bread soaked in a liquid such as water or milk. These poultices provide moist heat and are often used to soften and soothe the skin as well as draw out infections and promote healing. For example, bread poultices can be applied to boils, abscesses, or wounds to help soften the affected area and promote drainage.

15.10.3.4 Salt poultices

Salt poultices involve the use of salt, often combined with water or other liquids, for their antiseptic and drawing properties. These poultices are applied topically to the skin to help cleanse wounds, reduce inflammation, and relieve pain; for example, epsom salt and sea salt.

15.10.4 Composition of poultice

The composition of a poultice varies depending on its use, but generally, it consists of herbs, used according to medicinal purposes. Some commonly used ingredients include herbs (comfrey; anti-inflammatory and wound-healing properties, calendula; known for its soothing and healing effects on the skin, plantain; has drawing properties, making it useful for drawing out toxins, chamomile; known for its calming and anti-inflammatory effects, lavender; has soothing and antimicrobial properties), base, for binding purposes (common bases include clay, bread, and oatmeal), and liquid, to form a paste (commonly we used liquids are water, herbal teas, and vinegar).

15.10.4.1 Example

15.10.4.1.1 Kaolin poultice
Kaolin poultice is the only example of a poultice mentioned in B.P.

Take heavy kaolin, boric acid, and glycerin in a beaker. Heat them for 1 h at 120 °C with occasional stirring. After this, mix methyl salicylate and thymol in peppermint oil and mix both the mixtures thoroughly (Tab. 15.12). The glycerol, because of its hygroscopic nature, is believed to draw infected material from the tissues to which the poultice is applied. Methyl salicylate is analgesic, boric acid is weak bacteriocidal, and thymol is a powerful bacteriocidal agent, whereas peppermint is used to improve odor. Remember that heavy kaolin is cheaper than light because the preparation is thick enough, so the sedimentation rate is small, and hence, there is no serious problem with the preparation.

Tab. 15.12: The composition of poultice.

S. no.	Ingredients	Concentration
1.	Heavy kaolin	52.7 g
2.	Boric acid	4.5 g
3.	Methyl salicylate	0.2 g
4.	Thymol	50 mg
5.	Peppermint oil	0.05 mL
6.	Glycerine	42.5 g

15.11 Percutaneous absorption

The absorption of substances from outside the skin to the positions beneath the skin, including entrance into the bloodstream, is referred to as percutaneous absorption. Percutaneous absorption not only depends upon the chemical and physical nature of the drug but also upon the condition of the skin. The drug is absorbed from the skin via keratinized cells of the horny layer (stratum corneum) and hair follicles.

Via keratinized cells

In the stratum corneum, there is no intercellular space, so the drug is absorbed from the outermost layer and then diffuses slowly toward the living cells. As the horny layer contains both hydrophilic protein (keratin) and hydrophobic fatty substances (from sebum) hence both hydrophilic and hydrophobic substances are absorbed. However, the materials having both hydrophilic and lipophilic portions are more easily absorbed or penetrated.

Via hair follicle

Although the hair follicle occupies only a small area of the total epidermis, it provides a very important route of percutaneous absorption because they are extended from a relatively impermeable horny layer to the dermis and hypodermis, and also the hair follicular cells are not keratinized.

15.11.1 Factors affecting percutaneous absorption

Factors associated with skin

If hydration of the keratinized layer is done, then absorption is increased. The greater the thickness (palm and soles) less absorption, and the less the thickness greater the absorption. If the skin is broken (wound), then absorption is high.

Factors associated with medicament

If the drug is both lipid and water-soluble, then it is absorbed easily. The smaller the particle size more its absorption.

Factors associated with the vehicles

Due to sweating, pore size is increased, hence the drug is absorbed rapidly; therefore, oleaginous bases should be used for the hydration of the skin. Drugs should leave the vehicle easily. If it does not leave the vehicle easily, then percutaneous absorption is slow.

Other factors

Drug concentration should be high. The drug should be applied to a greater surface area. By the application of rubbing, there is greater absorption of the drug. The vehicle should be in unionized form. The longer the period for which the drug has contact with the skin, the greater the absorption. Increasing the frequency of application to the same site of the skin increases absorption. Use of percutaneous absorption enhancers; for example, DMS, dimethyl acetamide, dimethyl formamide, polyethylene glycol.

15.12 Transdermal drug delivery system

Transdermal drug delivery systems are designed to support the passage of a drug from the surface of the skin through its various layers and then into the systemic circulation. The field of transdermal drug delivery has experienced tremendous growth for the past few years. One of the driving forces behind this growth is the increasing number of drugs that can be delivered to the systemic circulation, in clinically effective concentrations, via the skin portal. The major advantage is the maintenance of the blood concentration of a drug at the desired therapeutic level using controlled permeation of the drug through the skin [35].

The system consists of two basic parts. That controls the rate of drug delivery to the skin and that allows the skin to control the rate of drug absorption. All this can be obtained by dividing the system into layers. So the rate drug is delivered to the skin by the system is less than the rate of drug absorption from the skin into the systemic circulation. Hence, the system, not the skin, controls the delivery of drugs.

15.12.1 Examples of different transdermal systems

15.12.1.1 Transdermal scopolamine system

As scopolamine has a narrow therapeutic index, its plasma concentration should be controlled; that is why the scopolamine transdermal system is used to counter the adverse effects and short duration of action that have restricted the usefulness of scopolamine when administered orally or parenterally [36]. It is used to counteract nausea and vomiting. The system is marked by CIBA by the name of transderm-scop. It is a circular, flat adhesive patch designed for the continuous release of scopolamine. It is a 0.2-mm-thick patch having four layers backing layer, a drug reservoir, a semipermeable membrane, and an adhesive layer. The backing layer consists of aluminized polyester film. In the drug reservoir, we have scopolamine 1.3 mg, mineral oil, and polyisobutylene. Semipermeable membranes consist of polypropylene as a membrane-making material. The adhesive layer has scopolamine 0.2 mg, mineral oil, and polyisobutylene.

This system delivers 0.5 mg of scopolamine daily for 3 days (72 h). Initially, 0.2 mg scopolamine is released from the adhesive layer and rapidly brings the plasma concentration to the required steady-state level. Then, from the drug reservoir, a continuous release of the drug occurs, maintaining the drug level. The microporous semipermeable membrane causes the release of drugs from the reservoir at a rate less than the ability of skin absorption. So here it is, a semipermeable membrane, not the skin, controlling the delivery of a drug into the circulation.

The recommended dosage is a single patch applied to the postauricular area at least 6–8 h before the anti-motion sickness effect is required. For faster protection, the patch may be applied 1 h before the journey in combination with oral scopolamine (0.3 or 0.6 mg). After 72 h, the patch should be removed and a new one applied behind the opposite ear. Its place in therapy is mainly on long journeys (6–12 h or longer) to avoid repeated oral doses, when oral therapy is ineffective or intolerable. It is not recommended for use in children (in some European countries, the minimum age limit is 10 years) and should be used with special caution in the elderly, in patients with pyloric obstruction, urinary bladder neck obstruction, intestinal obstruction, and impaired metabolic, liver, or kidney function. It is contraindicated in patients with known hypersensitivity to any of the components of the delivery system, in patients with glaucoma, and in pregnancy. Special care is required when handling the transdermal patch, and patients should be warned to wash their hands immediately after application since contamination of the fingers and subsequent rubbing of the eyes has already resulted in several cases of cycloplegia. If the system becomes dislodged during use, it should be removed and replaced by a new unit in a different postauricular area [37].

The adverse effects produced by the transdermal scopolamine system, although less frequent, are qualitatively typical of those reported for the oral and parenteral formulations of this agent. Dry mouth occurs in about 50–60% of subjects, drowsiness in up to 20%, and allergic contact dermatitis in 10%. Transient impairment of ocular accommodation has also been observed, in some cases, possibly the result of finger-to-eye contamination. Adverse CNS effects, including toxic psychosis (mainly in elderly and pediatric patients), have been reported only occasionally such as difficulty in urinating, headache, rashes, and erythema [37].

15.12.1.2 Transdermal nitroglycerine system

It is used widely prophylactically for the treatment of angina pectoris. This has a very short therapeutic index when taken sublingually and when taken by the oral route. It is rapidly metabolized by the liver. So it is given safely by this route. Currently marketed delivery systems for transdermal administration of nitroglycerin (GTN) provide fairly constant plasma levels over the 24-h application period, corresponding to a relatively constant GTN release in vivo. When higher GTN doses are employed, the constant drug input may lead to nitrate tolerance, possibly explaining the results of clinical studies that show decreasing anti-ischemic effects within 8–12 h of initial patch application and only minor clinical responses during repeated administration.

This problem associated with constant plasma levels led to the development of a GTN transdermal delivery system that provides a time-dependent, nonconstant drug delivery rate. The system was also designed to provide a more efficient delivery system, which would also reduce production costs. One approach in this field of technological development is represented by a new biphasic release system, which was developed to provide high GTN blood levels during the first 12 h after application to prevent exercise-induced anginal attacks and low GTN levels during the night to prevent the development of tolerance [38]. Commercially, it is marked by the names of three different products of three different companies: transdermal nitro, nitro-dur, and nitro-disc.

Transdermal nitro: Similarly to transderm scop, it consists of four layers and a protective peel. Four layers include a backing layer, drug reservoir, semipermeable membrane, and adhesive layer. The drug reservoir is made up of aluminized plastic, which is not permeable to nitroglycerine, hence preventing it from loss because it is a volatile drug. The drug reservoir consists of nitroglycerine absorbed by lactose, silicon fluid, and silicon dioxide. A semipermeable membrane is made up of ethylene/vinyl acetate. The adhesive layer is made up of silicon. Before use peel strip is removed. One-fifth of the total nitroglycerine in the system is delivered transdermally in 24 h. There are four strengths of transdermal nitro, releasing the amount of drug in 24 h. This product is applied to the shoulder, chest, and especially to these areas, which are dry and hairless. The drug is present in the blood after application within 30 min. The rate of drug release is 0.2 mg/cm^2 in 1 h from the system.

Nitro-dur: It consists of 2% nitroglycerine in a gel-like matrix composed of glycerin, water, lactose, polyvinyl alcohol, povidone, and sodium citrate. The drug release rate is 0.5 mg/cm^2 in 24 h. The system is available in various concentrations. Nitro-dur transdermal infusion system, 5, 10, 15, and 20 cm^2 for 20, 40, 60, 80, and 120 mg drug, respectively [39].

About 5 mg nitroglycerine/cm^2 of the matrix is present in the fluid phase of the drug reservoir on the lactose crystal [39]. When the system is applied to the skin, then some fluid is shifted to the skin, so equilibrium is present between the drug in the drug reservoir and the drug on the skin. When the drug from the solution on the skin is absorbed into the skin, then equilibrium is established, so the drug from the reservoir is passed to the skin, thus maintaining the drug level. Therapeutic effects are obtained within 30 min after the application of the system and remain for 30 min after the removal.

This drug product has been conditionally approved by the FDA for the prevention and treatment of angina pectoris due to coronary artery disease. In terminating treatment of anginal patients, both the dosage and frequency of application must be gradually reduced throughout 4–6 weeks to prevent sudden withdrawal reactions, which are characteristic of all vasodilators in the nitroglycerin class. Adverse reactions reported less frequently include hypotension, increased heart rate, faintness, flushing, dizziness, nausea, vomiting, and dermatitis [40].

Nitro-disc: It consists of a solid drug reservoir mixture. The mixture consists of nitroglycerine impregnated in the silicon polymer. Now nitroglycerine is mixed. Cool the mixture to obtain the core of the drug reservoir, which is solid in nature. The system is applied to the skin as a flexible, nonsensitizing foam adhesive bandage. Two strength systems are present: 5 and 10 mg. It released the drug in 24 h. The therapeutic effects are obtained within 1 h after application and remain for 30 min after the removal of the system.

15.12.1.3 Catapres-TTS

Catapres, 2-(2,6-dichlorophenyl amino ~ 24midazolinc hydrochloride), lowers blood pressure in humans after intravenous or oral administration [41]. Catapres transdermal therapeutic system is produced by Alza Corporation. It consists of clonidine, formulated in this form because of its high plasma availability. A transdermal therapeutic system for delivering clonidine (Catapres-TTS) is used for the treatment of hypertension. In a two-way crossover study comparing Catapres-TTS and oral [42]. One patch provides therapy for 7 days. The product is present in a layer system. The backing layer is made up of pigmented polyester film. The drug reservoir consists of clonidine as a drug and polyisobutylene and colloidal silicon dioxide. The permeable membrane is made up of microporous polypropylene. The adhesive layer consists of clonidine, mineral oil, polyisobutylene, and colloidal silicon dioxide. The system is available in three different brands. The system is 0.2-mm thick and available in three surface areas that are 3.5, 7, and 10.5 cm^2 with 0.1, 0.2, and 0.3 mg corresponding concentrations of API.

The drug from the adhesive layer is present on the skin, and when absorbed, a potential gradient exists between the adhesive layer and the drug reservoir. Due to this, the reservoir releases the drug. This drug is then absorbed from the skin surface into the skin layers and then enters the blood via blood capillaries. Therapeutic plasma levels are achieved 2–3 days after initial application. Drugs remain in the plasma for 8 h after the removal of the patch. If continuous therapy is required, apply a new on fresh skin with the removal of the old patch. It is used for hypertensive patients.

15.12.1.4 Estradiol transdermal system (estraderm)

Estraderm represents the third drug for which well-controlled transdermal delivery devices have been developed [43]. The estradiol transdermal therapeutic system is a cutaneous delivery device that delivers estradiol into the systemic circulation via the stratum corneum at a constant rate for up to 4 days. Physiological levels of estradiol (the major estrogen secreted by the ovaries in premenopausal women) can therefore be maintained in postmenopausal women with low daily doses because first-pass hepatic metabolism is avoided [44]. Estradiol is formulated in a transdermal dosage form by CIBA. It releases 17 β-β-estradiol on contact with the skin. It is metabolized orally, so given in this dosage form. It is used in the treatment of female hypogonadism, castration, primary ovarian failure, atrophic vaginitis, and kraurosis valvae. Treatment of vasomotor symptoms is associated with menopause.

Its layers consist of a backing layer of polyester film. The drug reservoir contains estradiol alcohol gelled with hydroxypropyl cellulose. The permeable membrane contains ethylene vinyl acetate. The adhesive layer contains drugs, mineral oil, and poly-isobutylene. The protective peel strip is siliconized polyethylene terephthalate.

The recommended initial dosage of transdermal estradiol for the treatment of menopausal symptoms is 0.05 mg daily, which may be increased in cases of inadequate response after 2–3 weeks of treatment, or decreased if breast discomfort or breakthrough bleeding occurs. For maintenance therapy, the lowest effective dose should be used. Treatment may be continuous or may be given in 4-week cycles (3 weeks on and 1 week off). Sequential progestagen treatment should be administered for 10–12 days per month to patients with an intact uterus.

The transdermal estradiol delivery system should be changed twice weekly. Contraindications to the use of estradiol include carcinoma of the breast or endometrium, leiomyoma of the uterus, endometriosis, vaginal bleeding of unknown origin, severe renal, hepatic, or cardiac disease, and active or previous thromboembolic disease [44].

A significant proportion of patients experience dermatological reactions to the transdermal delivery device. The most common systemic adverse symptoms are typical estrogenic effects, such as breast tenderness and spotting/bleeding, and general effects such as fatigue, abdominal bloating, and nausea [44].

15.12.1.5 Trans-ver-sal

Trans stands for transdermal, and sal stands for salicylic acid. It is used in the treatment of viral warts. Available in two sizes, 6 and 12 mm, these may act to get the required size. However, each patch releases the drug for 8 h. The system consists of 15% salicylic acid in a vehicle of karaya (a substance of nonirritating nature and also has self-adhesive properties). Each patch is covered with a polyethylene moisture barrier, which creates a desired occlusive effect. As salicylic acid is an irritant to the skin, it is fresh, so apply it only on infected areas. Cover fresh skin with some material. Wet the infected area with warm water and apply the patch for bedtime the next morning, after which the patch should be removed.

15.12.1.6 Testosterone transdermal system

The transdermal testosterone patch is a medicated adhesive device applied to the skin, designed to deliver testosterone directly into the systemic circulation. Its typical composition includes. At its heart, the patch contains testosterone stored in a special gel or polymer base. This is the medicine's source. To control how quickly the hormone is released, some patches use a built-in membrane that lets it seep out slowly and steadily. Others rely on the special material of the gel itself to manage the release. The patch stays in place with a gentle but strong adhesive layer to ensure that it has full contact with your skin for effective delivery. A protective backing on the outside gives the patch its structure and keeps the medicine from leaking out.

You typically apply a new patch once a day, often at night, to a clean, dry area like your back, stomach, upper arm, or thigh (4–6 mg/day). While earlier versions were designed for the scrotum because the skin there absorbs medicine more easily, the modern patches made for other parts of the body are now much more common. They are preferred for being more convenient and comfortable.

As testosterone therapy requires close observation by a health care provider, the testosterone patch also comes with the limitation of keeping an eye on treatment. Patch application may be followed by skin irritation leading to redness, itching, or small blister formation. Moreover, acne and gynecomastia may also present after testosterone patch therapy. Testosterone can sometimes cause blood clotting issues, increasing the risk of high blood pressure or blood clots. It can also stimulate the prostate gland, so regular check-ups are crucial to monitor for an enlarged prostate or prostate cancer. Other possible effects include fluid retention, which can be a concern if you have heart or kidney problems, and mood changes like increased irritability or aggression.

15.12.2 Technology of transdermal delivery system

Transdermal drug delivery systems may be of two types: monolithic systems and membrane-controlled systems.

15.12.2.1 Monolithic system

Transdermal drug delivery systems of the monolithic type are patches consisting of a polymeric adhesive layer in which the drug is directly dispersed or loaded in a solubilized form. The adhesive matrix has several functions, including skin adhesion, storage of the drug, and control over drug/enhancer delivery rate, and it also governs drug partitioning into the stratum corneum [45].

It consists of a drug matrix (drug and polymer) between the backing layer and the frontal layer. This polymer matrix controls the rate at which the drug is released. In this system, the second and third layers of the transdermal system are mixed and are not separate. The mixed second and third layer is called the drug matrix. Matrix is of two types: gel and solid (nitro-dur and nitro-disc, respectively), depending upon the concentration of the drug. Here we have two options. A system having an excess amount of drug will release the drug at a continuous rate for a prolonged time, so the plasma drug level is maintained for a long time without the removal of the patch. A system having a smaller amount of drug does not release the drug at a continuous level, and the drug release rate decreases with time, so we have to replace the patch again and again to maintain the drug level. Drugs and polymers are dissolved or blended and dried. Another way for a gel matrix is in which gel is produced on a sheet or cylindrical form and assembled between the backing and frontal layer. Examples are nitro-dur and nitro-disc.

15.12.2.2 Membrane-controlled system

The drug may be in gel or liquid form. The second and third layers of the membrane assembly are separated. All four layers are separate. Thus drug reservoir (second layer) and polymer (third layer) have a distinguished character. Here we have the advantage that the drug release is controlled by the membrane controlling it (third layer). A small amount of the drug is present in the adhesive layer to initiate the release of the drug and absorption of the drug via the skin. If the drug is delivered to the stratum corneum at a rate less than the absorption capacity, the device is the controlling factor. If the skin is a controlling factor, then the drug is released at a rate to saturate the skin. So in a transdermal system (monolithic and membrane-controlled; for example, transderm nitro and transderm scop), either skin or artificial layers control the rate of release of the drug.

15.12.3 Future of transdermal system

It is anticipated that the first-generation patch and gel-based technology will be continued for efficient delivery of small molecules with certain properties, particularly those therapeutics that have recently been administered through the oral and parenteral route and are approaching off-patent. The second-generation nanoparticles and chemical penetration enhancers will be persistently used for transdermal delivery of small molecules. The third-generation targeted delivery approaches (a combination of biochemical and physical enhancers) present more preferential enhancement, yet are in the early phases of development [46]. They have a uniform drug release over a long period. Whenever we apply these patches, the adhesive layer occludes the passage of skin, and there is a way flux (the passage is one way, not both ways, or secretion is blocked and only the drug enters the skin). There is no immediate drug release from the drug reservoir, but the drug is released immediately from the adhesive layer. The adhesive material is nonsensitized; that is, the material does not irritate the skin. Applied to a specific area and have a specific size for that area; for example, estradiol transdermal patches are applied to the abdomen.

15.12.4 Physicochemical properties

15.12.4.1 Physicochemical properties of the penetrant molecule

More than 70 years of research on the percutaneous absorption of drugs reveal that the solubility of the drug in the carrier system plays a fundamental role in the amount of drug permeating the skin. Since permeation is a thermodynamically driven event, saturated formulations ensure maximum transcutaneous penetration. The optimal K, partition coefficient, is required for good action. Drugs with high K are not ready to leave the lipid portion of the skin. Also, drugs with low K will not be permeated. Drugs possessing both water and lipid solubilities are favorably absorbed through the skin since very lipophilic molecules would diffuse more easily in the stratum corneum, but would have difficulty leaving it and migrating to the deeper layers of the skin. On the other hand, a hydrophilic substance will also find it difficult to melt with the lipids composing the sebum, compromising its absorption. The transdermal permeability coefficient shows a linear dependence on the partition coefficient. Varying the vehicle may also alter the lipid/water partition coefficient of a drug molecule. The partition coefficient of a drug molecule may be altered by chemical modification without affecting the pharmacological activity of the drug.

The permeation of the drug increases tenfold with temperature variation. The diffusion coefficient decreases as the temperature falls. Weak acids and weak bases dissociate depending on the pH and pK_a or pK_b values. The proportion of unionized drugs determines the drug concentration in the skin. Thus, temperature and pH are important factors affecting drug penetration. For molecular weight to permeate significantly through the skin, the molar mass of the permeating agents should be less than 500 Da. However, this property does not exclude that larger molecules can penetrate the stratum corneum and disorganize the supramolecular lipid structure, favoring the intentional or accidental entry of other molecules and/or nanoparticles. Transdermal permeability across mammalian skin is a passive diffusion process, and this depends on the concentration of penetrant molecules on the surface layer of the skin. The flux is proportional to the concentration gradient across the barrier, and the concentration gradient will be higher if the concentration of the drug is higher across the barrier [47].

The potential for the skin components to establish chemical interaction with the drug and the local enzymatic metabolism cannot be overlooked. Bonding with normal skin components will prevent further penetration, while metabolism can reduce the amount of drug molecules that can permeate it effectively. Skin metabolism normally exists for the conversion of endogenous substances (e.g., hormones, steroids, inflammatory mediators, and lipids) and for the removal of xenobiotics [48]. Diffusion coefficient of the penetrating molecule is between the delivery system and the skin. In homogeneously diluted solutions, the diffusion coefficient can be calculated using the Stokes-Einstein equation, provided that the limiting factor for absorption is not the dimension of the penetrating molecule. The diffusion coefficient undergoes a few changes for molecules with a molar mass between 100 and 1,000 Da. For smaller molecules, the diffusion coefficient is affected by changes in polarity [49].

15.12.4.2 Physicochemical properties of the drug delivery system

The affinity of the vehicle for the drug molecules can influence the release of the drug molecule from the vehicle. Solubility in the vehicle will determine the release rate of the drug. The mechanism of drug release depends on whether the drug is dissolved or suspended in the delivery system and on the interfacial partition coefficient of the drug from the delivery system to skin tissue. The composition of the drug delivery system may affect not only the rate of drug release but also the permeability of the stratum corneum using hydration. Release of the drug from the dosage form is less due to the dead nature of the stratum corneum. Penetration enhancers cause physicochemical or physiological changes in the stratum corneum and increase the penetration of the drug through the skin. Various chemical substances are found to possess drug penetration-enhancing properties [49].

References

[1]   De Villiers M. Ointment bases. Practical Guide to Contemporary Pharmacy Practice. 2009; 3: 277–90.
[2]   Madasamy S, Sundan S, Krishnasamy L. Preparation of cold cream against clinical pathogen using *Caralluma adscendens* var. attenuata. Asian Journal of Pharmaceutical and Clinical Research. 2020; 13(9): 120–23.
[3]   Bułaś L, Szulc-Musioł B, Siemiradzka W, Dolińska B. Influence of technological parameters on the size of benzocaine particles in ointments formulated on selected bases. Applied Sciences. 2023; 13(4): 2052.
[4]   Mayba JN, Gooderham MJ. A guide to topical vehicle formulations. Journal of Cutaneous Medicine And Surgery. 2018; 22(2): 207–12.
[5]   Yadav R, Thakur S, Parihar R, Chauhan U, Chanana A, Chawra HS. Pharmaceutical preparation and evaluation of cold cream. International Journal of Innovative Science and Research Technology. 2018; 8(5): 1069–72.
[6]   Bhowmik D. Recent advances in novel topical drug delivery system. The Pharma Innovation. Edinburgh, England, 2012; 1(9). 12–31
[7]   Hacker M, Messer WS, Bachmann KA. Pharmacology: Principles and practice: Academic Press; New York, USA 2009.
[8]   Farndon LJ, Vernon W, Walters SJ, Dixon S, Bradburn M, Concannon M, et al. The effectiveness of salicylic acid plasters compared with 'usual' scalpel debridement of corns: A randomised controlled trial. Journal of Foot and Ankle Research. 2013; 6: 1–8.
[9]   Weiss SC. Conventional topical delivery systems. Dermatologic Therapy. 2011; 24(5): 471–76.
[10] Sarathchandraprakash N, Mahendra C, Prashanth S, Manral K, Babu U, Gowda D. Emulsions and emulsifiers. The Asian Journal of Experimental Chemistry. 2013; 8: 30–45.
[11] Joy N. Calamine lotion. Journal of Skin and Sexually Transmitted Diseases. 2022; 4(1): 83–86.

[12] Lubowe II. Use of Hydrocortisone and 9-AlphaFluorohydrocortisone derivatives: Evaluation in the treatment of the pruritic dermatoses. AMA Archives of Dermatology. 1955; 72(2): 164–70.

[13] Christensen JM, Chuong MC, Le H, Pham L, Bendas E. Hydrocortisone diffusion through synthetic membrane, mouse skin, and epiderm™ cultured skin. Archives of Drug Information. 2011; 4(1): 10–21.

[14] Tiganescu E, Abdin AY, Razouk A, Nasim MJ, Jacob C. The redox riddle of selenium sulfide. Current Opinion in Chemical Biology. 2023; 76: 102365.

[15] Cohen PR, Anderson CA. Topical selenium sulfide for the treatment of Hyperkeratosis. Dermatology and Therapy. 2018; 8(4): 639–46.

[16] Hull CA, Johnson SM. A double-blind comparative study of sodium sulfacetamide lotion 10% versus selenium sulfide lotion 2.5% in the treatment of pityriasis (tinea) versicolor. Cutis. 2004; 73(6): 425–29.

[17] Alexander K. Dosage forms and their routes of administration. In: Pharmacology: Elsevier; Edinburgh, England, 2009. pp. 9–29.

[18] Garg T, Rath G, Goyal AK. Comprehensive review on additives of topical dosage forms for drug delivery. Drug Delivery. 2015; 22(8): 969–87.

[19] Kumadoh D, Ofori-Kwakye K. Dosage forms of herbal medicinal products and their stability considerations-an overview. Journal of Critical Reviews. 2017; 4(4): 1–8.

[20] Swathy B, Menaka M, Veerareddy PR. An overview of plant medicines in novel form. World Journal of Pharmaceutical Research. 2019; 8(7): 938–958.

[21] Shokri J, Adibkia K. Application of cellulose and cellulose derivatives in pharmaceutical industries. In: Cellulose-medical, pharmaceutical and electronic applications: IntechOpen; London, England, 2013.

[22] Rathod HJ, Mehta DP. A review on pharmaceutical gel. International Journal of Pharmaceutical Sciences. 2015; 1(1): 33–47.

[23] Şahiner A, Halat E, Yapar EA. Comparison of bactericidal and fungicidal efficacy of antiseptic formulations according to EN 13727 and EN 13624 standards. Turkish Journal of Medical Sciences. 2019; 49(5): 1564–67.

[24] Scardamaglia L, Nixon R, Fewings J. Compound tincture of benzoin: A common contact allergen?. Australasian Journal of Dermatology. 2003; 44(3): 180–84.

[25] Steiner K, Leifer W. Investigation of contact-type dermatitis due to compound tincture of benzoin. Journal of Investigative Dermatology. 1949; 13(6): 351–59.

[26] Geier DA, King PG, Hooker BS, Dórea JG, Kern JK, Sykes LK, et al. Thimerosal: Clinical, epidemiologic and biochemical studies. Clinica Chimica Acta. 2015; 444: 212–20.

[27] Costa R, Camelo S, Ribeiro-Costa R, Barbosa W, Vasconcelos F, Júnior JS. Physical, chemical and physico-chemical control of *Heliotropium indicum* Linn., boraginaceae, powder and tincture. International Journal of Pharmaceutical Sciences and Research. 2011; 2(8): 2211.

[28] Fiume MM, Bergfeld WF, Belsito DV, Hill RA, Klaassen CD, Liebler DC, et al. Safety assessment of nitrocellulose and collodion as used in cosmetics. International Journal of Toxicology. 2016; 35 (1_suppl): 50 S–9 S.

[29] Carnell PH, Cassidy HG. The preparation of membranes. Journal of Polymer Science. 1961; 55(161): 233–49.

[30] Inn K, Hall E, Woodward J, Stewart B, Pollanen R, Selvig L, et al. Use of thin collodion films to prevent recoil-ion contamination of alpha-spectrometry detectors. Journal of Radioanalytical and Nuclear Chemistry. 2008; 276(2): 385–90.

[31] Tsai J-C, Chuang S-A, Hsu M-Y, Sheu H-M. Distribution of salicylic acid in human stratum corneum following topical application in vivo: A comparison of six different formulations. International Journal of Pharmaceutics. 1999; 188(2): 145–53.

[32] Maqbool A, Mishra MK, Pathak S, Kesharwani A, Kesharwani A. Semisolid dosage forms manufacturing: Tools, critical process parameters, strategies, optimization, and recent advances. Indian American Journal of Pharmaceutical Research. 2017; 7: 882–93.

[33] Mansour MMA, Salem MZM. Poultices as biofilms of titanium dioxide nanoparticles/carboxymethyl cellulose/Phytagel for cleaning of infected cotton paper by Aspergillus sydowii and Nevskia terrae. Environmental Science and Pollution Research International. 2023; 30(53): 114625–114645.

[34] Mundy GS. A kaolin poultice heater. Occupational Medicine. 1955; 5(2): 68–.

[35] Krishnaiah Y, Bhaskar P, Satyanarayana V. Formulation and evaluation of limonene-based membrane-moderated transdermal therapeutic system of nimodipine. Drug Delivery. 2004; 11(1): 1–9.

[36] Trozak DJ. Delayed hypersensitivity to scopolamine delivered by a transdermal device. Journal of the American Academy of Dermatology. 1985; 13(2): 247–51.

[37] Nachum Z, Shupak A, Gordon CR. Transdermal scopolamine for prevention of motion sickness: Clinical pharmacokinetics and therapeutic applications. Clinical Pharmacokinetics. 2006; 45: 543–66.

[38] Wolff H-M, Bonn R. Principles of transdermal nitroglycerin administration. European Heart Journal. 1989; 10(suppl_A): 26–29.

[39] Key Pharmaceuticals I, Roche H-L. Nitro-Dur.

[40] Coate JD, Colburn JR, DiedMar O. Nitro Dur®. 1987.

[41] Constantine JW, McShane WK. Analysis of the cardiovascular effects of 2-(2, 6-dichlorophenylamino) -2-imidazoline hydrochloride (Catapres). European Journal of Pharmacology. 1968; 4(2): 109–23.

[42] Shaw JE. Pharmacokinetics of nitroglycerin and clonidine delivered by the transdermal route. American Heart Journal. 1984; 108(1): 217–23.

[43] Good WR, Powers MS, Campbell P, Schenkel L. A new transdermal delivery system for estradiol. Journal of Controlled Release. 1985; 2: 89–97.

[44] Balfour JA, Heel RC. Transdermal estradiol. Drugs. 1990; 40(4): 561–82.

[45] Rhee Y-S, Kwon S-Y, Park C-W, Choi N-Y, Byun W-J, Chi S-C, et al. Characterization of monolithic matrix patch system containing tulobuterol. Archives of Pharmacal Research. 2008; 31: 1029–34.

[46] Qindeel M, Ullah MH, Ahmed N. Recent trends, challenges and future outlook of transdermal drug delivery systems for rheumatoid arthritis therapy. Journal of Controlled Release. 2020; 327: 595–615.

[47] Mali AD. An updated review on transdermal drug delivery systems. skin. 2015; 8(9): 244–254.

[48] Reddy YK, Reddy DM, Kumar MA. Transdermal drug delivery system: A review. Indian Journal of Research in Pharmacy and Biotechnology. 2014; 2(2): 1094.

[49] Ramteke K, Dhole S, Patil S. Transdermal drug delivery system: A review. Journal of Advanced Scientific Research. 2012; 3(01): 22–35.

Imran Tariq, Muhammad Yasir Ali, Sana Inam, Sajid Ali, Saeed Ahmad, Muhammad Ijaz, and Ijaz Ali

16 Suppositories

16.1 Introduction

The word "suppository" is derived from the Latin word "supponere," meaning "to place under." Suppositories are solid dosage forms intended for insertion into body cavities or orifices, where they melt [1]. They are also defined as a solid form that consists of one or more active pharmaceutical ingredients (APIs) dispersed in suitable bases and then modified into various shapes according to the anatomy of the organ in which they are intended to be inserted for local and systemic effects. Suppositories soften or dissolve and exert localized or systemic effects. They are commonly employed rectally, vaginally, and sometimes urethrally [2]. Generally, they weigh between 1 and 2 g [3], and cocoa butter or glycerinated gelatin is commonly used as a base.

16.2 Uses of suppositories

Suppositories can be used to produce local effects like pain relief, treatment of hemorrhoids by decreasing itching, inflammation, and swelling. They can also be used for the treatment of vaginal diseases like bacterial and fungal infections. Suppositories produce a mechanical action on the lower bowel and facilitate evacuation in the treatment of anal irritation and constipation. Moreover, suppositories are usually preferable when other dosage forms, like tablets and capsules, are not easily swallowed by the patient, especially elderly patients and children, or when a drug cannot be tolerated orally. Anorectal diseases are usually treated with suppositories. They are used to produce systemic effects and incorporate or introduce various drugs into the body. The drugs are easily released after the melting of the base and show their local and systemic actions.

Remember that the suppositories may be for rectal, vaginal, urethral, nasal, and otic routes, but they are mostly prepared for rectal applications. Vaginal suppositories are called pessaries, while all others, especially urethral ones, are called bougies. Vaginal suppositories are also called as vaginal tablets [4–6].

Suppositories should be prepared not too distant from the pathway. Their route of administration is not a first-choice route, but they may be used for local effects, e.g., pain, inflammation, hemorrhoids (bleeding), itching, constipation, and contraceptives [7].

https://doi.org/10.1515/9783111438108-016

16.3 Types of suppositories

Based on their action, suppositories are classified into two types, that is, suppositories for local action and suppositories for systemic action.

16.3.1 Suppositories for local action

16.3.1.1 Rectal suppositories

These are described in USP as suppositories for adults as tapered on one end or both ends, usually in cylindrical or pencil-like form, weighing 2 g each. The length may be 3.2 cm; however, in some cases, pediatric suppositories are shorter and weigh 1 g [8]. Rectal suppositories are used as tranquilizers and for sedative effects. However, the largest single-use category is probably hemorrhoid remedies [9]. They are mainly used for the hemorrhoid condition, in which inflammation, irritation, constipation, and itching usually occur.

16.3.1.2 Vaginal suppositories (pessaries)

The USP describes vaginal suppositories or pessaries as a globular or oviform dosage form that weighs 2 g each. A special device is used for the insertion of these suppositories [3]. Some of the vaginal suppositories are contraceptives and antiseptics. These are employed principally to treat infections occurring in the female genitourinary area. Additionally, they are used to restore the vaginal mucosa to its normal state and for contraception. In combating vaginal infections, the usual pathogenic organisms involved are *Trichomonas vaginalis*, *Candida albicans*, and *Hemophilus vaginalis*. Examples include Nonoxynol (a spermicidal agent) employed for contraception, miconazole (an antifungal drug), tetracyclines (e.g., oxytetracycline) used as an antibacterial agent, tri sulfa drugs, for example, sulfa thiazole, sulfacetamide, and sulfa benzamide are available in form of vaginal suppositories and sulfanilamide used to treat infection due to *Candida albicans*.

The most commonly used suppository base for vaginal suppositories is polyethylene glycol (PEG). PEG is mostly used in combination with parabens (Tab. 16.1). As the pH of the vagina is constant at 4.5, this should not be altered by the administration of suppositories. Hence, a buffer system is also used. We have an example here:

The amount of progesterone may range between 25 and 600 mg per suppository. This suppository is used for the luteal phase defects and premenstrual syndrome.

Polyethylene glycol-based vaginal suppositories are firm enough to be inserted easily. However, certain manufacturers provide plastic devices with their products for easier, quicker, and more accurate insertion of this type of suppository [10].

Tab. 16.1: Composition of a typical progesterone suppository.

S. no	Ingredients	Concentration
1	Polyethylene glycol 400	60%
2	Polyethylene glycol 800	40%
3	Progesterone powder (q.s.)	100%

16.3.1.3 Urethral suppositories

These are not specifically described in the USP but are long, thin, and cylindrical in form, rounded at one or both ends. The diameter is usually 5 mm, and the length is 12.2 mm (for males) and 50 mm (for females). The weight of the suppository for females is 2 g, while it is 4 g for males. These can be used as local anesthetics and antibacterials.

16.3.1.4 Nasal suppositories

They are also known as nasal bougies or bouginaria. Their weight is about 1 g, and their length is 9–10 cm. They are always prepared in a glycerinated base [11].

16.3.1.5 Ear cones

They are also known as Arenaria. They are very rare in use. Generally, Theobroma oil is used as a base [12].

16.3.1.6 Anti-hemorrhoidal suppositories

They contain several components, including astringents, emollients, soothing agents, analgesics, vasoconstrictors, etc. The glycerin suppository, which is used as a laxative, is one of the common examples.

16.3.2 Suppositories for systemic action

Suppositories can be used to attain sufficient systemic drug concentration. Drugs administered rectally in the form of suppositories for systemic action are prochlorperazine, an antipsychotic also used for nausea and vomiting; chlorpromazine, an antipsy-

chotic used as a tranquilizer; oxymorphone hydrochloride, an opioid used as a nar-
cotic analgesic; indomethacin, an NSAID for analgesic and anti-pyretic effects; and er-
gotamine tartrate, which is used for the relief of migraines. Suppositories and other
rectal formulations in clinical trials are given in Tab. 16.2.

Tab. 16.2: Rectal formulations in clinical trials.

Drugs	Dosage form	Indications	Status
Ceftriaxone	Rectal suppository	Bacterial infections	Phase 1 completed
Nifedipine	Rectal suppository	Chronic and fissure	Phases 1 and 2 completed
Diclofenac	Rectal suppository	Carcinoma prostate	Phase 2 completed
Belladonna + opium	Rectal suppository	Nephrolithiasis	Phase 4 completed
Meloxicam	Rectal suppository	Ankylosing spondylitis	Phase 3 completed
Fecal microbiota	Enema	Severe acute malnutrition	Phases 1 and 2 completed
Maraviroc	Rectal gel	HIV, AIDS	Phase 1 completed

16.3.3 Some special types of suppositories

Morphine sulfate in slow-release suppositories, prepared with a base that includes a
material such as alginic acid, will prolong the release of the drug over several hours.

Speeding up or slowing down the release and prolonging the action of the incorporated drug in suppositories has been investigated using various coatings, emulsions, hydrogels, layering, matrices of different substances, nanoparticles, osmotic release, micellar solutions, and thermo-reversible liquid suppositories.

16.4 Advantages of suppositories over oral therapy

Drugs given orally may irritate and can be given as rectal suppositories. This route is
mainly preferable for older patients, children, unconscious patients, and mentally re-
tarded persons. Suppositories are not destroyed by the GIT enzymes. Furthermore,
larger doses of drugs are usually administered by suppositories as compared to the
oral route. Suppositories are devoid of bad taste and odor as compared to tablets and
capsules.

16.5 Disadvantages of suppositories

The main problem is patient acceptability. Moreover, patients with diarrhea cannot
be treated with suppositories. Additionally, as suppositories promote evacuation of

the bowel, this results in inadequate absorption of drugs. Narrow therapeutic index drug-containing formulations cannot be interchanged because of the toxicity risk. Due to the presence of microflora in the rectum, some drugs may be degraded in the suppository. Suppositories are expensive if made on demand.

16.6 Factors affecting the drug absorption from the rectal suppositories

The dose of a drug administered rectally may be greater or less than its dose for oral administration, depending on two factors: namely, physiological factors and physico-chemical factors of the drug.

16.6.1 Physiological factors

When systemic effects are desired from the administration of a medical suppository, greater absorption may be expected from the rectum. A drug is greatly absorbed from the rectum in the absence of fecal matter, but for local effects, there is no need for absorption. Other conditions, such as diarrhea, colonic obstruction due to tumors, and tissue dehydration, can all influence the rate and degree of drug absorption from the rectum [13]. The effect of pH cannot be overruled in the absorption of drug content, as the ionized drug is preferably absorbed over the unionized form [14].

16.6.2 Physicochemical factors of the drug

The most important factor affecting the absorption of a drug from rectal suppositories is the partition coefficient. The drug must be soluble in both water and lipids because, for better absorption, it must cross lipid barriers. The drug is also incorporated into the base, which may be lipid or fat. If the drug is water-insoluble, its absorption will decrease because its release from the base will be difficult and it will not be available for absorption. Therefore, for the release of the drug, it must be soluble in water to some extent [15]. Particle size also affects the absorption of the drug from suppositories. The smaller the particle size means the more dissolution and the more chances of rapid absorption are there [16]. The barriers separating the colon from the lumen are usually preferable for unionized drugs. Thus, the absorption increases as the number of unionized drugs increases. The concentration of the drug in a base also affects the absorption. The more the drugs are present in a base the more will be the absorption. The partitioning of the drugs from the base is the rate-limiting step in the absorption from the suppository. Surfactants and particle size are key factors affecting the release of drugs from the base.

16.6.3 Physicochemical factors related to the base and the adjuvant

The base is always the carrier of the drug; it has to be melted, softened, or dissolved to release the drug, so bases have a significant influence on the rate of absorption. For better absorption, there should be no interaction between the base and the drug. Before incorporating the drug into the base, it is first melted at some temperature, and then the drug is incorporated. Therefore, the release of drug content from such systems will also depend upon the choice of base. On the other hand, the presence and nature of adjuvants affect the dissolution, rheological properties, and drug absorption.

16.6.4 Pathophysiological conditions

The effect of different pathological conditions on drug release from different dosage forms has always been an apple of discord. Drug release from suppositories is also affected under different pathological conditions like irritable hemorrhoids, GIT motility issues, irritable bowel syndrome (IBS), and inflammatory bowel disease (IBD). Absorption will be painful and unpredictable in the presence of disease conditions with changes in mucosal integrity and inflammation. Such conditions include fissures, hemorrhoids, IBS, and IBD. Similarly, mucosal inflammation can enhance epithelial permeability and, consequently, increase the amount of drug absorbed across the colorectal mucosa. For instance, mucosal inflammation in IBD causes pathophysiological changes, including a disrupted intestinal barrier due to the presence of mucosal surface alterations, ulcers, and crypt distortions, as well as infiltration of immune cells that promote inflammation. Diarrhea or constipation, either drug-induced or in the presence of infection, may also decrease the absorption of drugs from the rectum. Drugs that may cause diarrhea are like antibiotics, magnesium salts, NSAIDs, and laxatives.

16.7 Factors affecting the action of suppositories

Dissolution depends upon the melting point of the base and the liquefaction of the base used. Diffusion depends upon the spreading capacity, solubility of the drug, particle size, retention of the active principle by excipients, and excipient viscosity at rectal temperature. Absorption depends upon the pK_a value of the drug, the presence of buffers, the additive effect on membrane permeability, and the partition coefficient of the drug.

16.8 Suppository bases

Suppositories play a significant role in the release of the drug they contain and, therefore, in the drug's availability. The suppository bases used to manufacture this dosage form are classified as fatty/oily bases, water-miscible/water-soluble bases, or miscellaneous bases [17].

16.8.1 Fatty bases

16.8.1.1 Cocoa butter or Theobroma oil

This has been used for 200 years. It is a yellowish-white solid with a chocolate-like odor obtained from the roasted seeds of *Theobroma cacao*. At room temperature, it is in a liquid state, readily liquefied on warming, and rapidly sets on cooling. It is miscible with many other substances. The fat extracted from *Theobroma cacao* seeds, often known as chocolate beans, can be extracted via a solvent extraction process or by squeezing the seed oil.

Chemistry-wise, triglycerides of saturated and unsaturated fatty acids, including stearic, palmitic, oleic, lauric, and linoleic acids, make up cocoa butter. It has a yellow hue and a strong, distinct smell. It is solid at normal temperatures but melts at body temperature.

The lack of emulsifiers in cocoa butter prevents it from absorbing a large amount of water. However, Tween-61, a solid, nonionic surfactant that is tan and waxy, can be added in amounts ranging from 5% to 10% if considered necessary. Cocoa butter's capacity to absorb water is increased by its addition; however, the nonionic surfactants cause the suppositories to become unstable while being stored. Cocoa butter is a gentle base that does not irritate delicate membranes. It is readily available. It is practical to utilize in the production of suppositories without the required equipment.

The impact of certain APIs on the melting point of cocoa butter is one of its main drawbacks. The melting point of cocoa butter decreases when phenol, thymol, or chloral hydrate is added to suppositories. By adding between 4% and 6% white wax or between 18% and 28% cetyl ester wax, this issue can be mitigated. Due to its low melting point, cocoa butter and products made with it should only be kept in the refrigerator. This base exhibits polymorphic forms with even lower melting points: 18 °C, 24 °C, and between 28 °C and 31 °C.

The production of suppositories containing cocoa butter is therefore challenging, as this base is prone to overheating and changing into a form with a lower melting point. This implies that if suppositories are prepared incorrectly, they may already melt at room temperature or when the patient tries to give themselves a dose. Adherence to the molds is yet another problem.

Because of this characteristic, the drug's manufacturing process must be strictly regulated. The base should appear slightly opalescent when it melts. The time at which the melted cocoa butter fully transforms into a clear, straw-colored liquid signifies that the melting point was surpassed, all stable crystals were obliterated, and the suppositories would melt at a temperature lower than the intended 34 °C [18].

16.8.1.2 Witepsol

Witepsol has a density of 0.95 to 0.98 at 20 °C and is a white, waxy, brittle solid substance that melts into a clear or yellowish liquid with almost no smell. There are about 20 varieties of Witepsol, which are divided into the series H, W, S, and E, but the most available for use in the pharmaceutical field is C15. It has a melting range of 33.5 °C to 35.5 °C, which is close to its pour point.

16.8.1.3 Hydrogenated oil

As a substitute for Theobroma oil, many hydrogenated oils, for example, hydrogenated edible oil, palm oil, pea oil, stearin, and a mixture of oleic and stearic acids are used. Overheating does not affect the solidifying point. They are resistant to oxidation [19]. The water-absorbing power of such bases is high; they do not stick to molds and produce colorless, odorless, elegant suppositories. However, on the other hand, they become brittle upon rapid refrigeration after heating. They are more liquid when melted than Theobroma oil [20].

16.8.2 Water-soluble bases

16.8.2.1 Glycerinated gelatin

This is a mixture of glycerin and gelatinized water (to make a stiff jelly). Glycerin is hygroscopic and can absorb atmospheric moisture, so if added to suppositories, it may soften and cannot be easily administered. If such a suppository is administered, dehydration of the cavity takes place. To solve this problem, suppositories must have some water content, that is, glycerin (20%), gelatin (60%), and drug solution (20%). The drug solution provides water for the glycerin when placed in the cavity and inhibits dehydration [21]. This is one of the oldest hydrophilic suppository bases, and in the British Pharmacopoeia 2009, the ratio of gelatin/glycerin/water is 14:70:16. These are hygroscopic because of the moisture-absorbing ability of glycerin. This effect may irritate the insertion site. To reduce this effect, it is moistened with water before use.

Glycerinated gelatin bases are used in vaginal suppositories and urethral suppositories. In the case of vaginal suppositories, they provide a prolonged effect of the medicament because they soften slowly in body fluids. Moreover, in the case of urethral suppositories, the composition changes (gelatin 60%, glycerin 20%, and medicament 20%). It provides a rapid effect with less irritation. The glycerol–gelatin-based suppositories have a soft, elastic consistency, which makes them suitable for vaginal use.

Besides all the usefulness of glycerinated gelatin, it is difficult to prepare and handle and is hygroscopic in nature. Gelatin is incompatible with many drugs, for example, tannic acid, gallic acid, $FeCl_3$, etc. They can support bacterial and mold growth [4], and one of the most significant disadvantages of such suppositories is that they are a good medium for the reproduction of bacteria and the rest of microorganisms. Therefore, it is compulsory to introduce preservatives while manufacturing (e.g., methylparaben 0.18% or propylparaben 0.02%). They do not melt but slowly dissolve in the mucous secretions of the vagina.

16.8.2.2 Polyethylene glycol

These are polymers of ethylene oxide and water. They are most widely used in extemporaneous preparations and commercially manufactured suppositories. As it is a polymer, the molecular weight increases as the polymer units increase. There are several molecular weight ranges for them. Polyethylene glycol 300, 400, 600, 1,000, 1,500, and 1,540 are predominantly used. The molecular weight of PEG, ranging from 200 to 1,000, is liquid in form, but above this range, they are solid or semisolid.

These are used when a suppository with desired characteristics and properties is needed. By increasing molecular weight, their hardness increased, and they changed from liquid to solid form. Additionally, by increasing molecular weight, the melting point also increases.

Polyethylene glycol is chemically stable, i.e., it can be easily stored without the fear of any chemical change. It is physiologically inert, i.e., it does not allow bacterial growth. Suppositories containing PEG are not easily melted if the patient has a fever. Additionally, suppositories with varying melting points and solubility can be prepared [22] using polyethylene glycol bases. In general, polyethylene glycol bases are made in such a way that in the future they will not melt at body temperature, but dissolve in body fluids, so they must be moistened with water before use. Some commercial polyethylene glycol bases are also available for suppositories that also contain additional components such as surfactants. Polyethylene glycol bases have a relatively high release of water and fat-soluble substances compared to hydrophobic bases. Suppositories' melting point is easily controlled by the appropriate mixing of certain PEGs with different molecular weights; these bases and suppositories based on them do not require a carefully controlled storage temperature [23]. Table 16.3 shows the properties of some polyethylene glycol polymers most commonly used in pharmaceutical practice.

Tab. 16.3: Physical properties of different classes of polyethylene glycol.

Class (M.W.)	M.W. range	Physical form	Melting point range (°C)	Solubility in water (%)	pH of 5% solution
300	285–315	Liquid	−15 to − 8	100	4.5 to 7.5
400	380–420	Liquid	4 to 8	100	4.5 to 7.5
600	570–630	Liquid	20 to 25	100	4.5 to 7.5
1,000	950–1,050	Semi-solid	36 to 40	80	4.5 to 7.5
1450	950–1,050	Semi-solid	43 to 46	72	4.5 to 7.5
3,350	3,000–3,700	Flaky or powder	54 to 58	67	4.5 to 7.5
4,600	4,400–4,800	Flaky or powder	57 to 61	65	4.5 to 7.5
8,000	7,000–9,000	Flaky or powder	60 to 63	63	4.5 to 7.5

*M.W., molecular weight.

16.8.3 Miscellaneous

Certain synthetic surface-active (miscellaneous) agents have been suggested for use as hydrophilic suppository bases that may frequently be used without the addition of other materials for the preparation of some formulations. Polyglycol 40 stearate (Myrl 52) and polyoxymethylene sorbitan monostearate (Tween 60) are two commercially available bases. These are safer than cocoa butter base because they melt at a higher temperature than that of cocoa butter oil [17].

16.9 Methods of preparation of suppositories

Suppositories are prepared by three methods, i.e., molding from a melt (fusion), compression (cold compression), and hard rolling and shaping.

16.9.1 Molding from a melt

The basic steps in molding are the melting of the base and the addition of medicaments. Pouring the melted mass into the molds. Allow the molds to cool down in the refrigerator. Removing prepared suppositories from the molds. The suppositories used in this method are glycerinated gelatin and polyethylene glycol.

16.9.1.1 Molds

Suppository molds are commercially available with the capability of producing individual or a large number of suppositories of various sizes and shapes. Some small molds have 6 or 12 suppository holes. But industrial molds can prepare hundreds of suppositories that form a single batch. These molds are of stainless steel, aluminum, brass, or plastic. Molds are reusable or disposable. These molds can be opened for cleaning before and after the batch has been removed. Pour melted material, let it cool, and then open the mold for the removal of suppositories.

16.9.1.2 Lubrication

Depending on the formulation, suppository molds may require lubrication before the melt is poured to facilitate cleaning and the easy removal of molded suppositories. Whenever cocoa butter or polyethylene glycol is used as a base, then lubrication is necessary. Also, when glycerinated gelatin is used as a base for the preparation of suppositories, then lubricate the mold [10]. Otherwise, these suppositories will not have a smooth surface due to the sticky nature of these substances. Remember that the lubricant must be non-irritant and should be applied before pouring (Tab. 16.4). Whenever a miscellaneous base is used then there is no need for lubrication. A thin coating of mineral oil is applied with a finger to molding surfaces. The lubrication can be done with the help of a brush or a cotton swab.

Tab. 16.4: Different lubricants used for suppository bases.

S. no.	Bases	Lubricants
1	Theobroma oil	Soap spirit
2	Glycerol-gelatin base	Almond oil, liquid paraffin
3	Synthetic fats	No lubricant required
4	Macrogols	No lubricant required

16.9.1.3 Calibration of molds

Sometimes, the calibration of molds is necessary. Each mold is capable of holding a specific volume of material in each of its openings. The pharmacist should calibrate each suppository mold for the usual base (generally cocoa butter and polyethylene glycol bases) to prepare medicated suppositories, each having the proper quantity of medicaments. From the top of the mold cavities, the volume is removed after the filling of the molds. Then, after removal, they are weighed. The volume of the suppositories remains the same, but the weight varies with the change of bases and medicament. This is due to the change in the density of the materials used [10]. It is a simple method. It is more elegant than hand-molded suppositories. Sedimentation of solids in the base is prevented. This method is suitable for heat-labile medicaments. On the other hand, air entrapment, weight variation, and oxidation of drugs and bases may be present in this method.

16.9.1.4 Displacement value

It is the quantity of a drug that displaces one part or 1 g of cocoa butter or any other base.

Supposed calculations:

Weight of six suppositories containing the base alone = 10 g

Weight of six suppositories containing 40% drug = 12 g

Weight of six suppositories containing 60% drug = 60/100 × 12 = 7.2 g

Weight of six suppositories containing 40% drug = 40/100 × 12 = 4.8 g

So, displacement of base by drug = 10–7.2 = 2.8 g

Displacement value = weight of drug/weight of displaced base

$$= 4.8/2.8 = 1.7$$

16.9.1.5 Preparing and pouring of melted mass

- Clean and lubricate the molds with a suitable lubricant.
- Melt the base and mix the medicament into it. If the drug is heat-labile, it should be added to the base near its congealing point.
- Fill the mold to the maximum extent.
- To avoid the larger bubble formation in the suppository melted mass should be poured continuously.
- Scrape off the extra portion of suppositories.

16.9.2 Cold compression method

On a small scale, a mortar and pestle may be used. Heating the mortar in warm water (then drying it) greatly facilitates the softening of the base and mixing. This method is usually used when a drug is heat-labile and when the drug is insoluble in the base. Here, suppositories may be prepared by forcing a mixed mass of base and the medicaments into special molds using suppository-making machines that apply pressure to the mass out of a cylinder into the molds. This is a commercially used process. Remember that this process is not useful for bases, e.g., glycerinated gelatin, where heating is necessary. In this process, mix the powdered drug with an equal amount of base (usually cocoa butter). Then, add the remaining amount of base. The mixed mass is forced into the cavities of molds by applying pressure.

In large-scale manufacturing, hydraulically operated cold compression machines are used, which are cooled by water jackets to prevent the heat of compression from making the mass too fluid [3]. On a large scale, a process employing mechanical kneading mixers and a warm mixing vessel may also be used. In small-scale manufacturing, a mortar and pestle may be used. If the mortar is heated in warm water before use and then dried, the softening of the base and mixing process is greatly facilitated [24].

It is time-saving and produces more elegant suppositories than hand-molded suppositories. The cold compression method is suitable for heat-labile medicinal substances, as no mold preparation or heating is required.

A special suppository machine is required, and there are some limitations to the shapes of suppositories that can be made. Additionally, manipulation requires considerable skill. The appearance is not elegant, and it is not commonly done.

16.9.3 By hand-rolling and shaping

The hand molding and shaping method is the oldest and simplest method for suppository preparation. It is of historical importance and is not used nowadays. It requires more skills and expertise [25]. It is a method of choice when only a few suppositories are to be prepared in a cocoa butter base. A plastic-like mass is prepared by triturating grated cocoa butter and active ingredients in a mortar. The mass is then formed into a ball in the palm and rolled into a uniform cylinder using a large spatula or a small flat board on a pill tile. The cylinder is then cut into an appropriate number of pieces, which are rolled on one end to produce a conical shape. This method avoids the necessity of heating cocoa butter; no equipment is required, and no special calculation is to be done. However, manufacturing is difficult, and the appearance is not pretty.

16.10 Formulation variables

Some of the formulation variables that should be considered during formulation are physical state, viscosity, brittleness, volume contraction, and drug release rates.

16.10.1 Physical state

A drug in its active form can exist as a solid, liquid, or semisolid. In the case of solids, the size of the drug particles plays a crucial role, especially if the drug is not very soluble in water. Decreasing the particle size can increase the surface area, thereby enhancing its effectiveness. When dealing with liquids, it is essential to incorporate the liquid into the suppository base using various methods, such as emulsification, the addition of a drying powder, or the inclusion of a suitable thickening agent when mixing with the base. As for semisolids or paste-type drugs, they can either be combined with a solid to thicken the drug before mixing with the base or mixed directly with the base along with a thickening agent.

16.10.2 Viscosity

Viscosity considerations play a crucial role in both the formulation of suppositories and the controlled release of the drug. In cases where the base has a low viscosity, it becomes necessary to incorporate a suspending agent, such as silica gel, to ensure uniform dispersion of the drug until solidification occurs. During the preparation of the suppository, it is imperative for the pharmacist to continuously stir the melt while maintaining the lowest possible temperature to achieve a higher viscosity. Once the suppository is administered, the drug's release rate may be hindered if the base has a very high viscosity. This is because high viscosity slows down the diffusion of the drug through the base, consequently delaying its absorption by the mucosal membrane.

16.10.3 Brittleness

Handling, wrapping, and using brittle suppositories can be quite challenging. However, cocoa butter suppositories are usually not brittle unless the percentage of solids present is high. Generally, brittleness occurs when the percentage of non-base materials exceeds about 30%. Synthetic fat bases with high stearate concentrations or those that are highly hydrogenated tend to be more brittle. Additionally, shock cooling can lead to the cracking of fat and cocoa butter suppositories. To prevent this, it is important to ensure that the temperature of the mold is as close as possible to the temperature of the melted base.

16.10.4 Volume contraction

When making a suppository, the pharmacist pours a molten substance into a mold and lets it cool down. As the substance cools, it tends to shrink in size. This shrinkage facilitates the removal of the suppository from the mold, but it can also result in a hollow space at the back or open end of the mold. This hollow space is undesirable and can be avoided by allowing the substance to reach its solidifying temperature just before pouring it into the mold.

16.10.5 Drug release rates

Table 16.5 shows the drug release rates of oil-soluble drugs and water-soluble drugs in oil bases or water-miscible bases.

Tab. 16.5: Drug release rates of oil-soluble drugs.

S. no.	Drug: base characteristics	Approximate drug release rate
1	Oil-soluble drugs: oily base	Slow release; poor escaping tendency
2	Water-soluble drugs: oily base	Rapid release
3	Oil-soluble drug: water-miscible base	Moderate release
4	Water-miscible drug: water-miscible base	Moderate release; based on diffusion; all water-soluble

16.11 Packaging

Extemporaneously prepared suppositories are usually placed in shallow, partitioned, rigid paperboard boxes, but packaging is based on the type of base used. Gelatin base suppositories are packed in tightly closed glass containers to prevent moisture penetration. Cocoa butter base suppositories are packed individually to prevent adherence. Suppositories containing light-sensitive drugs are packed in opaque materials [3]. Suppositories must be placed in containers in such a way that they do not touch each other. A suppository is foiled in polyethylene, tin, PVC, and aluminum. Staining, breaking, or deformation by melting caused adhesion can result from poorly wrapped packaged suppositories.

16.12 Storage

Storage of suppositories is very important in maintaining their normal shape; for example, suppositories stored in high-humidity environments may absorb water, forming sponge-like structures, while those stored in dry temperatures and dry environmental conditions may lose water content, resulting in brittleness. Cocoa butter base suppositories are stored in refrigerators. Glycerinated gelatin base suppositories are stored below 35 °F. PEG-containing suppositories are stored at room temperature [26].

16.13 Properties of an ideal suppository base

Ideally, the base should melt at or adjust below body temperature, or it should dissolve in body fluids. It should solidify quickly after melting and be miscible with the ingredients. It should be nontoxic and nonirritant, and stable in different storage conditions. It should be resistant to handling and stable to heating above its melting point. Base should release active ingredients readily and should be easily molded and removed from the mold.

The drugs to be added determine other desired qualities. For instance, bases with higher melting points can be utilized to create suppositories for use in tropical areas or to combine medications (such as volatile oils, camphor, chloral hydrate, menthol, phenol, and thymol) that often have lower melting points than the base. When adding substances that will raise the melting points or when adding a lot of solids, bases with lower melting points can be employed.

Selecting a suppository base that would be suitable in a particular situation, based on the physical and chemical characteristics of the APIs found in suppositories, is a prerequisite for developing pharmaceuticals. A more thorough investigation of the various suppository bases, their characteristics, benefits, and drawbacks is required for this reason. Of course, there are a lot of recognized surrogate bases available today, but more and more advanced carriers are emerging every day [1].

16.14 Displacement value

The quantity of the drug which displaces one part of the base (usually Theobroma oil) is called the displacement value. The volume of a suppository from a particular mold remains the same, but its weight varies due to the variation in densities of the medicaments and the base with which the mold was calibrated. To get a product of uniform and accurate weight, an alliance must be made for the change in density of the mass due to the added drug. For this purpose, the displacement value of the medicament is taken into consideration [27]. Displacement values with respect to fatty bases are given in Tab. 16.6.

There are three different methods of calculating the quantity of base that the active medicament will occupy, and the quantities of ingredients required can then be calculated.

Tab. 16.6: Displacement values with respect to fatty bases.

S. no.	Medicament	Displacement values
1	Aspirin	1.1
2	Bismuth subgallate	2.7
3	Chloral hydrate	1.4
4	Cinchocaine hydrochloride	1.0
5	Codeine phosphate	1.1
6	Hamamelis dry extract	1.5
7	Hydrocortisone	1.5
8	Ichthammol	1.0
1	Liquids	1.0
2	Metronidazole	1.7
3	Morphine hydrochloride	1.6
4	Paracetamol	1.5
5	Pethidine hydrochloride	1.6
6	Phenobarbital	1.1
7	Zinc oxide	4.7

16.14.1 Dosage replacement factor method

Here, we have a formula written as follows:

$$F = 100(E - G)/(G)(X) + 1$$

where E is the weight of pure base suppository (blank suppository) and G is the weight of the suppository with $X\%$ of the active ingredients.

16.14.2 Density factor method

Follow the following procedure:
- Determine the average weight of the blank suppository.
- Weigh the quantity necessary for the total batch.
- Weigh the amount of the given medicament.
- Calculate the amount of medicament for one suppository.
- Weigh the batch after the incorporation of the medicament.
- Determine the density factor

$$\text{Density factor} = B/A - C + B$$

- Find the replacement value by dividing the weight of the medicament required for one suppository by the density factor.
- Subtract this from the weight of the blank suppository.

- Multiply by the total number of suppositories to get the total amount of base.
- Multiply the weight of the drug/suppository by the total number of suppositories to get the total amount of the medicament required.

16.14.3 Occupied volume method

- Determine the weight of the blank suppository.
- Weigh the quantity of the base for the batch.
- The density of the active drug is divided by the density of the base.
- Divide the weight of the active ingredient for the whole batch by the result of step 3.
- Subtract the result of step 4 from the result of step 2 to obtain the weight of the suppository base required.
- Multiply the weight of the active drug per suppository by the number of suppositories in the batch.

16.15 Mechanism of action of suppositories

A suppository will melt on the mucous layer depending on its nature, such as hydrophilic or lipophilic. As the suppository melts and dissolves, the drugs it contains diffuse toward the mucosal epithelial surfaces. If the drug is not water-soluble, it needs to break free from the base of the suppository through gravity before it can dissolve in liquid. The softening and dispersion of lipophilic melting suppositories are not reliant on the presence of fluid. This method of medication administration remains consistent for suppositories that dissolve upon heating.

16.16 Quality control test for suppositories

Quality control procedures for manufactured suppositories involve the assessment of identification, assay, and, when necessary, water content, residual solvent, dissolution, and content uniformity. The quality control of suppositories encompasses both physical and chemical evaluations. Physical analysis comprises visual inspection (physical appearance), consistency of weight, consistency of texture, melting point, liquefaction time, melting and solidification time, and mechanical durability. Chemical testing involves the examination of potency and dissolution characteristics.

16.16.1 Physical examination

Physical examination, also known as visual examination, includes observing the shape, color, odor, surface condition, and melting ranges.

16.16.1.1 Shape of suppositories

You should always check suppository shapes. Make sure they are all the same, as consistency in form matters a lot. If they look different, something is wrong. This check is really important for proper use.

16.16.1.2 Surface condition

A good suppository has a smooth, unbroken surface. It shouldn't show cracks or gaps running through it. You won't find any splits or rough patches either. There should be no dull spots or visible air pockets. Holes are not acceptable, ensuring a solid form.

16.16.1.3 Color

The intensity, nature, and uniformity of the suppositories should be validated.

16.16.1.4 Odor

Confirming the odor can help avoid confusion when processing similar suppositories. A change in odor may also indicate a degradation process.

16.16.1.5 Weight

Suppositories can be weighed using an automatic balance to determine the weight of 10 suppositories. If the weight is too low, it is advisable to check if the mold is properly filled and if there are axial cavities or air bubbles, it is due to poorly adjusted mechanical stirring or the presence of an undesirable surfactant.

16.16.1.6 Melting point determination

The utilization of a U-shaped capillary tube for determining the melting point offers accurate data for monitoring excipients and ensuring consistency in the production of suppositories that contain soluble active ingredients. Another method to determine the melting point involves inserting a thin wire into the suppository melt within the mold before it solidifies. The mold is then submerged in water, held by the wire, and the temperature of the water is gradually increased (approximately 1 °C every 2–3 min) until the suppository slides off the wire; this indicates the melting point of the suppository.

16.16.2 Mechanical Strength

Suppositories can be categorized as either brittle or elastic based on their mechanical strength. Tests have been conducted to determine the maximum mass (in kilograms) that a suppository can withstand without breaking. A favorable outcome is achieved when the suppository can endure a pressure of at least 1.8–2 kg. During the test, the suppository is positioned upright, and weights are gradually added until it loses its form and collapses. The objective of this test is to ensure that the suppository can be safely transported and administered to the patient under typical conditions.

16.16.3 Content uniformity test

To guarantee consistency in content, it is necessary to analyze each suppository individually to obtain data on the uniformity of dosage from one suppository to another. This testing involves assessing the amount of drug substance present in each dosage unit to determine whether it falls within the specified limits.

16.16.4 Dissolution test

Dissolution testing is frequently necessary for suppositories to examine the hardening and polymorphic transitions of active ingredients and suppository bases. Various dissolution testing methods are available, such as the paddle method, basket method, membrane diffusion method (also known as the dialysis method), and the continuous flow (or bead) method.

16.17 Stability of suppositories

The ingredients or conditions that lead to any chemical or physical instability in suppositories should be avoided.

16.17.1 Physical stability

The major changes in suppository characteristics due to natural aging and their causes are given in Tab. 16.7.

Tab. 16.7: The major changes in suppository characteristics due to natural aging and their causes.

S. no.	Modifications	Causes	Examples
1	Odor	Fungal contamination	Suppositories with vegetable extract
2	Color	Discoloration due to oxidation	Suppositories with a tartrazine yellow-colored solution
3	Shape	Incorrect temperature during storage	Suppositories with essential oils
4	Surface condition	Whitening	Vegetable extract or caffeine with suppositories.
5	Weight	Loss of volatile substances	Suppositories with camphor or menthol, etc.

16.17.2 Microbiological stability

Most suppository formulations do not contain preservatives or antioxidants, as water is usually excluded from the formulations. However, if water is present or the formulation may support the growth of microorganisms, an appropriate preservative may be added.

16.17.3 Chemical stability

When working with suppositories, the majority of them are anhydrous, so the presence of water is not a significant concern. However, in certain cases, water may be included to aid in the integration of the drug into the base, or it may be present as part of the hydrated form of the drug components' crystalline structure. Additionally, if emulsions or suspensions are included in the suppositories, water may be present.

Lastly, certain suppositories may be hygroscopic and can absorb water from the surrounding atmosphere. This leads to chemical instability [10].

Example
Calculate the exact amount of cocoa butter for 100 suppositories from the following data.
Weight of blank suppository = 2 g
Density of drug = 1.6 g/L
Amount of drug/suppository = 0.2 g
As we have a 2 g suppository blank and we have to prepare 100 suppositories. So following steps are involved:
Weight of 1 blank suppository = 2 g
Weight of 100 blank suppositories = 2 × 100 = 200 g

Calculation for displacement value
Density factor of drug = 1.6
Amount of drug/suppository = 0.2 g
Displacement value = 0.2/1.6 = 0.125 approx.

Calculation for the medicament suppository
As the weight of the blank suppository = 2 g
Now, the weight of the base in one medicament suppository = 2 – 0.125 = 1.875
Amount of base for 100 suppositories = 100 × 1.9 = 190 g
Base required for one suppository = 2–0.125 = 1.875
Actual weight of each suppository = 1.875 + 0.2 = 2.075

References

[1]   Melnyk G, Yarnykh T, Herasymova I. Analytical review of the modern range of suppository bases. Systematic Reviews in Pharmacy. 2020; 11: 503–08.
[2]   Kumar A, Kolay A, Havelikar U. Modern aspects of suppositories: A review. European Journal of Pharmaceutical Research. 2023; 3(4): 23–29.
[3]   Shargel L, Kanfer I. Generic drug product development: Solid oral dosage forms: CRC Press; Florida, USA, 2013.
[4]   Allen LV. Suppositories as drug delivery systems. Journal of Pharmaceutical Care in Pain & Symptom Control. 1997; 5(2): 17–26.
[5]   Gupta P. Suppositories in anal disorders: A review. European Review for Medical & Pharmacological Sciences. 2007; 11(3): 165–170.
[6]   Yakabowich M. Prescribe with care: The role of laxatives in the treatment of constipation. Journal of Gerontological Nursing. 1990; 16(7): 4–9.
[7]   Iwobi S. Suppository solid provision technology. International Journal Papier Advance and Scientific Review. 2020; 1(1): 30–35.
[8]   Abd-el-maeboud K, El-Naggar T, El-Hawi E, Mahmoud S, Abd-el-hay S. Rectal suppository: Commonsense and mode of insertion. The Lancet. 1991; 338(8770): 798–800.

[9]   Havaldar VD, Yadav AV, Dias RJ, Mali KK, Ghorpade VS, Salunkhe NH. Rectal suppository as an effective alternative for oral administration. Research Journal of Pharmacy and Technology. 2015; 8(6): 759–66.

[10]   Allen L, Ansel HC. Ansel's pharmaceutical dosage forms and drug delivery systems: Lippincott Williams & Wilkins; Philadelphia, USA, 2013.

[11]   Dash A, Singh S, eds. Pharmaceutics: Basic principles and application to pharmacy practice. Elsevier, Edinburgh; England, 2023.

[12]   Garg R, Kaur S, Ritika Khatoon S, Naina Verma H. A complete and updated review on various types of drug delivery systems. International Journal of Applied Pharmaceutics. 2020; 12(4): 1–16.

[13]   Alharbi HM Suppositories.

[14]   Schanker L. Mechanisms of drug absorption and distribution. Annual Review of Pharmacology. 1961; 1(1): 29–45.

[15]   Hua S. Physiological and pharmaceutical considerations for rectal drug formulations. Frontiers in Pharmacology. 2019; 10: 1196.

[16]   Schoonen A, Moolenaar F, Huizinga T. Release of drugs from fatty suppository bases I. The Release Mechanism. International Journal of Pharmaceutics. 1979; 4(2): 141–52.

[17]   Srivastava P. Excipients for semisolid formulations. Excipient Development for Pharmaceutical, Biotechnology, and Drug Delivery Systems (vol. 1). CRC Press; Florida, USA, 2006. p. 197.

[18]   Noordin MI, Chung LY. Robustness of palm kernel oil blend in suppository preparation using acetaminophen as a model drug. Journal of Pharmacy Technology. 2007; 23(6): 339–43.

[19]   Shanmugam S, Kim Y-H, Park J-H, Im HT, Sohn YT, Kim KS, et al. Sildenafil vaginal suppositories: Preparation, characterization, in vitro and in vivo evaluation. Drug Development and Industrial Pharmacy. 2014; 40(6): 803–12.

[20]   Noordin M, Chung L. Thermostability and polymorphism of theobroma oil and palm kernel oil as suppository bases. Journal of Thermal Analysis and Calorimetry. 2009; 95(3): 891–94.

[21]   Dodou K. Rectal and vaginal drug delivery. Any Screen Any Time Anywhere. In: The Design and Manufacture of Medicines. Elsevier; Edinburgh, England, 2018. pp. 739–757.

[22]   Chen J, Spear SK, Huddleston JG, Rogers RD. Polyethylene glycol and solutions of polyethylene glycol as green reaction media. Green Chemistry. 2005; 7(2): 64–82.

[23]   Hua S. Physiological and pharmaceutical considerations for rectal drug formulations. Frontiers in Pharmacology. 2019; 10: 489933.

[24]   Woerdenbag H, Visser JC, Kauss T, Sznitowska M. Rectal and vaginal. In: Le Brun P, Crauste-Manciet S, Krämer I, Smith J, Woerdenbag H, editors. Practical pharmaceutics: An international guideline for the preparation, care and use of medicinal products: Cham: Springer International Publishing; 2023. pp. 405–37.

[25]   Yasmitha B, Pallavi A, Devaki D, Vasundhara A. A review on suppositories. GSC Biological and Pharmaceutical Sciences. 2023; 25(1): 186–92.

[26]   Thushara P, AR MI, Tk AB, Vigneshwaran L. Formulation and characterization of suppository. World Journal of Pharmaceutical Research. 2023; 12(16): 996–1012.

[27]   Oladimeji Francis A, Akinrinola Ibukun A, Dawodu Tolulope A, Ogundipe Omotola D. Quantitative evaluation of effects of drugs concentrations and densities on their displacement factors in suppository bases. Journal of Chemical and Pharmaceutical Research. 2018; 10(2): 32–39.

Index

https://doi.org/10.1515/9783111438108-017

www.ingramcontent.com/pod-product-compliance
Lightning Source LLC
Chambersburg PA
CBHW080123220326
41598CB00032B/4933